Mike Ellis

Pig production problems

John Gadd's guide to their solutions

PIG PRODUCTION PROBLEMS

John Gadd's Guide to Their Solutions

John Gadd

NOTTINGHAM
University Press

Nottingham University Press
Manor Farm, Main Street, Thrumpton
Nottingham, NG11 0AX, United Kingdom

NOTTINGHAM

First published 2003
© J.N. Gadd

British Library Cataloguing in Publication Data
Pig Production Problems: A guide to their solutions

ISBN   1-897676-34-4

Disclaimer

Every reasonable effort has been made to ensure that the material in this book is true, correct,
complete and appropriate at the time of writing. Nevertheless the publishers, the editors and the
authors do not accept responsibility for any omission or error, or for any injury, damage, loss or
financial consequences arising from the use of the book.

Typeset by Nottingham University Press, Nottingham
Printed and bound by Hobbs the Printers, Hampshire, England

CONTENTS

Dedicated to:

The 5,000 or so hardworking, dedicated, patient,
enthusiastic, generally under-rewarded people
in pig production world-wide whom I've met,
sometimes argued with, and always listened to
– all of whom helped make this book possible.

FOREWORD

Several people over the years have kindly suggested I write a 'textbook' on pigs. Tempted, I've always turned it down on the grounds that it would take me two years to write it and about another year to see it in print. This three year delay would run the risk of parts of it being out-of-date as soon as it appeared. That – and the hard work it involves – has so far turned me aside.

What seems to me to be a better and easier idea is to put down some experiences gathered together over a lifetime's work trying to get to the bottom of problems which confront the hard-working pig producer. The solutions arrived at – if they were seen to have worked – were less likely to be overtaken by events because so many were based on practical, hands-on remedial action which always will stand the test of time. In fact the passage of time itself is a benefit in this respect as the longer a problem is studied and dealt with, the more solutions appear due to technological progress over the years which help basic farm practice.

This is not a book which goes into the technical background of the main subjects which affect performances such as food conversion efficiency – there are plenty of excellent textbooks covering technical minutiae to be found in the 'Further Reading' Appendix.

You should find, instead, what are the target performances and suggested action levels of many common problems I have been asked to advise on over the years; how to recognise poor or uneconomic achievement; what action to take *and* why, and the evidence for it if this exists.

I confess that as an on-farm adviser – I prefer this old term to today's more fashionable 'consultant' – I am very lowly-qualified, just a couple of Scottish agricultural diplomas obtained fifty years ago. At the time I started my interest in farm livestock these were enough to get me an entrance ticket into the advisory arena – as a 'gofer', a junior in a commercial company who had to do all the mundane tasks the senior advisers didn't want to tackle. So I dipped 20,000 sheep, chemically caponized as many chickens and injected about 1,000 litters of piglets in my role as a (very) junior adviser in a drug company – all in one year.

Quite a start! I got my hands dirty, which set me on course to marry technology with mud-on-your-boots practice, something I've held to be paramount all my life.

But it was the pigs I fell for and identified with, and it was pigs I decided, happily, with whom to throw in my future lot. I loved the practical work involved in testing new pig products and gradually became confident enough to make suggestions of my own. When back in the company office I was given a deskload of enquiries and grumbles from real pig producers to answer, not being too proud to telephone my team of expert pig farmer friends for the answer if my colleagues in the advisory office were stumped or just disagreed with each other, as all experts do, and gave me contradictory answers.

My seniors, who were patient with me and very helpful, had enormous knowledge as specialists in nutrition, animal disease, chemotherapy and all the other scientific '-isms', and so were a tower of strength to a callow youngster. But I see now with hindsight that they may have lacked two things when giving advice. First, a practical knowledge of how solutions based on science might or might not work in a real farm situation from the practical point of view. Secondly, being experts and specialists to a man, they might not have been good enough at seeing how the problem might impinge on, or be linked together by several scientific areas. What the scientist today calls 'multi-factorial'. The broad canvas. As specialist experts, they weren't so good at the broad canvas which was something else I thought I'd better remember wherever I went.

In my days as a salaried commercial adviser making sure that my employers' products, first in the drug trade and then in the feed trade, worked as they were designed to do – or finding out why they didn't if complaints arose – I learned two vital pre-requisites necessary to help come to the right answer in either situation, and then a third, which was how to make the advice 'stick'!
All of them are also germane to the suggestions I make in this book. These are:–

1. Are you being told the truth ?

Of course farmers are not liars – but they, or their stockpeople may be mistaken. If the adviser works from inaccurate information, it may be impossible to achieve the right answer to the problem. Worse, working from incorrect assumptions, he may go on to give the wrong advice!

The following chapters describe some ways of finding out what really happened which I have found useful so as to sidestep this pitfall.

2.   What do the records reveal ?

Most people keep records these days. Some people keep good records. But both are not necessarily the same. Of course good records make the problem-solving easier and quicker, but *any* records can be of help too. A good adviser must ask to see what records are available – if necessary to do with the problem but preferably across-the-board – *well before he arrives on the farm*. Too many consultants are encouraged to tour the farm, then migrate to the office for a cup of coffee and essentially a quick look at the records. This misses a major opportunity to get at the truth. Records first, pigs second!

It takes *time* to assess records properly, or so I find after 30 years of studying them.

Why? Quite often the problem can be seen from a careful study of the records. In theory anyway. But it is vital to put what the records may suggest to the test of visual inspection. Let the eye (and/or the other senses) confirm or otherwise what the records suggest. If it is positive then the chance of a correct solution moves up a gear; if negative, one may need to look elsewhere, or realign your preconceptions.

3.   How to convince the client to do what's necessary ?

In my early days conviction and enthusiasm got me a long way. Today it is different. Farmers need to be convinced, or better, to be shown how to convince themselves! Putting everything you say into a cost-effective syntax is a good persuader. It must be, as mathematics are precise, accurate, unarguable. Econometrics – the measurement of cost-effectiveness – is a valuable persuader.

THE ECONOMETRICS OF
PROBLEM SOLVING

One of the essentials of sound farm advice is to know what things are going to cost. Most advisers – and all scientists except a select few – tend to be performance-oriented. This was fairly satisfactory when profits from pig production were reasonable, and there was room for fairly easy improvement on physical performance. These days profits are there, but tend to be of a hairline nature. Improved performance is still possible but tends to be constrained by cost/investment. So we need to be ***profit-oriented*** in our guidance. How do we maximise income at minimal cost?

Because I started my advisory career working in agricultural commerce, I had forced on me, initially unwillingly on my part, the need to keep value-for-money uppermost in my mind. There was always a competitor who could do it cheaper. And there was usually more than one way to achieve maximum performance – but which method or product could achieve it most cost-effectively? Not always the same thing as choosing the cheapest product, or the easiest method, by any means.

As margins got tighter and as technology became more expensive I found the old favourite performance technology terms (FCR, ADG, Cost/kg gain, % mortality, etc.) becoming less useful than new profit-oriented terms (MTF, REO, PPTE, etc) which I had to devise myself so as to give my customers the best economic advice to suit each farm's individual circumstances. And just as important – to convince them to adopt the better idea. In some cases the old (existing) terms were actually misleading the farmer into adopting an incorrect solution! Today this is happening more and more often because everything is more expensive in relation to income. With computers making the assembling and assessment of information easier and quicker, (but possibly making it more confusing to implement) the need for a new look at how we measure performance becomes even more essential.

I thus devote a section of this book to describing how the New Terminology works and why you will need to adopt it – or your own version of it – sooner or later. Pig producers are not good with figures (like myself for that matter), so don't be tempted to skim over this chapter. Once you get the hang of it – and it is not difficult – you'll see its benefits. Many of the remedial suggestions I make are reinforced by it, and some depend on it. I have used both old and new terminology in this book, but because you will come across REO and MTF and PPTE and AIV in the pages, you should read the New Terminology section, and how to use it, as soon as you can.

You'll be a much better pig producer thereby!

CONVINCING YOU TO DO IT !

Problems cost money – a lot of it. Solutions can also cost money. Looking at both in terms of cost-effectiveness puts both financial expenditures into perspective. Any farm adviser can tell you that finding the likely solution to a performance fall-away is one thing – but persuading the client to then do what is necessary can also be another equally tough hurdle to clear!

Looking at things and presenting them in New Terminology terms, being based as it is on cost-effectiveness, I have found to be a valuable persuader in this respect because it keeps the bottom-line in the front of the pig producer's mind. What it costs in income or profit terms if he doesn't follow the advice, for example. Then, just as important, to compare the extra cost of putting it right with what it costs if it is left unattended or half-solved. This comparison in New Terminology language is Return on Extra Outlay (REO). Nearly all the suggestions made in this book have provided, from experience, REOs of 3:1 or more, and some a lot more.

One of my seniors in my early days – Stephen Williams, who was Boots the Chemist's farm manager at Thurgarton, Nottinghamshire in the 1950's – told me "True economy is not money *not* spent, but the *correct amount* of money spent in the *right place*". The New Terminology helps this truism forward, that's all.

THE ORIGINS OF THIS BOOK

Have I always been successful in my advice? No, not always. Of course not. I know this because I've always followed the visit up. "Did it work? What do you think of things now?" Again I was trained to do this by my business colleagues. In my early days as an on-farm adviser supporting a range of products, complaints had to be investigated promptly. They said "A complaint is a potential customer for life". How true this is! I also did it because I'm a writer-holic – I write everything down. Always have done. It helps keep track of things when you are busy, and I've a rotten memory anyway! I have kept a farm diary since 1950, which subsequently became a general (social) diary in the 1960's and has been what I call an "Omnium Gatherum" – in which anything which interests me is recorded and, if needs be, photographed or copied – for the past 23 years. There are over 22,000 photographs and 3 million words in it now, many to do with pig production – a feat which astonishes me as much as anyone else, but it is surprising what you accumulate over time if you keep at it. Eccentric maybe – but useful!

It is useful in the present context of writing this book because, annually indexed and cross-referenced as it is (I take 10 days off every Christmas to do this) it is a databank of experience to draw upon – to refresh memory and correct the distortions of time.

And it is because I never have been able to remember things too well that I soon found that consulting a written-down check-list of my own was a useful thing to carry with me when a pig problem had to be solved. After all, even the pilot of an aircraft has his checklists to ensure he doesn't miss something before he leaves *terra firma*, and we advisers have just as many things to take into account, even if our feet should be firmly on the ground, or in the muck in my case.

Some of the experiences I've gathered in this way are in the following pages, presented as far as I am able in a checklist format, to which I've added the probable reasons why, and the evidence for them if I can locate it, as well as the economics – costs and benefits. Quite rightly, the modern pig producer wants evidence; a lifetime's experience is all very well, but it needs to be leavened with fact, and what has gone before.

There are bound to be – only a few, I hope – errors and omissions. These will be entirely my fault and I apologise in advance. No man knows it all, especially me! I learn something new about pigs every day, something important once a month, and something of revolutionary importance once a year. And you learn more as you get older; but then become increasingly disturbed and ashamed at what you *don't* know about the subject of your life's work. One good thing with age is that you are then quite happy to confess you don't know, and are content to pass the enquirer on to a specialist who probably does know. As I said, I'm a broad-brush man, which is why I've found my checklists so useful to keep me straight on the detail.

I hope you do, too.

METRIC AND IMPERIAL

The publishers and I have decided to put as many figures as possible in both metric and imperial terms, accepting that it is sometimes a little unwieldy and may even irritate some readers. In my travels across the world there are still many pig producers who think in non-metric terms, and I hope those of you converted to metrication will bear with me on their behalf.

This is something of an experiment in a textbook, and I hope it works.

SPONSORS

We are indebted to the following organisations for their financial support.
We hope that their generosity will enable the book to reach a
wide and relevant readership.

ABNA Ltd
PO Box 250, Oundle Road
Peterborough
Cambs PE2 9QF

Alltech Inc
3031 Catnip Hill Pike
Nicholasville
KY 40356
USA

Antec International
Chilton Industrial Estate
Sudbury
Suffolk CO10 2XD

Big Dutchman International GmbH
P.O. Box 11 63
D-49360 Vechta
Germany

Elanco Animal Health
Kingsclere Road
Basingstoke
Hampshire RG21 6XA

Farmex
Pingewood Business Estate
Pingewood
Reading RG30 3UR

SCA
Maple Mill
Dalton Airfield Industrial Estate
Dalton
Thirsk
Yorkshire YO7 3HE

ACKNOWLEDGEMENTS

Although one man has written this book, it wouldn't have been possible without a huge number of people who have been so patient and helpful to him across 50 years.

Chronologically from the present to my student days, I owe much to ....

Barbara, my wife, for the use of our dining room table for 18 months, and innumerable cups of coffee or something stronger when the writing goes badly.

To Laura and Janeen, who type at the speed of light and also play arpeggios on a battery of word processing computers.

Next, the wise counsel and hard work of the Nottingham University Press team who have produced such an attractive and readable volume.

Then, in the years I've been self-employed as a pig consultant since 1984, 30 loyal and persevering British and overseas pig producers, most of whom prefer to remain anonymous, who have generously allowed me to use their pigs and facilities to do farm trials on products and problems on which there seemed to be no published information. We all owe them a debt of thanks for their trouble and patience.

Next, it must be around 3000 pig producers and their stockpeople who, across the years, have opened their premises to a visitor who criticised too much and praised too little. How much all of you have taught me over four decades about pigs – and people!

Then there are my erstwhile colleagues in the feed and allied industries. While we had shared interests on our employer's behalf, we didn't always agree on how best to attain them, but we have, I hope remained friends as our careers went their separate ways. And I include my peers in 'the competition' too. One of the pleasant memories of my days in commerce was how you could ring up a competitor and just ask a technical question of their adviser or nutritionist. The answer was nearly always friendly and quietly given – I'm sure it is no longer like that in these more flurried intensively competitive and – yes – less gentlemanly days!

Then there are the many academics and scientists who know so much more about matters technical than I do, some of whom I have had the temerity to disagree with or even upset from time to time, and I trust will continue to do – but still in a constructive way! Where would we be – and where would this book be – without them. I certainly owe them a debt, especially to those whose advice or work are gratefully acknowledged in the references, and in the 'further reading' appendix.

Then there are about 40 commercial firms who have given me work to do. As an ex-commercial man myself I have been able to appreciate their problems, and this econometric background has influenced and I think has also improved much of the advice given in these pages, centred on the bottom line as it should always be.

We now arrive at my early, formative years, thus names can be mentioned either because they are sadly no longer with us, or must have retired, so this will not cause them any embarrassment.

Stephen Williams, that original thinker on farm management. Very wise, a bit cynical but always right, none the less. My early mentors, veterinarians Norman Black and Ollie Murch, who convinced me of the continuing value of the pig veterinarian to us all. Another original thinker in pigs and pig farmer extraordinary, David Taylor, MBE; maybe difficult to work for, but a marvellous practical tutor and so generous with his knowledge to a young agriculturist finding his way among real pig producers.

We are all salesmen these days. Producers sell pigmeat while I sell information about producing pigmeat effectively. A couple of individuals may not have noticed it, but I have learned much from travelling to pig farms with two superb professional pig-feed and pig-product salesmen still actively working – Reg Hardy (Wiltshire, UK) and Mike King (Iowa, USA). If I get some messages across in this book it is partly due to them and their exemplary persuasive techniques.

And three college tutors – my 'prof' at Aberdeen, New Zealander Professor Neil Cooper. As he was a sheep man we actually *argued* about pigs – and also we shared an interest in mountain climbing, my other lifelong passion. Then Dr. R.V. Jones, fresh from his secret work on Radar Beams in WW2, who got me interested in physics – how air moved and about thermodynamics, a vital groundwork subject for anyone finding himself in pig advisory work. R.V. Jones shot real bullets into a sandbox to wake up us students after lunch – he was that sort of teacher. And in contrast Dr. Tom Dodsworth at the College's Craibstone Farm, a quiet, friendly cattle researcher who taught me about the lessons and the pitfalls of achieving accurate trial work.

Finally farmer Bob Milne, up in Laurencekirk, Scotland, my first employer. Another original thinker. It was he who ordered me (aged 19) to keep a farm diary and demanded it be initialled every month during my 18 month farm apprenticeship and thus started me writing professionally. I still have his pencilled 'No!' on many pages! Thankfully my editors didn't continue the practice.

I have written a monthly column for four successive editors of the UK 'Pig Farming' magazine over a period of exactly 35 years. That's 413 articles in all, just for one journal (among the half dozen or so others I write for) which is, I suppose, quite an achievement and a tribute to the editors' patience in allowing me the space to write about pig production as I see it for the last three and a half decades.

I thank you all because every one of you has contributed to this book. Without you I couldn't have written it.

LITTER SIZE 1

The number of piglets born. Normally expressed as 'total-borns' i.e. the number of fully-formed individuals expelled at farrowing, alive and dead. Also sometimes classified as 'born-alives' i.e. those piglets known to have drawn breath immediately after expulsion.

The former figure is preferred, as in an examination of poor litter size, the number of foetuses carried to full term could be important in establishing causes.

TARGETS

Are of course variable across the world, hence a range of targets are given:–

Table 1 LITTER SIZE TARGETS

Total piglets born and piglets born alive per litter
(It is advisable to sample a minimum of 50 to 100 litters)

	Poor	*Typical worldwide*	*Good/Target*	*Target (Hyperprolific genes)*
Total-born	9.5	9.9	11.3 – 11.8	13.1
Born-alive	9.0	9.4	10.75 – 11.25	12.5

Generally, target for no more than 5% piglets born dead.

THE PROBLEM AND ITS COST

Low litter size is a major problem world-wide. The breeder is usually alerted when periods occur where only 8.3 pigs per litter or less are weaned and/or where the numbers weaned per sow per year drop below 18.5 at any weaning age.

It is impracticable to estimate what the financial penalty is for a drop in litter size, as costs, incomes and margins vary across the world, and within one pig industry across, say a five year period, or even among one farm's financial picture over as little as one year's output.

However, as it costs an appreciable amount just to get one piglet born, whether dead or alive, my experience of the financial cost between the 'Typical world wide' and 'Good/Target' figures cited in Table 1 of 1.6 piglets total born per litter reduces sow income by 20% and has affected my clients' gross margin by 18% to as much as 45%

In many pig industries a 20% gross margin on sow output is considered a minimal baseline, then litter size must become a major influence on breeding efficiency.

The problem of reduced litter size can be due to many factors, confounded by possible interactions between them.

A list of primary factors encountered by the author is:–

CHECKLIST - PRIMARY FACTORS ALLIED TO LOW LITTER SIZE

Gilts
- ✓ Serving too light, grown too fast
- ✓ Stress rather than stimulation
- ✓ Lack of flushing

Sows
- ✓ Lactation nose-dive
- ✓ Incorrect lighting pre-service
- ✓ Adequate feed weaning to service
- ✓ Boar presence
- ✓ Poor AI technique

Sows	✓	Poor and late heat detection
(contd)	✓	Clean floors and rear-ends
	✓	Rest and quiet after service
	✓	Bullying
	✓	Litter scatter not analysed & acted on
	✓	Disease
	✓	Poor checking for returns
	✓	Herd age profile
	✓	Stress, discomfort, anxiety
	✓	Genetics
	✓	Short lactation
Boars	✓	Poor AI technique
	✓	Not monitoring boar records
	✓	Overuse of favourites
	✓	Lethal genes in some boars
	✓	Not sanitizing boar's sheath
	✓	No separate insemination/breeding area
	✓	Nutrition
	✓	Lack of exercise
General	✓	Lack of the 'Feel Good Factor'
	✓	Poor use of pig specialist veterinarian
	✓	Too few man-hours/sow/year

TACKLING A LOW LITTER SIZE PROBLEM

Good records are vital, particularly on sow and boar use. While mixed semen can be an aid to improved AI results, the inability to identify problems associated with individual males is a distinct drawback.

GENERAL FACTORS

The 'feel good' factor

A difficult term to define, but any pigs which seem harried or stressed rather than comfortable and contented are prone to poorer performance.

In litter size problems the Feel Good Factor is particularly important in the lead-up to breeding gilts, and in the post service period in sows. Producers should do a stress, anxiety and comfort audit of their breeding stock especially at these times.

Stockmen should distinguish between stimulation and stress before and during breeding, and do everything they can to keep the sows calm and contented in a restful atmosphere once mating is over.

Too few man-hours per sow per year

Experience suggests that stockmen (especially experienced stockmen as distinct from trainees) who spend more time with their breeding stock enjoy better litter sizes. It is difficult to suggest a minimal number of man-hours devoted to the breeding sow per year, but a threshold of around 20 is suggested. Some massive units, due to economy of scale, are run as low as 10 or less, but the performance of such units is not high even if the profit is considered 'adequate'.

The major labour problem among medium to small farms with reduced litter size is always the time spent on urgent repairs & precautionary maintenance at the expense of breeding & farrowing time or adequate hygiene. Some specialist help/contract labour can help.

Lack of pig specialist veterinary advice and presence

Disease, particularly viruses at a low level like PRRS, can be a cause of poor litter size. All farms, whatever their size, should have a regular input from a pig specialist veterinarian. In the author's records there are several cases where the disease level fell, adding generally more than 20% to the net income. However the veterinarian's time only increased vet/med costs by 4% and increased preventive protocols (drugs, vaccines, hygiene products) by another 5%, an REO of 20÷9 or 2.2:1.

In these cases litter size based on born-alives rose by an average of 0.9 pigs/litter.

GILTS

Gilts at a 37% - 41% replacement rate (which is too high but commonplace) can constitute a large part of the modern breeding herd.

Modern gilts have the capability to give large litters (over 11 total-borns) but too often give 8 or less. Even so a gilt (*i.e.* 1st parity sow) should not be culled on poor litter size. [*See Culling Checklist.*]

FACTORS AFFECTING GILT LITTER SIZE

Serving too light

125 kg (276 lb) is the current advised threshold, though some producers prefer 130 kg (287 lb). 135 kg (298 lb) has been suggested for some ultra-lean genotypes. Consult your supplier, and then a nutritionist, see below.

Grown too fast

The modern gilt can grow over 900-950 g/day (2-2.1 lb), and can be in danger of her hormone system falling behind her precocious growth. Thus she may be heavy enough and even display enthusiasm alongside or in with companions showing vigorous signs of oestrus, but still give you a poor conception rate or low first litter size due to immature sexual hormones.

Table 2 gives a suggested weight-for-age table to suit most genotypes. Check with a nutritionist that his gilt developer feed will achieve this level of weight for age.

Table 2   GILTS : SUGGESTED TYPICAL WEIGHTS FOR AGE FOR MODERN HIGH LEAN GAIN EUROPEAN BREEDS*

Gilt growth rate		*Aim to achieve 100 kg (221 lb) in 170-180 days, gilt growth-rate at 550 g/day, rising to 600 g/day towards puberty (1.2 to 1.3 lb/day)*		
100 kg	(221 lb)	180 days	25th or 26th week	6½ months+ old
104 kg	(229 lb)	187 days	week 27	
108 kg	(238 lb)	194 days	week 28	7 months old
112 kg	(247 lb)	201 days	week 29	
116 kg	(256 lb)	209 days	week 30	
120 kg	(265 lb)	216 days	week 31	
125 kg	(276 lb)	223 – 225 days	week 32	8 months old

* *Consult your seedstock supplier for actual targets.*

Table 3 shows the effect of this steadier approach to puberty in the modern gilt.

Table 3   GILT LITTER SIZE BY AGE AT FIRST SERIVCE

Age at 1ˢᵗ Service (days)	200-210	215-225	230-240	245-255	260-270
Percentage of gilts	28	27	21	16	8
Total borns	10.58	12.27	12.92	12.87	10.44

Source : Easton Lodge Pigs (UK) March 2000

WEANER OR 'JUNIOR' GILTS

Buying the gilt at 35–40 kg (77-88 lb) and raising it yourself avoids any possibility of the multiplier 'forcing' the gilt before delivery in order to save costs. It has other significant beneficial effects apart from that of a possible rise in gilt litter size, but the higher cost/lower selection/ increased failure rate should be taken into account. *See 'Choosing a Gilt' section.*

FLUSHING

Flushing is particularly valuable if gilts are delivered rather lean and light, as is common today, because multiplying farms want to get their finished gilts off their hands as quickly as possible. Table 4 gives a programme I've found valuable for many years. Table 5 gives the sort of results it can give when allied to later mating.

Gilts can arrive light, lean, stressed (nervous) with some of them quarrelsome. Along with capturing their interest (e.g. straw), giving them a good high-nutrient, specialised diet settles them down quickly into the 'feel good' factor which is so necessary for successful first litter formation.

We now have to keep them quiet for quite a time before eventual service at their 3rd oestrus at least, and just grow them steadily and quietly across the second heat period on a diet of modest nutrient density. How long this modest plane of nutrition should last depends on the gilt's weight increase and condition. If they are thought to be still thin, light, cold or stressed, it is better not to continue for the full 3 weeks in Table 4. Shorten it to a few days before the sudden increase 7-10 days before the oestrus chosen for mating.

Economics

In the UK the extra cost of food is about £9 per gilt (6-8% of purchase cost) but the total cost/gilt, including labour and housing costs, plus other overheads, can reach £20/gilt (+14%). But this is a lifetime cost, and

Table 5 suggests that the resultant 5-6 more weaners per lifetime can raise margin by £50 - an REO of 2.5:1. Money spent at the gilt rearing stage is always a good investment. The younger and lighter the gilt on arrival, the more flushing will pay.

Table 4 TYPICAL GILT FLUSHING REGIME

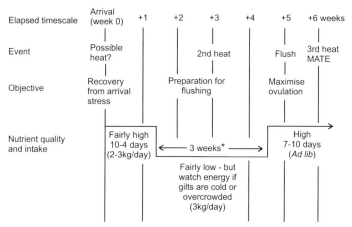

Note: If (a) virus disease immune build-up or (b) low arrival weights (<100kg) then for (a) buy weaner gilts and for (b) delay final service by one heat.

I have found the above programme to be particularly valuable for young or lean gilts. Even so *this period of relatively modest nutrient intake should be shortened to 10-14 days if the gilts are still lightish or thinish at 2-3 weeks after arrival. Be observant; be flexible. Nothing in gilt management is carved in stone!

Table 5   RESULTS OF THE ABOVE TECHNIQUE (SAME GENOTYPE, SAME FARM)

	Flushing / later mating *Cost (from previous table) £20/gilt.* *Served at 3ʳᵈ heat and flushed.*		*Old system* *Gilts rested for* *2 weeks, fed well* *(to appetite).* *Served at 2ⁿᵈ heat.*	
Results	*Pigs born alive*	*Weaned*	*Pigs born alive*	*Weaned*
Parity 1	10.8	9.4	10.3	9.4
2	11.1	10.0	9.2	8.2
3	11.7	10.4	10.7	9.4
4	11.6	10.8	10.9	9.4
5	12.1	10.9	10.8	9.6
TOTALS		**51.5**		**46.0**

Extra Cost – £ 20   Extra Productivity 5.5 weaners worth £50 margin in 1997
REO (1997) $50 \div 20 = 2.5{:}1$

Source : Authors' Records (1997)

STIMULATION, NOT STRESS

Hormones encouraged by fear and anxiety tend to neutralize the pro-oestrus hormones and affect litter size. Stockmen also tend to approve noisy mounting and chattering behaviour in the gilt pens during the run-up period to service, especially if boars are nearby to encourage this.

However some of this noise can be due to fear and protest. So check for:–

* Adequate space, especially fleeing space (3.5 m²/gilt, approx 38ft²/gilt)

* Pen shape – as square as possible

* Numbers together. This depends on space adequacy but normally 6 is enough.

* Evenness. Excited heavier gilts can be dominant and aggressive. Also it is a cardinal error to run gilts with multiparous sows; even to pen them adjacently at oestrus is unwise.

* Adequate food and water access.

SOWS

The Nose Dive – a term given to muscle and body-fat fallaway through lactation (Figure 1).

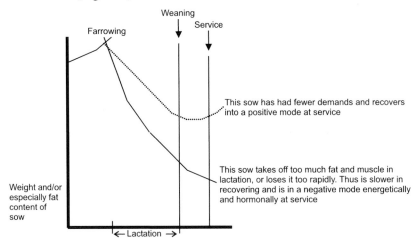

So how you feed and manage the sow way back in pregnancy, and especially in lactation, can markedly influence re-breeding

Figure 1 The 'nose-dive' effect. How it could affect ovulation - returns to service and litter size

Sows in lower than normal condition at weaning can often give poorer subsequent litter performance as well as re-service problems.

Don't let the sow lose too much condition down through lactation and/ or (new suggestion), lose less condition **overall** but suddenly lose it badly towards the end of lactation. For each 10 kg (22 lb) weight loss in lactation, subsequent litter size can be reduced by 0.5 pigs/litter, and for each 1 kg (2.2 lb) less food eaten, 23 day weaning weight can reduce by 400-500 g (approx 1 lb) per weaner, dependent on litter size and sow condition.

A great deal of future sow productivity depends on the stockperson's skill in defending the sow from the depredations of a voracious and, these days, larger litter.

The following checklist gives a list of factors which can help.

CHECKLIST – HOW TO AVOID THE 'NOSE-DIVE'

✓ **Body Condition Score*** so you can quickly detect when it is occurring and its degree of severity.

✓ Don't breed gilts too soon, too light, too thin.

✓ Flush all gilts (new advice on feed intakes available – see Table 4).

✓ Buy breeds/strains with good appetites.

✓ Keep sows cool (21°C/70°F maximum) in lactation.

✓ Feed a special lactation diet (altered for hot weather).

✓ Adequate water and accessibility especially in lactation.

✓ Water by bowl, not by bite drinker.

✓ Feed wet (by pipeline).

✓ Don't overfeed in pregnancy, especially 7 days before farrowing.

✓ Take the load off the sow by fostering/piglet swapping/ "weaning by weight/not date" etc (but take care if PMWS is about).

✓ Feed 3 times a day, last main feed at night. Maintain feed freshness.

✓ Avoid all stressors, especially discomfort.

✓ Be especially careful in hot weather.

*   Body Condition Scoring has its detractors among academics, due to its undoubted imprecision and subjectivity. However, many years of advisory work has convinced me that its value lies in encouraging stockpeople to really examine their sows by feeling as well as looking. It is especially valuable in monitoring the start of the nosedive phenomenon (Figure 1) and in determining fat cover and fleshing post-weaning. Academics, while right to point out its disadvantages, should ease up on their public denunciations for fear of damaging a useful sharp-end stockmanship tool.

PRACTICAL ADVICE

- Check carefully with your nutritionist that his diet contains sufficient protein/amino acids and energy to satisfy the weight of progeny to be suckled daily.

- Many successful breeders follow the U.K. 'Stotfold' lactation feed scale. (Tables 6a and 6b)

- Try to buy females with an ample appetite potential. Ask the breeder for evidence of likely appetite, then contact his customers for their opinion on how easy or otherwise it is to get sufficient daily food intake in lactation.

CORRECT LIGHTING PATTERNS

Light, like water, is relatively cheap but its importance is poorly appreciated in the breeding unit. The mating area and the areas where the newly weaned sow or new-entrant gilt are held prior to service need really bright light – at least 350 lux (lumens). This is bright enough to read a newspaper easily. Bright summer sunshine in a green field is between 500-600 lux. Figure 2 suggests that in some breeding units white fluorescent tubes are needed, and suggests the spacing and height needed of 100w fluorescent tubes. Remember, the light needs to be taken in by the sow's eye so the tubes need to be over stalled sows' heads, or just forward of them, not over the back.

Table 6a LACTATION FEED SCALE (metric)

First 10 days (All sows/gilts)		Sow Identification
Day	Kg Fed	Total Fed:
1	2.5	Date Farrowed (Day 1)
2	3.0	NOTES: *This dietary scale is now widely used in Europe.*
3	3.5	*Liaise with a pig nutritionist to formulate a diet density which*
4	4.0	*will satisfy the published daily intakes (cf Close & Cole*
5	4.5	*'Nutrition of Sows and Boars', NUP 2000). Total litter weight*
6	5.0	*at weaning can be on some farms over 30% higher than for average*
7	5.5	*herds.*
8	6.0	
9	6.5	
10	7.0	

Gilt<10 piglets Sow<9 piglets			Gilt 10 piglets Sow 9 piglets			Gilt 11 piglets Sow 10 piglets			Gilt 12 piglets Sow 11 piglets			Gilt 13 piglets Sow 12 piglets		
Day	Kg	Fed	Day	Kg	Fed	Day	Kg	Fed	Day	Kg	Fed	Day	Kg	Fed
11	7.0		11	7.5		11	7.5		11	7.5		11	7.5	
12	7.0		12	7.5		12	8.0		12	8.0		12	8.0	
13	7.5		13	8.0		13	8.5		13	8.5		13	8.5	
14	7.5		14	8.0		14	8.5		14	9.0		14	9.0	
15	8.0		15	8.5		15	9.0		15	9.5		15	9.5	
16	8.0		16	8.5		16	9.0		16	9.5		16	10.0	
17	8.5		17	9.0		17	9.5		17	10.0		17	10.5	
18	8.5		18	9.0		18	9.5		18	10.0		18	10.5	
19	9.0		19	9.5		19	10.0		19	10.5		19	11.0	
20	9.0		20	9.5		20	10.0		20	10.5		20	11.0	
21	9.5		21	10.0		21	10.5		21	11.0		21	11.5	
22	9.5		22	10.0		22	10.5		22	11.0		22	11.5	
23	9.5		23	10.0		23	10.5		23	11.0		23	11.5	
24	9.5		24	10.0		24	10.5		24	11.0		24	11.5	
25	9.5		25	10.0		25	10.5		25	11.0		25	11.5	
26	9.5		26	10.0		26	10.5		26	11.0		26	11.5	
27	9.5		27	10.0		27	10.5		27	11.0		27	11.5	

Developed by the UK Meat and Livestock Commission – Stotfold PDU

Notes on how to use the feed scale

(1) Assess piglet and sow condition on Day 10
(2) Select appropriate scale consistent with piglet number and rearing ability of sow (*eg* a highly productive sow with 10 piglets may require the feed scale for a sow with 11 piglets)
(3) Where deviations from the scale are appropriate (either up or down) record the amounts consumed in the 'Fed' column
(4) Cross off the days in the 'Day' column as lactation progresses – allowing relief stock persons to refer to and maintain correct feed intake levels
(5) Record alterations to piglet numbers and change to the appropriate scale
(6) Feed lactating animals at least twice per day
(7) Two diet feeding system is recommended; the lactating sow requiring higher energy and lysine levels than the pregnant sow
(8) Ensure an adequate water supply. Drinkers should flow at least 1.5 litres per minute
(9) Ensure correct room temperature. As sow feed intake increases, room temperature should reduce from 20° to 16°C. Maintain at 16°C for the last 10 days of lactation
(10) When day time temperatures are high, feed one third of the daily requirement am and two thirds pm.

Table 6b LACTATION FEED SCALE (Imperial)

First 10 days (All sows/gilts)			Sow Identification											
Day	lb	Fed	Total Fed:											
1	5.6		Date Farrowed (Day 1)											
2	6.6		NOTES: *This dietary scale is now widely used in Europe and is very.*											
3	7.7		*successful. Because sow genotypes are now fairly similar in N. America*											
4	8.8		*to those in Europe, this valuable guideline is transposed into the*											
5	9.9		*Imperial system.*											
6	11.0		*Liaise with a pig nutritionist to formulate a diet density which will satisfy*											
7	12.1		*the published daily intakes (cf Close & Cole 'Nutrition of Sows and*											
8	13.2		*Boars', NUP 2000). Total litter weight at weaning can be on some*											
9	14.3		*farms over 30% higher than for average herds.*											
10	15.4													

Gilt<10 piglets Sow<9 piglets			Gilt 10 piglets Sow 9 piglets			Gilt 11 piglets Sow 10 piglets			Gilt 12 piglets Sow 11 piglets			Gilt 13 piglets Sow 12 piglets		
Day	lb	Fed	Day	lb	Fed	Day	lb	Fed	Day	lb	Fed	Day	lb	Fed
11	15.4		11	16.5		11	16.5		11	16.5		11	16.5	
12	15.4		12	16.5		12	17.6		12	17.6		12	17.6	
13	16.5		13	17.6		13	18.7		13	18.7		13	18.7	
14	16.5		14	17.6		14	18.7		14	19.9		14	19.9	
15	17.6		15	18.7		15	19.9		15	21.0		15	21.0	
16	17.6		16	18.7		16	19.9		16	21.0		16	22.1	
17	18.7		17	19.9		17	21.0		17	22.1		17	23.2	
18	18.7		18	19.9		18	21.0		18	22.1		18	23.2	
19	19.9		19	21.0		19	22.1		19	23.2		19	24.8	
20	19.9		20	21.0		20	22.1		20	23.2		20	24.8	
21	21.0		21	22.1		21	23.2		21	24.8		21	25.4	
22	21.0		22	22.1		22	23.2		22	24.8		22	25.4	
23	21.0		23	22.1		23	23.2		23	24.8		23	25.4	
24	21.0		24	22.1		24	23.2		24	24.8		24	25.4	
25	21.0		25	22.1		25	23.2		25	24.8		25	25.4	
26	21.0		26	22.1		26	23.2		26	24.8		26	25.4	
27	21.0		27	22.1		27	23.2		27	24.8		27	25.4	

Developed by the UK Meat and Livestock Commission – Stotfold PDU

Notes on how to use the feed scale

(1)   Assess piglet and sow condition on Day 10
(2)   Select appropriate scale consistent with piglet number and rearing ability of sow (*eg* a highly productive sow with 10 piglets may require the feed scale for a sow with 11 piglets)
(3)   Where deviations from the scale are appropriate (either up or down) record the amounts consumed in the 'Fed' column
(4)   Cross off the days in the 'Day' column as lactation progresses – allowing relief stock persons to refer to and maintain correct feed intake levels
(5)   Record alterations to piglet numbers and change to the appropriate scale
(6)   Feed lactating animals at least twice per day
(7)   Two diet feeding system is recommended; the lactating sow requiring higher energy and lysine levels than the pregnant sow
(8)   Ensure an adequate water supply. Drinkers should flow at least 0.33 Imp. gallons per minute
(9)   Ensure correct room temperature. As sow feed intake increases, room temperature should reduce from 68° to 61°F. Maintain at 61°F for the last 10 days of lactation
(10) When day time temperatures are high, feed one third of the daily requirement am and two thirds pm.

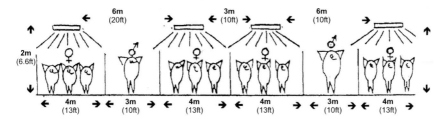

100 watt fluorescent strip lights, white
Aim for 16 watts per m (1.6 watts/ft)

Place the lights so that the majority of the light falls via the eyes

Lighting pattern

On maximum 16 hours/day
Off for 8 hours/day

Figure 2 Lighting a mating/breeding house

LIGHT PATTERNS

Most people – but not all academics – think that a 1/3 / 2/3rd pattern of off : on is best, say 8 hours of near darkness (10-12 lux) followed by a maximum of 16 hours of bright light (350 lux+), maybe 14 hours for gilts. I myself am quite convinced that this pattern is correct from following up too-dark breeding areas advice. A definition of lux (and also lumens, not the same) can be found in the Glossary.

LITTER SCATTER, AN UNDER-USED LITTERSIZE CHECK ON MANY FARMS

Litter scatter is a good indicator of infertility disease arising or already present, problems with ovulation/implantation or poor boar service. These are all pointers to litter size problems.

Litters with <8 : Target 10%. Action level 15%.

HERD AGE PROFILE – DON'T GET CAUGHT OUT

Because many herds, due to non-specific infertility and in hot weather, leg problems, are forced to cull prematurely (Table 7), the ideal herd age profile can quickly distort (*see Culling section*). This can affect litter size by up to 2 pigs a litter for a period and 1 pig/litter is very common.

Table 7   NUMBER OF YOUNG FEMALES CULLED FOR POOR REPRODUCTIVE/PERFORMANCE REASONS 1985 V. 1998 (%)

		1985	*1998*
Overall herd replacement rate		37%	45%
Reasons:–	Reproductive Failure	32%	48%
	Health & Losses	30%	27%
	Legs & Feet	18%	10%
	Other	20%	15%

Source : MLC Yearbooks (UK)

This can also allow virus disease to gain a hold because of reduced herd immunity from too few established 3 to 6 litter sows. This alone can reduce litter size substantially.

Practical advice : Keep an ongoing graphical account of your herd age profile. Have ample gilts available ***suitably acclimatized*** in a gilt pool; buying weaner gilts can assist here. Consult with a pig specialist veterinarian if your enforced culling is more than 33% due to non-specific infertility. Choose gilts with strong leg bone structure and good spring of pastern; this should not unduly affect a tendency towards coarse/ heavy bone structure in the finished progeny.

Remember – it is enforced early culling which can alter your herd age profile which eventually affects litter size.

SHORT LACTATION LENGTH

The pressures of economics favouring the reduction of lactation length are well-known. European producers, at the time these notes are written, have stabilised at 23/24 days (even though a move to 28 days has been proposed under further Welfare constraints), but the move down to 16 days seems to be preferred in the US, especially on larger and newer units. A fall-off of 0.5 pigs litter at least can be expected between 23 days and 16 days, but this could reduce as management and re-breeding skills improve among those permitted to wean early outside the Welfare limitations now in place in some countries.

REPEATS (RETURNS TO SERVICE)

Check repeats like this : –

Regular repeats (21 ± 3 days)   Target 10% Action level 15%

Irregular repeats (>24 days)   Target 3% Action level 6%

Attention to better pregnancy diagnosis and thus reducing repeats by 33% has improved litter size by 0.3 pigs/litter.

DISEASE

If disease is an influence on litter size it will strike mainly post-service. Check that gilts are vaccinated against parvovirus and test for PRRS. With your vet's help determine if causes are infectious (especially stillborns, together with mummies and size of mummifieds) or non-infectious causes (*e.g.* born alive but suffocated/weak).
The routine use of a pig specialist veterinarian is an important defence against low litter size.

BIOSECURITY

Follow a **proper** biosecurity protocol. Most used today are now outdated in technique and products used. (*See Biosecurity section.*)

Mycotoxins. Various mycotoxins at very low levels may have a bearing on low litter size.

- Always include a mould inhibitor in the feed or stored grain.

- Consider also (and it is especially advisable after wet warm harvests) adding a modern mycotoxin absorbent (not clays).

- Sanitise (*i.e.* steam clean and dry out) bulk bins regularly, probably twice a year.

- Agitate with bin manufacturers for 'bulkhead door' access ports at mid-height and swing-away food auger boots. It is invariably too onerous and dangerous to gain access to a bulk bin via the top inspection hatch.

- Feed hoppers can be a source of mycotoxins.

- Damp or mouldy bedding is a source in grouped sow yards.

- Fodder maize is also a potential source.

(This is discussed in detail in the Mycotoxin section, page 383)

GENETICS

All body functions implicated in litter size are under genetic control to a greater or lesser extent. However as a large number of genes are involved in the physiological processes which determine litter size, it is unlikely that any two females would be identical, thus considerable variation is likely to exist between groups of females.

Secondly, different females respond differently to stresses, further complicating the situation, especially regarding ovulation and implantation.

Various experts have reviewed the complexities of genetic improvement of litter size in pigs. The difficulties revealed show :

- The low heritability of the trait (<0.1)

- The low repeatability of the trait (<0.15)

- A very large sample is required to measure differences.

- The influence of heterosis.

- The influence of environment and management at all stages of the female's life.

The author fully accepts the scientific wisdom thus expressed, but has evidence of at least three cases of a dramatic improvement in litter size (range +0.92 to 1.86 b/a) where a batch of gilts from a different breed was tried – an average improvement of 21%. In all three cases the indigenous breed was a high lean hybrid, while the replacements were a cross based heavily on female traits (LW).

The general expert opinion is that the boar has little or no influence on litter size. *However*, individual boars (or AI) within a breed can have a very significant influence on litter size if semen concentration or quality is low, and some boars can produce lethal genes which can result in some embryo deaths or chromosome abnormalities rendering them infertile. In future these defects could be screened out before boars and/or AI semen is used.

A careful examination of boar/semen use is important in any investigation into low litter size.

A BOAR PERFORMANCE CHECKLIST

✓ No more than 15% of young boars under 9 months should be 'on duty'.

✓ Litter size varies between boars; keep a check on this.

✓ Check boar's success rate:–

For each boar, multiply the farrowing rate of his sows x the total born average litter size across 100 of his services or inseminations.

Target : 1000   ***Action / exploratory level*** : 800

e.g. Farrowing rate of his sows 85%   Across the last
Litter size of his sows & gilts 9.1   100 services

85 x 9.1 = 773.5 Action level.

✓ For herds using natural service, check on use of 'favourites'. This is quite a common, if understandable, weakness of stockpeople. Use a check-board sited in a prominent position. (Figure 3). If put in a passageway it can be viewed sideways (at an angle) and any over or irregular/sporadic use of a boar can be quickly noticed.

✓ ***Quality of service***. Hurried services can reduce litter size (Table 8).

Table 8 CONCEPTION RATE AND LITTER SIZE IN RELATION TO QUALITY SCORE AT FIRST SERVICE

Quality score	Duration of intromission (mins)	N°. of first services (%)	Conception rate (%)	Mean total borns
1	<1.5	7	86	7.67
2	1.5 – 3.0	28	75	10.11
3	>3	49	91.8	11.46
4	>3	16	75	11.50

Source : Bell R., *et al* (1994)

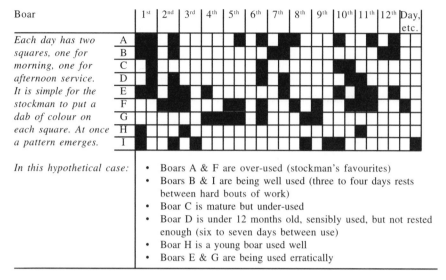

Boars A & F are over-used (stockman's favourites)
Boars B & I are being well used (three to four days rests between hard bouts of work)
Boar C is mature but under-used
Boar D is under 12 months old, sensibly used, but not rested enough (six to seven days between use)
Boar H is a young boar used well
Boars E & G are being used erratically

Figure 3 How graphics can make complicated things clear - to reveal pattern of boar use, look at an angle laterally. Thus siting the board on a passageway is useful to encourage this sideways view.

✓   ***Multiple Services*** : have an effect on litter size. Two being better than one, and three being better than two (Tilton & Cole 1982) in this case the two latter on consecutive days. However in the last 20 years breeders have paid more attention to correct heat detection and the placement of services with more even 12-hour spacing. In this case two ***correctly timed*** service inseminations will maximise litter size.

✓   Your heat detection timing.

✓   Your subsequent service timing.

✓   Your quality of service. Supervision & patience is essential.

REFERENCES

'Litter Size'. On Our Farm (report) Farmers Weekly March 2000.
Tilton & Cole (1982) Effect of triple v double mating on sow productivity Anim. Prod 34 279-282

BIRTHWEIGHTS 2

Birthweights are – on the working farm – defined as the weight of those born alive. The definition of a 'born-alive' can be taken as a neonate which took at least one breath after being born. The 'bucket-test' to help in determining those born alive from those born dead which never drew breath is given at the end of this chapter.

For their own purposes research workers may need to record both born-alives and born-deads, but in this section we deal with the birthweights of born-alives only.

TARGETS

These are changing year on year, but 10% under 1.2 kg (2.64 lb) and 50% over 1.45 kg (3.2 lb), leaving 40% between 1.2 and 1.45 kg gives a good start to the growth performance of the finished pigs.

BIRTHWEIGHTS ARE IMPORTANT

• You should know your birthweights even though it adds a further task to a busy time. You cannot estimate birthweights by eye. 'Baby-scales' are cheap and effective. My clients tell me that the extra time taken is under 15 minutes per sow per year (+1.25% extra labour). This is recouped if 33% of the slaughter pigs get shipped 2 days sooner.

• Does this happen? It seems so as Table 1 shows that careful attention to birthweights improved days to slaughter by 2.7 days for the whole herd, paying back about four times the extra trouble taken.

Table 1   RESULTS AFTER BIRTH WEIGHT AUDITS, ACTION BEING TAKEN ON SEVERAL OF THE AREAS DESCRIBED BELOW

	Before			*After (+ 14 months)*		
	Under			*Under*		
	1.0 kg	*1 - 1.3 kg*	*1.3 kg +*	*1.0 kg*	*1 - 1.3 kg*	*1.3 kg +*
	(2.2 lb)	*(2.2-2.86 lb)*	*(2.86 lb+)*	*(2.2 lb)*	*(2.2-2.86 lb)*	*(2.86 lb+)*
	13%	45%	42%	9%	28%	63%
Av. days to slaughter (birth - 88kg)	156	151	142	157	151	141

Av. days saved to slaughter 2.7, whole herd.

Source: Clients records

- Evenness of birthweights is more important than averages. Too many piglets at, say, 1.1 kg (2.42 lb) or below can markedly affect average days to slaughter, especially if the litter number is high (12.5+). First, my records suggest that a 1.0 kg (2.2 lb) birthweight takes a week longer to get to slaughter than one at 1.45 kg (3.19 lb) and the pigs were up to another 2 kg (4.4 lb) lighter at 28 days.

- Next, 25% of my clients' birthweights are between 1.1 (2.42 lb) to 1.3 kg (2.86 lb) and these small (but not over-small) piglets took 2.7 days longer to reach slaughter than the 30% between 1.5 (3.3 lb) and 1.65 kg (3.6 lb).

Of very small neonates (800 to 900 g) (1.75-2 lb), 62% died, with many being kicked to judge from their injuries. Farms with over 5% under 900 g (2 lb) neonates had a 2% higher pre-weaning mortality than those with only 2%. Target for as few as 10% under 1.20 kg (2.64 lb), and as many as 50% over 1.45 kg (3.19 lb). That's a good and achievable birthweight scenario, I find, profitwise.

THE PROBLEM OF 'AVERAGE BIRTHWEIGHT'

Average birthweight *conceals much larger differences in the numbers of smaller and larger piglets* born over a production/costing period.

Is this important?   Seems so. Table 2 gives some figures published by a well-managed European commercial/demonstration farm, and another very similar unit, nearby.

Table 2   ACTUAL BIRTHWEIGHTS / MORTALITY RATIOS FROM TWO FARMS

Birthweight category (kg)	(lb)	Distribution of born-alive birthweights (%)		Pre-weaning mortality (%)	
		Farm A	Farm B	Farm A	Farm B
<0.5	<1.1	0.5	1.8	80	78.2
0.5 – 0.74	1.1-1.63	2.2	1.4	62.4	63.1
0.75 – 0.99	1.64-2.1	6.2	11.8	24.7	25.2
1.00 – 1.24	2.2-2.7	16.5	20.9	13.4	13.0
1.25 – 1.49	2.75-3.2	24.1	29.1	6.6	6.2
1.50 – 1.74	3.3-3.8	27.9	24.3	3.7	3.5
1.75 – 1.99	3.9-4.3	15.1	6.4	2.5	2.6
>2.00	>4.4	6.9	3.8	1.7	1.7
Average pigs born alive		*11.7*	*11.1*		
Average birthweights (kg)		*1.48*	*1.37*		
(lb)		*3.26*	*3.01*		

Comment :Despite a seemingly small difference in average birthweights (under 8%) Farm B had 0.6 fewer pigs born alive; more than three times the piglets born alive under 0.5 kg, which involves hard work to keep them alive; nearly twice the percentage of pigs under 1 kg (2.2 lb) (the ones that got crushed/chilled/scour); and half the percentage over 1.75 kg (3.9 lb) (the ones which get to weaning 2 kg (4.4 lb) heavier or 5 days quicker)

I've been around looking at new born litters for some 40 years now, and I still find it difficult to single out litters which recorded *average* (and good) birthweights, say, of 1.4 (3.1) and 1.5 kg (3.3 lb) – as in Table 1 for example. The temptation is to say "Big enough – everything's OK".

Now this difficulty of mine may well be due to my own incompetence, sure, but I don't think so! Knowing his *spread* of birthweights alerts the stockperson to a problem – either there already or developing. So, back to "Is this important?"

Table 1 was from a breeder/feeder client of mine with not especially good birthweights (which was why I was called in) and shows the advantage achieved from improving the birthweights all up the scale – overall nearly 3 days faster to slaughter at 88 kg (194 lb).

TWO SUGGESTIONS WORTH EXPLORING

1.   I wonder if the global problem of pre-weaning mortality, still stubbornly over 10% of born-alives when it is consistently under 6% on some farms, is linked to the fact that stockpeople regard weighing pigs at birth to be a pain. Well, it *is*, but I find fully half of those farms which get well down into single figures now make time to do this. They know their lower birthweight litters and do something to improve them, or tighten up on culling.

2.   And as mentioned in the 'Mixing Pigs' section, some expert breeder-feeders are taking birthweights into consideration as well as evenness for size when batching and matching weaners. Research is needed on this suggestion – that pigs of similar *birthweights* tend to do better as a group to slaughter – the spread of close-out time is shortened by 3 to 4 days per group. An interesting hypothesis and important too, as batch production depends on minimal close-out variation to achieve one major advantage of the concept.

A BIRTHWEIGHT CHECKLIST

Start early

Success has come from convincing the stockperson that what is done *very early in the cycle* can influence birthweights on the commercial farm.

✓   Good implantation at 12-24 days is vital

✓   Provide rest and quiet at this time

✓   Freedom from stress and sexual excitement is beneficial

✓   Try not to mix sows at this time

✓   A relatively high nutrient intake between weaning and service seems to be beneficial. For sows with a birthweight problem, try feeding 1.5 kg (3.3 lb)/day more than you currently allow them. This may help follicle release synchrony – more follicles are released closer together in time – and/or it enables the womb surface (endometrium) to regain receptivity sooner. This is covered in the Litter Size section.

✓   Don't feed too good a diet in pregnancy. See your nutritionist for a low lysine (0.55% total) pregnancy food – a bit lower than currently favoured. However, just increase feed allowance

(1.8 to maybe 3.0 kg, approx 4-6.6 lb) of this diet as pregnancy progresses rather than keep flat as the text-books advise for sows in reasonable condition.

✓ Never let a sow nose-dive in lactation. Here we are looking at the effect on the next litter of what was done in the *preceding* breeding cycle.

✓ Don't deliberately feed more and suddenly just before farrowing *if birthweights are low*. This could be acceptable for other reasons.

✓ Don't worry too much about big litters affecting birthweights - you can have big litters and big newborns, up to about 14 total borns anyway.

✓ Several multisite farmers in the US, after changing from farrow-to-finish monosite, do not report lower birthweights next litter when weaning at 16 days old (but their litter size seems to be one pig less than in developed pig industries in Europe).

✓ Don't use prostaglandins too early, as foetuses are growing at 90 g (3.2 oz)/day just before farrowing.

✓ Latest research could be indicating that transferring to organic trace elements rather than inorganic (from rocks/soil) will be the future of trace element nutrition in the breeding animal.

✓ Genetics are not likely to be involved. Maybe only if the super hyperprolific strains are much in evidence, which outside specialised breeders in the Far East is not likely.

SMALL PIGLETS PROFIT MOST

Having done many cost-calculation exercises for farmers in my lifetime, I find time and time again that the extra cost of raising the smallest surviving pig in the litter tends to cancel out most or all of the profit of the biggest pig in the litter. Surprisingly, while the average to slightly smaller pigs in the litter don't convert food to slaughter much worse, or at all, compared to the larger piglets which get there faster, I find the very small piglets definitely do convert worse by 0.3 or more. One of the reasons why, in the broiler-like batch-production conditions of SEW/multisite in the USA, these 'very-smalls' may be sacrificed at birth. They are just not "profit potential"; at best only break-even. And why I said earlier that with birthweights, averages can be misleading and we need to study individual weight sectors to see which remedies are best.

PERSUASION DIFFICULT!

The problem has been to persuade farmers to go the extra mile and start weighing born-alives as routine.

If you take a modest average birthweight of 1.25 kg (2.75 lb), which is a typical figure presented to me by problem farms, then an average birthweight 0.25 kg (9 oz) lower is likely to reduce saleable meat/tonne of feed (MTF) at 100 kg (220 lb) slaughter by 31 kg (68 lb), and 0.25 kg (9 oz) above it will increase MTF by almost 36 kg (80 lb). This is from speedier, more efficient growth from those that live, not from lower mortality. In this latter respect, however (the value of reduced mortality) turn back to Table 2. In birthweight terms Farm A had 0.6 more weaners/litter from their lower number of 'smalls' and larger number of 'heavies'. Thus Farm A sold about 5.4% more finished pigs out of the yearly sow and boar's food share of, shall we say 1.4 tonnes. Let's also assume the farm sells, out of each sow, 22 x 100 kg (220 lb) live pigs year, or at 75% KO, 1650 kg (3638 lb) saleable meat. 5.4% more saleable meat is another 89 kg; and spread over 1.4 tonnes this is another 64 kg (141 lb) of meat sold off each tonne of breeding food sow per year. This is added to the benefits of faster growth already quoted. So what's one kg of saleable meat worth to you? Each extra kg of monetary income therefore reduces the sow food cost/tonne by that same figure. Work it out. Quite an eye-opener isn't it?

A BONUS WORKS WONDERS !

This is why these clients have started recording birthweights despite all the extra hassle involved. Sure, the farrowing house stockpeople still have the extra work to do, but explained this way, some owners have agreed to a 50% bonus based on an improved MTF over the current achievement. The bonus is fixed for 3 years and then reviewed again, up or down.

The breeding section heads use my checklist to explore the suggested avenues of improvement. Even so, some hard thinking had to be done by the manager to ensure the time was made available *and that other jobs around farrowing didn't suffer*. Three of my clients work this system. The bonus has meant that their stockpeople's take-home wage has risen by about 10%, so wage costs accordingly have increased

1.4% but as far as we can tell, increased productivity probably linked to the concept has improved by 30 kg (66 lb) MTF/tonne fed – although this was about half what was expected. ***Nevertheless this is enough to pay for the bonus three times over***.

THE BUCKET-TEST FOR TRUE BORN-DEAD

This helps distinguish between a true born-dead and a neonate which was born alive, but died soon afterwards. When trying to establish the possible causes of low litter size, the factors which led to pre-farrowing and post-farrowing deaths are different, and need different remedies.

Fill a bucket with water and lower the eviscerated lungs into it gently. A true born-dead's lungs will sink relatively rapidly, while those from a born-alive will sink more slowly, if at all.

This is because even if the neonate took just one breath some of it will remain in the lungs, while a true born-dead never drew breath at all.

Once you have observed a few of both, you will easily detect the difference.

THE POST-WEANING CHECK TO GROWTH 3

The slow-up in growth rate seen immediately after weaning. Correctly defined as the period in days the newly-weaned pig takes to recover the degree of daily gain achieved in the last 24 hours on the sow.

THE EXTENT OF THE PROBLEM

The post weaning check can be a matter of hours only to as much as 18 to 24 days, with 7 to 9 days being commonplace on typical farms and double that on the poorer-run units. We shall see what effect this has on profitability later in the chapter.

In keeping with high pre-weaning mortality, poor growth rates to slaughter in relation to what can be achieved with today's good genetic material, and high 'empty' or non-productive days (NPDs), the post-weaning check is a fourth area where a disappointing lack of progress at farm level has been seen across the last 30 years.

Primarily – maybe exclusively – a nutritional problem

In the author's opinion this is mostly due to stockpeople and owners failing to appreciate that the problem has been – and to a disappointing extent still is – due to *incorrect nutrition* over the crucial transitional period from the time when the piglet is on the sow to its acceptance of solid food.

Problem 1 – failure to invest sufficiently in the design of the post-weaning food

Much progress has been made by nutritionists on the design of diets to make the dietary transition as easy as possible for the piglet's digestive system. The trouble is that these specialised foods can be expensive – around three times more than what farmers have been used to or are offered for sale from manufacturers over-keen to get their business. There is no escaping the fact that to lower the immediate cost of post-weaning diets compromises the specialised diet design needed to avoid indigestion. The growth check problem then emerges, especially when the piglets are weaned 'early' *i.e.* 16-21 days. It is a tough discipline the seller of baby pig food has to adhere to.

Problem 2 – dirty feed receptacles

The second contributory factor is the farmer's failure to provide these expensive, high-quality transitional feeds in a clean enough manner. Troughs get fouled too quickly and too often. The contaminants further compromise the animal's delicately balanced digestive system at a challenging time digestively and immunologically, when both these systems (along with its thermodynamic system) are underdeveloped and need all the help science and stockmanship can give them.

The following pages are an attempt to persuade those who are responsible for raising weaners to . . .

* Pay sufficient for the right food and

* Feed it properly.

WHAT DOES THE POST-WEANING CHECK TO GROWTH COST ?

My survey of a substantial sample of British breeders revealed that 96% thought that some form of post-weaning interruption to growth was bound to occur and that, on their own farm . . .

88% thought it lasted 2 days or more

52% thought it lasted 4 days or more

36% thought it lasted 7 days or more

22% thought it lasted 10 days or more

What is quite serious is that, even in a sophisticated pig industry like ours in the U.K. more than one-fifth of us questioned were resigned to a 10 day check at weaning as being 'normal'.

In fact the best of us limit it to 2-3 days.

At Dean's Grove farm, even 20 years ago, we measured growth rate and often we estimated we had got it down to a 2 day check. We had a few pigs growing at 923g/day (2.03 lb) at 25 kg (55 lb) – which is close to the magic 1000g (1 kg or 2.2 lb)/day by 25 kg which the nutritionist/geneticist says is possible and is a figure which many producers disbelieve. I can remember lifting (with difficulty!) one or two 9½ week (66 day) monsters out of the nursery weighing 33.5 kg (74 lb). These must have achieved a daily gain of 485g/day (1.06 lb) from birth.

This is almost double what many people obtain.

And thereby hangs a tale. .

A minimal post-weaning check has a maximal effect on growth rate to slaughter, and thus the amount of saleable lean meat (MTF) produced for each tonne of grow-out food purchased.

This is confirmed by some figures from the UK (Table 1) where a reduction of 9 days in the check (12 days down to 3 days) saved the producer just under £20 per tonne of all food fed to slaughter, or 14%.

Table 1   LOSING TYPICAL GROWTH IMPULSION AT WEANING RAISES THE COST OF ALL GROWING FOOD BY OVER 10%

Liveweight	Length of check	Days to 94kg (207 lb)	Daily gain	First graders	Carcase lean	Lean per tonne of food
5.8 kg (12.78 lb)	12 days	156	567g	72%	52.3%	166 kg (366 lb)
5.8 kg	3 days	142	621g	86%	53.1%	182 kg (401 lb)

16 kg (35.28 lb) more LEAN for every tonne food

Worth, at 102p/kg* retail meat = **£16.32**

And 14 days fewer overheads at 20p/day = **£2.80**

Total £19.12

* UK / Dec 2001   *Source : Based on A1 Feeds data*

Expressed another way, 16 kg (35.28 lb) more meat/tonne feed (M.T.F.) at 5 pigs to the tonne of feed is worth £3.57/pig or 5% more income.

Remember, the average weaned pig checks for 7-10 days, thus for every day the pig slows down over 3 days adds another £1.88 to the cost of a tonne of growing/finishing feed he will eat to slaughter. On a world pig feed price this is about 1.5% feed price increase for each day's slowdown in post-weaning growth.

RESEARCHERS RARELY GO FAR ENOUGH IN BABY PIG EXPERIMENTS

I have read many worthy research trials which indicate statistically significant differences in performance *to weaning or to the end of the nursery period* – and then stop there!

Some even have gone on to give the financial benefit of the product or technique used (as distinct from the performance improvement) for that portion of the growth curve. However, smart farmers say "Yes, but the treatment barely paid back, despite the improved physical performance demonstrated ... *at the conclusion of the nursery trial!*

Had the trial been continued to slaughter, however, the cost-effectiveness picture could have changed markedly for the better even if the eventual physical performance improvement percentage could have slipped a bit by then.

I believe no young pig trial is satisfactorily concluded until the econometrics have been assessed on both groups of pigs raised under similar conditions to slaughter.

One may not be able to be as statistically certain because of subsequent variables, but any negative or 'not-worth-it' *economic* conclusions early in the growth curve could be altered by slaughter weight especially if the treatment tested had a positive effect on the post-weaning check. I understand the reasons (extra money, lack of facilities and staff needed) which hamper research departments carrying on young pig trials to slaughter in this way, but it is a weak point in their current approach which at least needs consideration at the trial design stage. After all it is the improved profit which comes from better physical performance which matters. Generally we don't sell weaner pigs, but we do (or someone does) sell the finished animal – and the finished pig incurs a lot more food cost/day than does the weaner.

WHY IT PAYS TO SPEND, SPEND, SPEND ON THE LINK FEED

The interim specially-designed pre-starter feed has been called the Link Feed – a better description than 'pre-starter' which could also refer to a creep feed. Many breeders refuse to pay the £600-£800 asked for even a moderately well-designed Link Feed (I have seen costs of £1,000/ tonne for a really top class feed) when they are paying £300/tonne "without too many problems".

In world pricing terms that is two to three times more.

"No feed can be worth *that*," they exclaim in disbelief.

Let's look at the situation coolly and dispassionately in terms of payback.

First, that statement "without too many problems". The problems referred to are digestive ones, such as scouring stall-out or inappetance. But the real problem is the underachievement at slaughter which goes unrecognised, like this:-

The Americans are adding data to those I've quoted in Table 1. The University of Minnesota has quoted a financial loss of 10 cents per lb gain from 11.5 lbs to 50 lbs (5.2 – 22.7 kg). Note that this loss had already occurred by 23 kg (51 lb). Table 2 cites the University of Georgia where 8 days were lost to slaughter on a conventional post-weaner diets. And while the food cost in the 7 days post-weaning was double on the more expensive diet, this was recouped three-fold by slaughter weight.

Table 2 DO POST-WEANER LINK FEEDS PAY ? EXTRAPOLATED FROM AMERICAN DATA

	Conventional post weaner diet	High digestible diet
Weaner growth rate/day 0-7 days post weaning	100 g (3.52 oz)	200 g (7.04 oz)
Days to 105 kg (232 lb)	171	163
Relative cost of food eaten in 7 days	100	199
Relative savings in costs to slaughter (food & overheads)	-	513
Relative value of highly digestible food	(513-199 = 314)	314

While the post-weaning food cost twice as much, the net income at slaughter was a third more

1. Link feeds pay best in the first 5 to 10 days after weaning

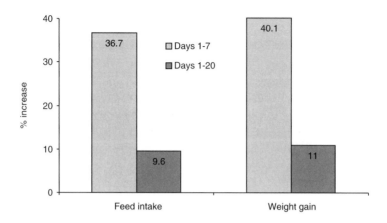

2. A little link feed after weaning shortens time to slaughter

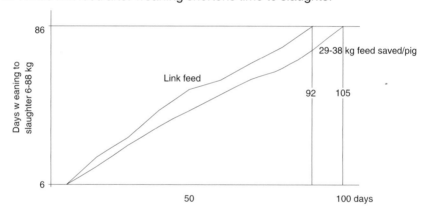

Figure 1 Improvement of higher cost link feed over standard starter feed (British Data). Source: SCA (1994)

Table 3 shows some data collected when I was working in Iowa and Minnesota just as the Link Feed concept was being tried out – hesitantly due to the ×3 increase in the cost per tonne asked for the whole feed! The table shows what each farm could have afforded to pay for the Link feed before the extra investment was eroded, based solely on food saved, not on reduced "overheads" as well. Including overheads would have given a 2 to 2.5:1 payback on extra feed cost. Incidentally, the standard of nursery feeder hygiene was not good on farms 2&3, reinforcing my point about trough cleanliness.

Table 3 BREAKEVEN COSTS AND PAYBACKS FROM FEEDING LINK FEEDS ON 4 US FARMS

	Farm 1	*Farm 2*	*Farm 3*	*Farm 4*
Days used post-weaning	5	10	7	12
Breakeven cost per ton ($)	2,631	1,840	2,135	3,100
Price actually paid per ton ($)	1200	1200	1500	1350
Pay-off ratio	2.2:1	1.5:1	1.4:1	2.3:1

Clients' Records : 1999-2000

A POORLY UNDERSTOOD EQUATION

Some farmers do not seem to realise how little the post-weaner pig eats in the 7 days post-weaning when the Link feed does so much good.

- A typical pig weaned at 21 days eats about 3.25 kg to 7.5 kg (7.2 - 16.5 lb) food in the 5 to 14 day post-weaning period. This means that one tonne of Link Feed will feed about 300 down to 133 weaners. Let's say 250 to 100 weaners to be on the 'safe' or pessimistic side. From this, take an average of 175 weaners.

- But because of the post-weaning check ('stall-out' in the U.S.A.) on low grade (conventional) post-weaning feeds, each of these weaners is costing 5 to 8 days in extra food by slaughter – 10.5 to 17 kg of food (23 - 38 lb). At 14p/kg this is £1.47 to £2.38/pig. Let's take £2/pig as a fair mean, on food costs alone – no saved overheads. A conventional post-weaning diet will cost from £250 - £350/tonne.

- So if a special high-cost Link feed avoids this waste of finisher food you can afford to pay 175 pigs x 2 = £350/tonne *more* before, even at the most pessimistic, customer-friendly scenario, the extra cost is eroded. About £400/t with saved overheads incorporated.

- Thus you can afford to pay more than double the cost of a conventional starter/grower feed – even using pessimistically-weighted figures – before any advantages are eroded.

DO THE SUMS

All I ask is for you to do your own sums based on this approach. You can always ignore my assumptions and substitute your own. Also, you should convince yourself to do a nursery trial with an expensive Link feed against your current choice and use the benefits as your base performance matrix. My experience is that only very rarely will a good Link feed fail to succeed econometrically at slaughter if your nursery stockmanship and housing is good.

The interesting exercise is how far up the price scale asked/tonne do you need to go? It does not seem to be a question of 'the cleaner and better the nursery environment the lower the price is required'. Rather the converse; very good nurseries seem to do much better the better the design of the post-weaning feed they use. This suggests that the geneticists are right in constantly telling us that at the sharp end we are nowhere near exploiting the genetic potential – in this case growth rate – already locked into the genetics we have purchased.

And it reinforces my plea that post-weaning research is never finished until the pig is shipped at slaughter.

"It's never over until the fat lady sings!"

WHAT HAPPENS IN THE GUT AT WEANING?

I hope I've convinced you in economic terms to pay more for a well-designed Link feed.

But what do I mean by 'well-designed'? First, we must understand what happens digestively when we wean a piglet. Nature never intended for it be removed suddenly from its dam – it evolved a gradual process taking at least 16 weeks, and more commonly 20 weeks, which allowed the gut to become accustomed to digesting solid food little by little. The bacterial and chemical pathways had time to adjust and change from milk to dealing with plant roots, acorns, mast, grass and weeds, apples and the soil ingested with them. And there was always time for a quick milk suckle to help level out any inconsistencies even late in the process.

Apart from feral pigs, and to a much lesser extent modern outdoor pigs, all that has gone. By weaning abruptly from 17 to 35 days we put an impossible strain on the piglet's digestion. If we don't help it counteract the suddenness of the changes, then the post-weaning check inevitably occurs. This is what happens . . .

Look at Figure 2. This is a simplified, diagrammatic representation of a very complicated – but extraordinarily elegant – chain of events best understood like this:–

- In the 3-week-old weaner the stomach is both a reservoir and a pre-digestive mixing tank holding about 0.2 litres – say a wineglassful.

- Milk from the sow arrives every 35 to 45 minutes or so in carefully measured amounts. The sow does this in response to the suckling stimulus by releasing milk from the udder cisterns, and switching it off again after around 17 to 30 seconds. However long and vigorously the piglet suckles it only gets its hourly 'ration' of about 150-200 cc.

- The stomach of the unweaned piglet is not very elastic and can only hold a certain volume of contents – as we've seen, about 200 cc or 0.2 litre.

- Cells in the stomach walls liberate both digestive enzymes and hydrochloric acid to start pre-digesting proteins (proteases, etc) and carbohydrates (amylases, etc.) especially. The acid helps disable pathogenic bacteria which are involuntarily eaten along with the food. The contents – sow's milk in the unweaned piglet – already contain nutrients in the right form for this to happen easily and within the 35 minutes or so needed for these pre-digestive (enzymes) and sanitation (acids) processes to take place.

- After this time the stomach contents are then passed into a short pipe-like channel, the duodenum, where fats are pre-digested. It also holds about 0.2 litre.

- At the third gut movement (each one of these instigated by the call to suckle) the duodenal contents, now largely ready prepared for absorption, enter the small intestine. They are also relatively free by now of potentially damaging organisms the piglet may have eaten as it scampers about and investigates life around it.

- The food (sow's milk) is now properly predigested and made safe for absorption in the small intestine.

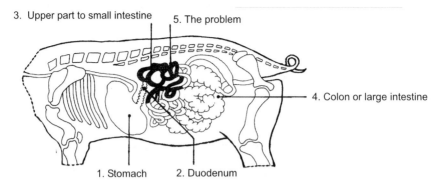

Figure 2 Digestive system of a 5.5-6.5 kg (12-14 lb), 3 week old weaner.

Understanding what will happen digestively when the piglet is weaned helps a great deal in solving food, food intake and growth check problems. This is what happens.

1. **Stomach** Holds 0.2 litre, does not expand. Food needs to remain here for nearly 45 minutes so as to be infused with acid to knock out harmful organisms and be washed with enzymes which get starches and proteins ready for digestion later on. Engorgement does not give sufficient time for either to happen. The stomach then refills the duodenum and is itself replenished by fresh sow's milk.
2. **Duodenum** 2" thick 9" long, also holds a little under 0.2 litre. Cells in wall wash stomach contents, once they arrive, with fatsplitting enzymes so that fats in the food are made ready for digestion and absorption in the next part of the digestive tract. Too much food from the stomach, too soon, causes the digesta to be pushed on only partially prepared.
3. **Upper part of small intestine** 5 to 7 yards long, convoluted with a huge surface area equivalent to half a football pitch, due to thousands of millions of tiny villi, or little microscopic fingers which absorb food. Billions of surface cells absorb the predigested nutrients. Insufficient predigestion – no absorption. This part of the tract cannot cope with poorly processed food.
4. **Colon or large intestine** 2 yards long, 1" thick. Water and fibre absorption. Not involved to much extent in the post-weaning check to growth.
5. **The problem** Piglet is weaned. Stops regular ingestion once every 45 minutes. Gets hungry… then overeats (engorges). Food not long enough in stomach or duodenum, so is insufficiently "sanitised" or pre-digested in the small intestine. Blockage occurs, bacteria breed causing villi to truncate. Water is liberated and piglet scours / dehydrates
6. **The solution is to:**
    - To accept that piglets *will* engorge after weaning;
    - And *will* therefore overload the small intestine with food;
    - So provide a post-weaning prestarter food which is so "pre-digested" that overloading will not harm absorption.
    - Only restrict feed such a diet for 12-36 hours, and then only marginally

- Then blend into a normal grower food once they have been on the starter for 5-7 days.
- Have electrolye solution available and plenty of clean fresh water.
- Alternatively wean later and accustom the piglet to sufficient daily creep feed.

Crude, indigestible fibre *is* indigestible so sow's milk is virtually fibre-free.

WHAT OFTEN HAPPENS WHEN WE REMOVE THE SOW AT WEANING ?

- The piglet gets a little hungry as is quite normal after 45 minutes to an hour, and looks for a feed. But mother is nowhere to be seen.

- After an hour or two the stomach is empty, the duodenum is empty and even the fore-end of the small intestine has moved its contents further down to other adsorption sites and additional processing by beneficial bacteria further on down the gut.

- "Yet some idiot has put down this solid but quite pleasant-smelling creep pellet/meal," thinks the piglet. "But it is not wet, it is not warm, it is gritty, it doesn't taste or feel like milk and I suspect it contains more of that fibre stuff than is good for me. I'll pass it by in the hope that mother will appear soon."

- By now three or four hours have passed. The piglet is ravenous. Moreover some of its bolder or hungrier penmates are beginning to eat the solid food provided. "Perhaps I could try a bit too," it thinks. It does, and while the solid food is a poor substitute for the real thing, it eventually overeats to remove its hunger pangs. This is called 'engorging'.

- But its stomach is inflexible. It cannot handle the volume of solid food which the piglet throws down, and there are only two ways the ingesta can go – either back up again and the piglet is sick, or through the more natural route into the duodenum and on further to the intestine – which is calling for replenishment and thereby activating the hunger response.

- So the ingested solid food does not remain sufficiently long in the stomach or duodenum for proteins, carbohydrates and fats to be prepared for absorption in the small intestine. Neither has it been sufficiently washed by acid to eliminate the hostile bacteria which nature has made susceptible to a natural, high acid level.

- The food arrives too soon in the small intestine with the wrong chemical signatures for absorption, and also loaded with damaging bacteria.

WHAT HAPPENS THEN ?

- The ingesta forms a traffic jam – a blockage – in the forefront of the small intestine. It cannot be sufficiently absorbed, so it stays there. It is a serious form of indigestion. But it is now an ideal breeding ground for the bacteria which have free-loaded in with it. These quickly proliferate and their toxins aggravate the delicate absorptive structures – the villi – which are covered in cells which recognise and absorb *properly digested* nutrients but refuse to accept those not predigested sufficiently.

- The bacteria cause the villi to reduce their length defensively (called truncation) so the huge absorptive area (about half the size of a football pitch in each piglet!) can be reduced to no more than the goalmouth area. Nutrient processing is drastically reduced.

- This villous reduction process stimulates cells (crypt cells) at the base of the villi to exude water. This liquifies the ingesta and stimulates bowel movements to flush the blockage down the gut to be voided. This is what scouring (diarrhoea) is – it is a lavatory-flushing operation to help cleanse the gut of potentially lethal material.

- This is why post-weaner pigs are prone to scour. It is a defensive mechanism - all that they have to try to put things right.

INSUFFICIENT WATER

- Trouble is the 6-kg piglet only has a limited amount of water in its bloodstream and body cells to 'flush the lavatory'. When this runs out, the blood thickens unless the water can be quickly replaced. Blood both conveys nutrient energy to the muscles (arterial blood sugars) and then removes toxins via the venous system to the organs (liver, kidneys) which can then process them for excretion as urine and also in the faeces.

  With thickened blood the piglet is starved of muscle energy (gets sluggish) and also gets cold (shivering). It already starts to poison itself with the

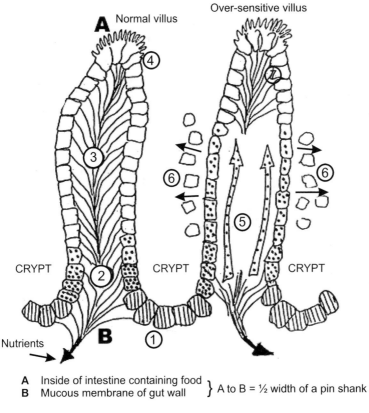

A Inside of intestine containing food
B Mucous membrane of gut wall } A to B = ½ width of a pin shank

1. Cells which secrete enzymes and in the scour of scour - water.

2. Replacement (germinative) cells which are non-absorptive at this stage. Within 3-5 days these become...

3. Absorptive cells which absorb amino-acids, sugars, water and minerals from food. (An ideal ratio is 1 germinative to 5 absorptive).

4. Microvilli on villus tip further increase absorptive surface area.

5. Germinative cells are speeded up if certain foods are given too early.

6. Healthy absorptive cells are pushed off the villus surface.

7. Absorptive area is drastically reduced - piglet cannot digest sufficient food - bacteria multiply, invade and scour occurs.

Figure 3 Microsection of villi

accumulating toxins (feels very ill). Thus an immediate source of specially treated water helps the piglet avoid these traumas. Treated in the form of added electrolytes which are simple minerals allowing the crypt cells to **insorb** water at the same time as **exsorbing** it. In other words it can now take on water while continuing to expel it by the scouring process.

• This is why as soon as looseness is noted, an electrolyte solution should be provided either as an additive to the normal water supply (in the early stages) or as a replacement for it when scouring is acute. The important deciding factor is never to affect the piglet's ability to drink clean water. Allowing an electrolyte container to run dry is disastrous and accustoming them to a new source of water needs to be borne in mind. So some farms, especially in hot, dry climates, provide an electrolytic solution as routine after weaning.

Here is an electrolyte formula recommended many years ago by a specialist pig veterinarian. (Table 4) However several commercial products exist and are less trouble to make-up.

Table 4   HOMEMADE ELECTROLYTE SOLUTION

To 2 litres (3½ pints) of water, add …	
Pure Dextrose BP	45g
Sodium Chloride (salt)	8.5g
Citric Acid	0.5g
Glycine BP	6.0g
Potassium Citrate	120 mg
Potassium Dihydrogen Phosphate	400 mg
Scouring pigs – full strength for 2 days for all pigs in room	
Post Weaning Depression – half strength for 10 days	

HELPING THE WEANER THROUGH THIS DIGESTIVE IMPASSE

Now we know what happens in the gut of the newly-weaned pig, we can do something about it.

1. We can pre-digest the solid food to such an extent that it doesn't need to remain in either stomach or duodenum for the necessary 35-40 minutes for normal processing. Some ingredient raw materials are already largely preconditioned/pre-digested and are essential at this time.

2. We can additionally add the essential enzymes (preferably from natural sources) needed to pre-digest proteins, carbohydrates and fats.

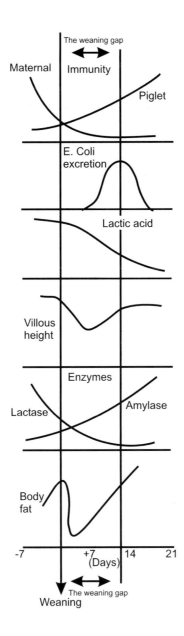

Figure 3 Why 7 days after weaning gives so much trouble. Extrapolated from Animal Talk, Cole and Close (2001)

3.   We can add minerals and trace elements in a far more absorbable form.

4.   We can condition and control, often by heat treatment, what fibre must be present so as to make it more easily digested.

5.   We can add extra acid to the food (or drinking water – moderately acidified water is surprisingly palatable) to help pre-sanitize the gut contents as they arrive in the stomach.

6.   We can make the food exceptionally palatable (texture, flavour and smell) to dissuade holding off eating solid food and lessen the resulting overeating (engorgement).

7.   Judicious creep feeding can accustom the piglet to solid food while on the sow. This will be of vital importance if, on welfare grounds, weaning is not allowed before 28 days.

8.   We can, if we wish to adopt the system, pre-ferment the diet or individual cereals/soya by adopting FLF (Fermented Liquid Feeding).

The upshot is that all these things – the raw material feeds, the additives, the processing and the precautions taken in manufacture and storage shelf-life *are much more expensive than with conventional foods*. By suggesting you should use a link feed at three or even four times your current cost does not mean the manufacturer is ripping you off. On the contrary, these are usually genuine on-costs and any reputable firm will be glad to explain things to you and answer your questions however doubtful you may feel.

FEEDING METHOD MUST BE CORRECT

Generally speaking the younger and lighter are the piglets you wean, the better Link feed they need. This does not always follow, as the *way you feed them* is of great importance to success. This is best described by a check-list.

THREE CHECKLISTS FOR GOOD POST-WEANING FEEDING PRACTICE

A.   Checklist – The food itself :

✓   What degree of post-weaning check do you suffer? See Target Growth Rate section (page 229).

✓   Have you considered a specially-designed Link feed? The degree of sophistication – and thus the cost – will largely depend on the magnitude of check your pigs suffer.

✓   Have you discussed a suitable Link feed with a nutritionist experienced in the design of these diets? Also, some pig specialist veterinarians and other consultants are a useful source of advice on how your housing and management will influence the quality of diet you need.

✓   Do not be overconcerned about the cost per tonne – this will repay itself in improved performance and thus dietary cost savings, *by slaughter*. Find a good Link feed and stick to it.

✓   If you sell end-of-nursery weaners, make sure the buyer appreciates the trouble taken and extra expense you have incurred in helping your pigs to reach his slaughterweight quicker. You are entitled to a premium for this to offset your higher feed costs. To give you an idea, the investment of one monetary unit before 30 kg is worth at least 2.5 monetary units at slaughter.

✓   Farms cannot make their own Link feed. They haven't the plant to make it, and many vital ingredients are only available in bulk lots, or unobtainable outside the feed trade due to restricted supplies. Leave it to the experts.

✓   But you should ask the supplier about his turnover of stocks both of raw materials and finished goods, especially in summer/ hot weather. These should preferably be days, not weeks.

✓   In this respect, never hold more than 14 days' Link feed yourself. Order frequently and often; accept small-load charges reluctantly if you have to, and store the food in a dry, cool space. An old ice-cream or frozen goods container is excellent. Never store any bags in the nursery. Because of the small quantities required (5 x 200 pig nurseries totalling 1000 weaners from 6-12 kg (13-26 lb) will eat – at 350 g/day (12 oz) growth rate and with a food conversion of 1.2:1 – over 10 days on the Link feed only a little over 4 tonnes of food). So bags could be considered rather than bulk. For one 200 pig nursery, the amount needed will be around 1 tonne only – a 1.5 tonne fortnightly order, maximum. But beware, some nurseries can require up to double this.

✓   Water adequacy, cleanliness and accessibility is essential to adequate uptake of a Link feed, which by its nature tends to be thirst-making. This itself is no problem if water supply and management is good – indeed it will increase feed intake which the design of a good Link feed encourages, without digestive kickbacks.

✓   The water problem is made worse by piglets drinking less once weaned. At this stage liquid intake can fall dramatically from as much as 800 ml/day on the sow to only 200 ml from a waterer (Figure 4) Until the weaner learns how to get all its needs from water, feed intake drops, its ability to digest food reduces and performance suffers.

This may be the fault of drinker design. The Japanese (Zennoh) have an excellent tongue or leaf drinker in aluminium or bright metal specially for 5-12 kg (11-27 lb) weaners – easy to maintain and keep clean.

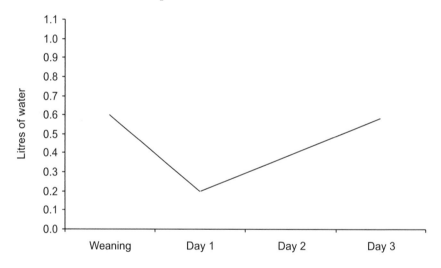

Figure 4 Effect of weaning on water intake. Source: Tibble (1992)

✓   Learn how to 'drive' electrolyte provision. While electrolytes can be put in the feed, the water route is far preferable. Auto-dispensing devices are available, but many smaller nurseries use dedicated canisters.

B.   Checklist –the way you feed the link feed

✓   Troughs and hoppers must be spotlessly clean and dry – an essential part of any AIAO process (*see Biosecurity section*).

✓   Allow *at least* 25% more trough space than the conventional allowance (70 mm, 2¾ inches, per pig at 5 kg) until the pens have settled down. In other words calculate the space needed – but remember, the weaners will have come from feeding all at one time on the sow. I myself prefer allowing each piglet their shoulder width plus at least 25%. A temporary extra trough or hopper, suitably 'barred' to deter nosing the food to one end and stepping right into it, is useful.

✓   The floor under any trough should be a solid 'comfort board' arrangement, even if it is used as a temporary slab, cover or board, or even a permanent tray as is popular in 'big pens', (nurseries on solid floors) to contain peat or shavings.

✓   The trough should be opposite to the drinker/elimination area etc. but as far away from it as possible.

How much food to allow in the first few hours ?

This depends on several things – the design of the food; the weight and fleshing of the weaners; trough adequacy, group size and stocking density; stress levels on arrival; creepfeed consumption before the day of weaning; and watering facility.

The ideal is to feed ad-lib and while this has been known to be successful, it is not usually undertaken in piglets weaned early (under 18 days, say 4 kg, 9 lb) but is more likely to be feasible in pens containing weaners of over 5.6 kg (12 lb) (21 days) and quite possible with weaners of 6 kg (13 lb) and over (24 days and over).

First : Contact your manufacturer. He will know how weaners under differing conditions, when fed his food, respond to various feed allowances and timings in the first critical 2 days. So seek his advice.

Second : Check that all the environmental desiderata (see Checklist C) are in place.

Here is a pelleted feeding plan I have used successfully for the past 6 years (see Table 5). It is especially valuable for light-weight batches which have had to be removed from the farrowing house, and/or if creep feeding has not been adequate, and/or the farmer is reluctant to pay enough for a really good Link Feed.

The idea is to avoid postweaning scouring by not pushing such vulnerable piglets too hard.

As baby pig nutrition progresses and the acceptance of the expensive link feed concept becomes universal, then immediate ad-lib feeding after 17 day weaning will become standard practice. Until then a cautious 'trial and error' approach similar to that outlined above is often needed, because both the suitability of the foods the farmer has chosen and the degree of investment and care in the housing and management of the weaners is, I find, still very variable.

C. A checklist of the environmental issues involved

✓   All weaners should be weaned within an AIAO regime.

✓   As well as the piglet having an undeveloped digestive system at 4 to 6 kg (9-13 lb) liveweight, its thermo-regulation and immune defence systems are also rudimentary. We must do all we can (as we have with the feed) to compensate for their lack of development.

✓   **Hygiene**. Check that the feed troughs are clean, disinfected and dry before first use, and then kept clean and 'sweet' thereafter. Stale accretions must be removed several times a day.

✓   Check that the in contact surfaces are warm *and dry* before entry. Up to twelve hour pre-heating is wise.

✓   **Temperatures**. Above-back temperatures should be 29°C (84°F) for well-fleshed 3.5 kg weaners, 28°C (82°F) for 4.5 kg weaners and 27°C (81°F) for 6 kg, 13 lb, weaners. For 'thinnies' allow 1°C (2°F) warmer. With ample dry strawed pens, 'thinnies' temperatures up to 4°C (8°F) warmer may be permissible in still air conditions without affecting appetite. A 'thinnie' is a standard-sized weight-for age weaner but lacks fleshing.

✓   Supplementary heating is essential/advisable in cold/temperate climates.

See Ventilation section for details, but in general airspeed over the newly-weaned piglet's back at thermoneutral temperatures should not exceed 0.15 m/sec (about 7 seconds to cross one metre or yard). Fans should begin to accelerate when the temperature is 0.5°C (1°F) above the correctly-set temperature and switch on when the temperature is 0.5°C below set temperature.

Table 5 POST-WEANING FEEDING – A SUGGESTED SCHEDULE FOR A HOT NURSERY WHERE THE WEANING SKILLS ARE ONLY AVERAGE

Post Weaning	Time	Weaning at 17 to 21 days – pigs at 5 kg (11 lb) or under Expected consumption	Weaning at 4 wks (Pigs 6 to 6.5 kg) Expected consumption
Day 1	10 am – wean. Do not feed for 2 hours. 12 noon – Place about ¼" (6 mm) of food in base of hopper or ½ round trough. 2 pm – Inspect. If clean add same quantity. 4 pm – Inspect. If clean add same quantity. 6 pm – Inspect. If clean add same quantity. Last thing: Inspect, tidy up, add ½" (12 mm) food. Leave light on over hopper.	60 to 70 g/pig over the day (2.1-2.5 oz). Certainly not more than 0.7 kg per 10 pigs (25oz/1.5 lb).	You can probably allow about 33% more on Day 1.
Day 2	8 am – Inspect. Tidy up, add similar quantity *i.e.* 12 mm approx. 11 am – Inspect. If clean, add similar quantity. 3 pm – as for 11 am 7 pm – Add appreciable quantity, enough to last the night on the basis of looseness-free consumption up to now. Leave light on.	70 to 90 g/pig (2.5-3.2 oz) over the day. Not more than 0.9 kg (2 lb) over 10 pigs.	Careful! Some pigs will stand 100g (3.5 oz) but others won't; keep to a lower level if so.
Day 3	Check and inspect. Check food eaten, looseness. If looseness is apparent, you are overdoing quantities, or the feed is not digestible enough. If OK feed to appetite or x2 or x3 times daily, as you see fit.	100 to 120 g/pig (3.5-4.25 oz) over the day. Do not exceed 1.2 kg/10 pigs (2.7 lb).	To appetite x3 day
Day 4	Ad-lib under x3/day supervision	Ad lib	Ad lib

- Spreading the food down the trough is essential at all times.
- Batches will vary in acceptance, thus each pen may have to be treated individually. Some can be ad-libbed from Day 2 evening onwards.

- You will find more variation in 4 to 5 week weaned pigs than in 21-22 day weaners in the amount they can eat in days 2 and 3 without looseness.
- Watering needs checking/cleaning at every feed.

REMEMBER: This is a cautious feeding table which should avoid digestive overload on many 'average' farms. If you are weaning later; have got a good creep feed intake by weaning (500g/day+, Varley, SCA, 2002); have correct trough cleanliness; the environment right and a well-designed link feed, you can increase the quantities offered substantially and quickly within 4 hours of weaning.

✓   A very common fault is chilling at night – even in the tropics.

✓   Draughts disturb airflow patterns. Check for draughts at night – use your wetted arm or back of hand and, or better, use a small smoke pencil.

REDUCING STRESS AFTER WEANING

Chilling and draughts raise stress (anxiety, worry) and generate low-level hormone reactions which dampen down both appetite and digestive competence.

So learn how to use door and window tape sealers and/or simple air deflectors.

•   When setting a temperature, allow for the smallest pig in the batch.
        The others will do no worse for being a little warmer.

•   Always check the lying pattern of the pigs both at the warmest and coldest time of the day. This will mean the occasional night-time inspection in cold or windy weather. Don't switch lights on, take a torch and move quietly to detect resting patterns and satisfactory breathing.

•   Never assume in your nursery that the temperature corresponds to that set on the control panel. Check, check, check! Call the electrician if you suspect an error. Over 1 in 5 nurseries I visit have got it wrong by 2°C or more. This is enough to cause low-level stress, slow growth rate and raise FCR (Table 6).

Table 6 FLUCTUATING TEMPERATURES COST MONEY

Effect of temperature variation in the first 2 weeks after weaning (6 kg) (13.23 lb)

	Variable *More than ± 2°C from* *set temperature*	*Steady* *Within ± 2°C of* *set temperature*
Daily feed intake	443g (15.6 oz)	404g (14.2 oz)
Average daily gain	306g (10.8 oz)	344g (12.1 oz)
FCR	1.45	1.17
Extrapolated extra weight at 9 weeks at +47g/day (1.65 oz) overall		+2.33 kg (5.13 lb)

Source: NAC Pig Unit (1989). These figures are fairly elderly now, but are frequently found on average farms across the world today.

• Reduce house temperature progressively so that at 11 kg (24 lb) the air temperature is 24°C (75°F). This is hotter than most people expect for this weight-for-age, but a weaner loses a lot of fat cover after weaning if a growth rate check occurs (Figure 5).

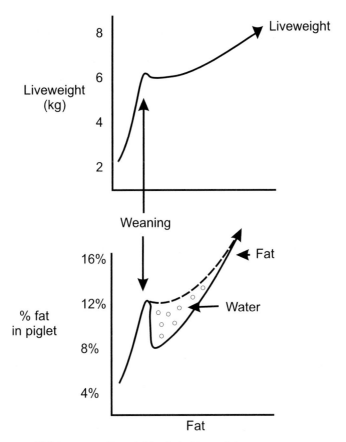

• Piglet recovers its weight gain but loses fat reserves immediately after weaning.
• Some or much of this is replaced by water in the tissues.
• This shows we must keep the piglets **warmer** after weaning, be very careful about **draughts and chilling at night**, and ensure the **water supply is accessible**.

Figure 5 The fat situation after weaning.

• Trough hygiene is vital. To a certain extent clearing up the food allocation will 'polish' a trough – but don't assume so. The best nursery stockpeople use a cleaning stick for trough corners, a garden trowel for removing stale uneaten food and a bucket sponge/swab and cloth for drying out corners. This sort of attention pays in the first 3 days, at least – it is not 'unnecessary'! We wash up food receptacles for baby humans – you must do the equivalent for baby pigs. *Our present standards are far too low!*

• Once on to *ad lib* feeding, trough and feed 'sweetness' (*i.e.* freshness) is vital. Too many *ad lib* troughs/hoppers are dirty and single-space feeders especially so. As well as food spoilage pathogens, mycotoxin poisons are a danger to young pigs, especially.

• Newly weaned piglets must not be without any food for more than 2 hours. This means constant monitoring and supervision – a reason why dedicated nurseries in a three-site arrangement are so successful – the staff have time to attend to routine inspection.

MORNING OR EVENING WEANING ?

Weaning the pigs in the quiet of the evening or the bustle of the day? 8 pm and 8 am have been tried.

American work suggests that the pigs settle together more quickly with a night's rest to come, and by 28 days feed intake was 5% higher than those weaned during morning hours and their weight gain 6% better. Little food was offered to both groups in the first 24 hours, however. Is the lack of weaner supervision in the first 8 to 10 hours of darkness compared to daytime weaning a drawback? It doesn't seem so. However, weaned sow movement is made more awkward and most have to stay put, without their litter, until the morning of the next day. Plenty of time for them to fret with little to take their minds off things as would be the case when they are moved to the service house straightaway. Would stress have affected conception rate? The American work doesn't say.

Me? I still favour morning weaning as I feel I need to be there during those first critical hours!

IN CONCLUSION

Three criteria are vital in the nursery environment:

- Good planning.

- Good observation and meticulous checking

- Elbow grease!

  Make time for all three. Many don't!

  Finally, Fermented Liquid Feeding (FLF) can put a new dimension into the weaning procedure. This is covered in the Fermented Liquid Feeding section.

REFERENCES

Data assembled from SCA Nutrition technical data sheet (1994).

Cole. D.J.A. and Close, W.H. 'Bridging the Weaning Gap' Pig Talk 8 No 5 May 2001.

Tibble, S 'Clearing the Confusion' 'Feeds & Feeding' May/June 1992 pps 9-11.

Proceedings 'Understanding Heating & Ventilation' Course NAC Pig Unit, Stoneleigh, UK March 1989.

Varley, M. 'Piget Nutrition: The Next Five Years'. Procs. J.S.R. Genetics Annual Technical Conference, Nottingham, UK. Sept 2002.

EMPTY DAYS 4

(Non productive days; open days)

Any day in the productive cycle on which the sow or gilt is not pregnant or lactating. Expressed as empty days per litter or empty days per sow per year.

An extremely important yardstick. Just like trucks, airplanes and ocean liners, sows only earn money when they are 'carrying', not when standing empty.

TARGETS

3 to 5 week weaning
(not including any
gilt induction period)

	Poor	Typical	Good	Exceptional	Action Level
Per litter	28	15-33*	12-13	4-7	16
Per sow per year	53	35-66*	28-30	10-15	35

One open day costs at least £2.65 ($4) sow/year and about £90 ($135) for each sow per year.
Approx guideline: Every 15 days empty days per sow per year reduces litters born sow per year by 0.1.

*From clients' records, 1993-2000

A BASIC EMPTY DAY CHECKLIST

✓   Length of time gilts enter the herd from date of effective service.

✓   Gilts not showing oestrus.

✓   Weaning to service interval

✓   Returns to service   – regular, *i.e.* up to 22 days
                              – irregular, *i.e.* over 23 days

✓   Sows showing anoestrus

✓   Sows proving to be not-in-pig

✓   Farrowing rate

✓   Delayed culling

THE EFFECT OF THE 'HERD ENTRY TO SERVICE DATE' TIME PERIOD ON EMPTY DAYS

This will be of growing importance if we adopt longer on-farm induction periods for bought-in gilts to acquire better immune status. However the trend to buy in weaner or 'grower' gilts at 30-35 kg (66-77 lb) will tend to remove this disadvantage.

Table 1 HOW TO CALCULATE EMPTY DAYS PER LITTER

1. Add gestation (usually 115 days) to lactation interval.
    Example : 115 + 28 = 143.
2. Multiply this by your farm's litters/sow/year
    Example : 143 x 2.2 = 315 productive days.
3. Subtract from days of the year (365)
    Example : 365 – 315 = 50 empty days/sow/year.

Empty days per litter is this figure divided by the Farrowing Index on the number of litters achieved per year.
e.g. Farrowing Index 2.3   50 ÷ 2.3 = 21.7 empty days per litter.

Targets

13 days	3 week weaning
12 days	4 week weaning
12 days	5 week weaning

Typical breakdown of target open days per litter is:–

1.	Weaning to breeding	6.0 days					8.0 days
2.	Breed to rebreeding	2.4 days	}	Good	Bad	{	4.0 days
3.	Breed to culling	2.5 days		(Target)	(typical)		8.0 days
4.	Weaning to culling	1.0 day					3.0 days
	(Litters/sow/year 2.35)	12 days		(Litters/sow/year 2.2)			23 days

Breaking down your empty days per litter will help you find your weak areas of which returns to service (1 & 2) is commonplace, but farmers often cull late (3 & 4) which affects open days/litter considerably. Many do not include the gilt's herd entry to effective service date period.

THE INFLUENCE OF A GILT'S RUN-IN (INDUCTION) PERIOD

If induction periods are increased from 21 to 49 days this increases empty days per sow year by 9 days:–

Assumptions

One-third of sows replaced per year. 100 sow herd.
Old system 3 weeks induction; new system 7 weeks induction
Old : 33.33 new gilts/year x 21 days before service = 700 days.
New : 33.33 new gilts/year x 49 days before service = 1633 days.
Divide by 100 sows = 7 days *v*. 16.33 days.

On a herd target of 30 empty days sow/year this is 9.33 days increase or 31% more, so it is quite an important future factor to remember when compiling your target empty day figure. Even so, it has been omitted from Table 1 as it is a voluntary, fixed figure, while the others listed are remediable.

How much does one empty day cost ?

How to work it out per sow

Add together your ...

·   Feed cost/sow/year *

·   Operating costs/sow/year *

·   Variable costs/sow/year *

·   Gilt purchase price less cull sow price received.

·   Divide by 365

    * Including unmated gilt costs.

Taking recent MLC figures, the figures reveal:–

	MLC Av.	*MLC Top Third*	*MLC Top 10%*
Empty day cost/sow/year	£93 ($140)	£61 ($92)	£48 ($72)
Average herd size (sows)	218	217	238
Empty days/sow	35	23	18

(MLC Signet Recording Scheme)
Source : Richardson (1999)

PUTTING EMPTY DAY COSTS INTO PERSPECTIVE

Thus the difference between a Good Target (23 days) and a typical 35 days on a typical world-wide 217 sow herd is a serious extra cost of £7000 (over $10,000) per herd, or over £32/sow ($48), the equivalent of not selling one 25kg (55 lb) weaner/sow/year.

So work out what your empty days really are and know their cost. If, as many producers have discovered, they turn out to be nearer 50 than 35 per sow per year, then that is well over 2 weaners' income foregone per sow per year compared to the average MLC-recorded producer.

REDUCING EMPTY DAYS CHECKLIST

If your empty days are increasing, check the following:–

✓   Are not-in-pigs (NIPs) rising? Past 16% ?

✓   Is sow pregnancy mortality rising – past 2% ?

✓   Are your gilts slow to get mated? >15% not served 10 days from commencement of service?

✓   Are returns-to-service rising ? Regulars over 6%. Irregulars over 5%.

✓   Is your oestrus detection programme up-to-scratch ?

How does your farrowing rate look? Is it below 85% over 100 samples?

✓   Are abortions over 2% ?

✓   Have you got enough gilts available for service at any one time? At least 6%?

✓   Are your average days from weaning to first mating passing through 6 to 7 ?   Check individual sow's history, as averages can be misleading here.

✓   Is your culling policy getting undisciplined? Refer to the Culling Section in this book.

As can be seen, keeping the necessary records to inform you of your current status on all these influencing factors is essential.

EMPTY DAY ACTION CHECKLIST

✓   Are you *stimulating* (exciting) sows before service rather than *stressing* them? A lot of noise may not be a good thing!

✓   Are you detecting heat early enough?

✓   Are you supervising your matings and recording quality of service?

✓   Have you brushed up on the latest AI techniques?

✓   Are you preg-testing?

✓   Have you a firm grip on picking up NIPs (non-pregnant sows) if not.

✓   If outdoors are you aware of the latest mating paddock techniques?

✓   Have you considered using hormone injections as per PG600?

✓   Are you culling efficiently?

✓   Have you a seasonal infertility problem?

✓   Did the affected sows 'nose-dive' in lactation?

THE PROBLEM OF THE 'SECOND LITTER FALLAWAY' – HOW IT AFFECTS EMPTY DAYS

Modern prolific gilts often give a large first litter. This, if allied to low body size or insufficient weight for age (both muscle and fat) at farrowing and insufficient feed intake during the first lactation, can result in the gilt being unable sufficiently to mobilise her rebreeding hormones to achieve a prompt and fertile second conception.

Such animals – and this includes any sow which has similarly been dragged down in condition – may need assistance with their endocrine (hormone) system. The supplementation of 400 i.u. serum gonadotrophin and 200 i.u. chorionic gonadotrophin (PG 600, Intervet) the day after weaning can assist such peri-fertile gilts and sows by stimulating the release of ovae, usually within 3 to 5 days of injection.

Farm trials have shown over 5 fewer empty days/litter together with a day saved between weaning and first service – in this case achieving the '5 day threshold' which seems to be critical in many modern herds (Table 2).

In the author's experience checking for heat should commence within 24 hours of injection as the response in some females these days can be surprisingly rapid.

Table 2 TYPICAL EFFECT OF PG600 HORMONE SUPPLEMENTATION

	Control	*PG600 treated*	
Number of sows	105	115	
Mating Performance			
Conception rate to 1ˢᵗ service (%)	84	86	
Average weaning to 1ˢᵗ service interval (days)	5.19	4.08 ***	Quicker oestrus onset
Average weaning to successful service interval of repeat breeders (days)	58	38	
Average overall weaning to successful service interval (days)	13.2	8.1	5.1 fewer empty days
Sows, not mating 7 days post-weaning (%)	10	0	Service area cleared in a week
Litter Performance			
Average total litter size	11.44	12.69**	
Average n°. born alive/litter	11.11	12.09*	Increased n°. born alive of 0.98 piglets/ litter

* P = 0.05   ** P = 0.01   *** P = 0.001
Source: Intervet UK (1999)

THE ECONOMETRICS OF THE HORMONE INJECTION TECHNIQUE (EXAMPLE: PG600/INTERVET)

Farmers can be put off by what they might consider a high cost per dose (£5 ($7.50) in the UK on average). This can be false economy. . . . From Table 2, 5.1 fewer empty days per litter per 100 sows is worth nearly £3200/year ($4800). When PG600 is used on just the 1ˢᵗ litter weaned sows to lessen the 'second litter fallaway', at a 45% sow replacement rate/year – too high but common enough, influenced by the 20% of 'gilts' which fail to produce 2 litters - then the typical damage done to performance of 100 sows at 2.3 litters/sow/year at 5.1 empty days/sow x £2.65/day x 45 first litter sows/year is £608 ($912).

But the cost plus labour for these 45 sows at £5/dose is only £225 ($339).

Thus the REO is 2.7:1 – a good bargain, and thus not expensive at all.

Hidden Benefits

In addition to the attractive payback when the hormone addition technique is used in this way there can be other advantages:

- Less time spent on heat detection (-18%)*

- Better use of AI semen (10% more effective)*

- Less difference between batched individuals and easier to establish batch production.

- This itself allows the several advantages of synchronized and supervised farrowing to be used, especially in lower neonatal mortality and heavier, more even weaners.

- Helps address seasonal anoestrus problems.

- Provides positive action to sort out those sows which really are NIPs rather than sitting back and waiting to see what happens.

(* = *From clients' observations*)

For those sows detected as anoestrus or NIP, Intervet recommend …

1. Re-group these sows with newly-weaned sows in the service area.

2. Give each problem sow an injection of PG600.

3. Commence oestrus detection the following day, for seven days.

4. Since the last oestrus of these sows is unknown, some may be in a stage of their hormonal cycle where PG600 simply cannot work. Give any sows who fail to show oestrus a second injection of PG600, 10-12 days after the first, and detect oestrus for a further seven days.

5. Any sows still not showing oestrus on the seventh day should be pregnancy tested, and – if negative – culled. In this way you can be certain that any decision to cull was the correct one.

REFERENCES

Richardson, J. (2000) *Low Cost Pigmeat Production*. Intervet UK Ltd, discussion paper. BPA Autumn Symposium.

Intervet UK (1999) *How to Improve Breeding Performance on the Pig Farm*. Promotion literature.

CULLING STRATEGY 5

Every breeding herd needs an established culling policy on when sows will be removed and replaced so as to maintain and maximise economic productivity. The intention is to remove any animal from undue suffering where alleviation is impossible or economically not worthwhile. Enforced culling is the removal of a sow due to circumstances beyond the stockperson's control.

TARGETS FOR CULLING

These will vary according to replacement rate (35 to 47% per annum). In the table below I have taken 40% - still too high, but commonplace. They refer to a reasonably healthy herd.

If your records reveal a substantially higher individual incidence, then these are the areas to examine.

Reasons for culling	% of the sow herd in any one year's records
Age alone	7
Infertility	12
Lameness	6
Other poor performance	4
Disease	3
Deaths	3
Abortion	1
All udder complications	2
Others (*e.g.* prolapse, refusal to suckle etc)	2
	40%

Summary: Generally sow culling is not done well. 'A bit of mathematics, some statistics and a lot of experience' – a view put forward by one expert, is not good enough. Much more examination of the records, careful forward planning and less gut-feel are involved these days. Good culling strategy involves three broad areas:–

- Have an idea of your target output as much as 3 years ahead. This is especially important if herd expansion is in mind. After all, 3 years is but a target sow lifetime so you must plan this far ahead.

- Realistically review your breeding records across the past 3 parities and use this performance as a base-line for your likely replacement needs.

- Review your culling needs, enforced or voluntary, every month, using a herd age profile as a guide, constantly reviewing your replacement targets and using the checklists provided here.

MORAL: Think 3 years forwards, 3 parities back, *every* month

COMMON MISTAKES IN CULLING POLICY

- Removing all sows at a predetermined life-span, often the sixth litter, especially on the larger, industrial units. This is a trade-off between waste of sow productivity and convenience.

- Removing sows where insufficient gilts are ready for mating, or are not fully acclimatized. *The latter is a very common error and can account for much of the enforced culling due to disease/poor reproductive performance later on.*

- Culling first litter sows on poor reproductive performance only.

- Poor monitoring of the empty days history of individual sows, thus keeping unprofitable sows.

- Holding on to unproductive sows in a price crisis.

SOW LONGEVITY

With increasing age the sow's performance will decrease primarily due to increasing disease and lameness, followed by milking problems, smaller litters and increased returns.

But if we cull too soon – unnecessarily – this raises overheads and inflates replacement charges.

Table 1, from a feed additive trial, Sowpack (Alltech) reveals how keeping sows longer can reduce the cost of producing a weaner.

Table 1 APPROXIMATE PROPORTION OF LITTERS (%) UPTO EACH PARITY

Parity	4	5	6	7	8	9	Average
Treatment	57	48	54	57	50	30	49%
Controls	52	29	20	19	15	16	25%

Source : Fehse & Close (2000)

Bearing in mind that it costs about £40 ($58) each time a gilt replaces a cull sow (what I call the topping-up charge) I calculate that the value of keeping the sows in productivity across and beyond 6 parities was £1.61 ($2.34) more income per weaner in this trial and gave a cost:benefit ratio (*i.e.* REO) of 6.4:1.

You've got to cull/got to pay that topping up charge sooner or later, but the longer you can effectively delay it by having more older and productive sows – the more income you secure per weaner.

Across 5 years' economic results, the £1.61 per weaner bonus represents an increase of gross margin/weaner of 37% – all from keeping appropriate sows on longer.

Producers often complain about the rising cost of gilts, yet fail to recognise how much they save on gilt costs, however 'expensive', by keeping sows longer.

Table 2 WHY SOW LONGEVITY IS IMPORTANT – YOUR GILT INVESTMENT IS HALVED

Cost of getting sow to first litter	*Sow lasting 3 litters more (40 pigs)*		*Sow lasting 6 litters more (70 pigs)*
		(per pig)	
£268	£6.70		£3.82
Plus empty-day lag in replacement female in 4th parity 25 days at £2.50/day ÷ 40 =	£1.56		Nil
Total per pig	£8.26		£3.82 (around 50% less)

UK figures only taken, but other countries will find the 50% reduction in gilt costs is about right.

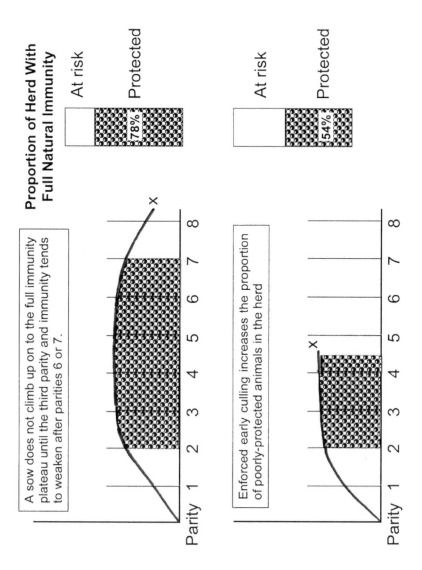

Figure 1 The danger of a short herd life

WHY THE 'FASHION' TO CULL AT THE 6th PARITY OR BEFORE?

1. Large herds tend to operate on a 'proforma basis' where replacement stock are 'ordered up' or prepared in gilt pools for automatic entry at a parity which the unit finds convenient to fit its pigflow plan.

2. It is thought that naturally-acquired immunity tends to fall off at around the 6th & 7th parities (being re-established again later) and a viral disease peak can occur at this time, which can also threaten younger stock, *e.g.* PRRS. This is called 'back-tracking'. (Figure 1)

3. The harsh and exercise-restricting conditions for sows in large herds tend to cause culling from leg and physical surface body debilitation by the 6th parity, even though these sows may have been otherwise healthy and productive enough.

4. Breeding companies encourage rapid sow turnover in order to maximise the benefits of genetic improvement in their latest replacement gilts for sale.

TARGET HERD AGE PROFILE

The Resting Lion (Lion Couchant) shape of Herd Age Profile

The percentage of viable sows remaining in production varies a bit, but most of us agree that at present it is something like *Figure 2*. With knowledge (and skills) improving all the time on how to make a sow productive for longer, this 'lion couchant' shape could well change, becoming longer but lower as it would be in the upper example in Table 1.

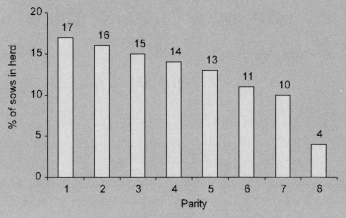

Figure 2 Ideal 'resting lion' shape of herd age profile (Source: Carroll, 1999)

Gilt numbers can be ideally anywhere between 15% and 20% (dependent on farrowing index and replacement rate). If down towards 15% the Lion does not have a 'head' and his back is straighter! Either profile is productively efficient.

However, from the trial cited earlier, maybe our lion of the future could have a rather lower but much longer profile? No loss of productivity, but much lower replacement costs?

I hope it does, because it fits into the correct economic theory of SLC or producing the Same at Less Cost – see Business Section. SLC doesn't flood the market, and pockets the cost savings as profit.

CULLING CHECKLIST

While textbooks cover culling quite well, invariably they are rather long-winded about it and usually leave out some considerations.

A good culling strategy is flexible, reviewed constantly and specific to your conditions only.

A sow should be culled if:

✓   She is unhealthy, lame (after consulting the vet) and has a history of lameness.

✓   Low numbers born more than once, but only after the second parity.

✓   If not in-pig more than once (consecutive oestrus).

✓   If over 10 litters – keep an eye on herd age profile.

✓   There are too many negative comments on her sow card.

✓   Your herd age profile slips from the 'Resting Lion' shape towards the 'Reverse J' shape. *See Figure 3*, the 'n' shape is an intermediate stage and is a warning sign.

✓   If you've met your weekly breeding targets; using your herd age profile graph to cull any sows you suspect are unhealthy or lame. If you have none, cull based on a history of low-litter size twice in a sow, excluding parities one and two.

✓   Those that fail to come on heat after hormone treatment, by the 18th day from weaning and/or seven days from the treatment. Get your vet to advise you on procedure.

Consider culling the sow after the next farrowing if:

✓   The sow's performance falls below the herd average, especially after litter number six.

✓   The sow falls below the target thresholds given in Table 3.

✓   Failure to breed on two consecutive oestrus cycles, after seeking veterinary advice.

✓   If a seven-litter sow performs worse than your gilt average.

✓   Two litters have had three piglets born alive under one kg (2¼lb).

✓   When high stillbirths have occurred twice consecutively, and especially in a sixth-litter sow onwards.

A "DON'T CULL" CHECKLIST

Don't cull if:

✗   Weekly breeding is below target. Try to stiffen up your culling resolve, only culling the essentials.

✗   Her sow card shows she is an asset – for example if 7th parity sows have farrowing rates above 86%.

✗   Her problem is likely to be due to a boar. If so, don't cull.

✗   The problem is stockmanship or management. If so, don't cull.

✗   Don't cull a gilt or second parity sow on few livebirths. This is nearly always mismanagement. Genetically the chance of successive litters the same size is low. Only about 15% of the apparent inferiority of a gilt's single record is likely to be typical of her future performance.

✗   Genetically only one out of four sows performing badly at first parity is likely to give a small litter again due to genetics, so do not cull solely on this basis.

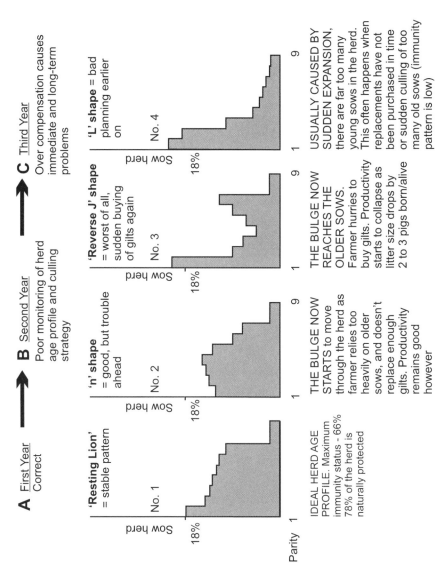

Figure 3 An example of unplanned culling strategy revealed by herd age profiles

Table 3 SUGGESTED CULLING THRESHOLDS ON HISTORY OF PIGS REARED PER LITTER

	Parity	*Pigs reared cumulatively*	*Pigs reared in previous 3 litters*
A sow may be a culling candidate when below these standards in the U.K. only. Different standards exist overseas, and these numbers should be compared to the herd's average. Seek local advice.	2	16	-
	3	28	-
	4	39	-
	5	50	-
	6	60	Under 26
	7	70	Under 26
	8	79	Under 25

How long to keep her ?

Sow longevity is related to high fertility levels and low non-productive days.

If you try to lower culling rate to increase lifetime litters per sow, always keep low sow non-productive days as a priority in your mind. Monitor these closely. See Table 4 for non-productive day targets.

Table 4 NON-PRODUCTIVE DAYS

	Poor	*Typical*	*Good*	*Action Level*
Per litter	28	15 – 23	12 – 13	16+
Per sow per year	53	35 – 66	28 – 30	35+

Lifetime litters per sow is a trade-off against the cost of empty days per sow, so you must monitor non-productive days on the sow record card and balance this cost against the lower cost per finisher of keeping her longer. Too many people don't work out these two vital figures and so penalise themselves.

A FINAL CHECK

To cull efficiently, producers must have answers to the following questions:

Across all sows ...

✓   Do you know your herd cost of one extra empty day? See *Empty Days Checklist* for a guideline.

✓   Do you know your savings per pig reared for one extra litter per lifetime?

   If you don't, you can't cull efficiently.

✓   Finally – do you know how much you can save per weaner by keeping a productive sow on for two more parities, rather than replacing her automatically at the sixth litter? Work it out – it will surprise you.

REFERENCES

Carroll, R. (2000) Culling Guidelines. *Pig Interational.* **30**, 2.

Fehse, R. and Close, W. (2000) 30 Piglets per Sow per Year - Fact or Fiction? *Biotechnology in the Feed Industry.* Nottingham University Press. 309-326

PROLAPSE

RECTAL, VAGINAL, CERVICAL

6

The tissues supporting the rectum and birth canal are elastic but tough. When these connective tissues fail the organs supported by them can invert and protrude, in the case of the rectum through the anus. In the case of the cervix/vaginal prolapse the cervix (neck of the womb) protrudes into the vaginal orifice, or outside it or, in the worst example, the uterus protrudes outside.

TARGETS AND INCIDENCE

Rectal prolapse occurs in both growing pigs and sows, and seems to be on the increase. It sometimes occurs in 'waves' although individual cases are common in growing pigs. Target is nil.

Vaginal/ cervical prolapses occur normally in about 1/200 pregnancies (Muirhead, 1998), usually in older sows, especially if overweight and/or heavy in pig with a large potential litter. Target therefore 0.5%.

COMMENT

Prolapses are a veterinary matter, especially if serious, but as management and feeding could play a part in the disorder, the condition is covered in this book.

IMMEDIATE ACTION IN THE CASE OF THE SOW

Vaginal and cervical prolapses may show early on between the lips of the vulva. These can revert when the sow stands, but the animal must be watched carefully thereafter, especially as pregnancy progresses.

Rectal prolapses in sows can also revert, but often only for a while. The sow should be removed to be on its own and a veterinarian called if the protrusion persists.

More serious vaginal/uterine protrusions also call for the same action; the veterinarian can carry out/demonstrate suturing techniques which will help save life.

IMMEDIATE ACTION IN GROWING PIGS

The main problem time is from 10-20 weeks, and the onset is rapid. If small *e.g.* 10-15 mm (½-¾ in) protruding, the rectum can revert to normal – larger than this *i.e.* with 50-80 mm (2-3 in) everted this is unlikely, so immediately remove affected pigs to isolation so as to prevent cannibalism.

Your veterinarian can demonstrate to you gentle reversion by manipulation and then various suture repairs. Once done, the affected pig can be returned to the pen.

Bitten prolapses need hospitalisation. Consult your veterinarian.

VAGINAL/CERVICAL PROLAPSE A CHECKLIST

There is no proven genetic weakness for birth canal prolapse, but there is evidence in the rectal variety. However, selecting gilts with over-small vulval size may be implicated, but that is not really genetics, just bad selection.

Check : Incidence in old, overweight sows, known slow farrowers / large litter sizes.

Also lack of appetite in very hot weather.

Likely causes are :–

Mycotoxins
- ✓ Check bins and feed outlets / troughs for cleanliness. Sanitize bins during summer; allow to dry thoroughly.
- ✓ Check mouldy grain, especially for zearalenone mycotoxin.
- ✓ Add a modern mould inhibitor (*e.g.* Mold-Zap/Alltech).
- ✓ Add a modern mycotoxin binder (*e.g.* *M*ycosorb / Alltech). This is much preferableto the cheaper clays.
- ✓ Check for hyperventilation after farrowing,poor milking ability, both of which can also occur when mycotoxins are too high.

Constipation
- ✓ Especially pre-farrowing.
- ✓ Ensure lead-in diet contains 4 to 6% crude fibre.
- ✓ Check water availability, accessibility and flow-rate (up to 2 litres, 3.6 pints/min).
- ✓ Add a cupful of dried sugar-beet pulp to the diet 2 days before and over farrowing and for 1-2 days subsequently if needed. Watch stool softness – it should deform when hitting the ground.
- ✓ Make dietary ingredient changes as gradual as possible, especially wet to dry / dry to wet.

Lack of water
See above and the final paragraph of this checklist.

Flooring
- ✓ Sows lying over a lip (as in solid-floor gestation rooms).
- ✓ Floor too flat. Solution is to raise the sow on a 'farrowing board' which raises the rear-end 2" (5 cm). Also a slope to the rear *greater* than 1:20, *i.e.* more than 4" (10 cm) fall in a dry sow stall,is suspect.

Slow farrowing	✓ See the vet re possibility of judicious oxytocin injections. Do *not* attempt this without veterinary advice.
Other suggestions	✓ Lack of exercise.
	✓ Over-fat sows.

✓ Clumsy pulling during assistance with the preceding farrowing, or torn tissues due to a bad farrowing / over large / badly-presented pigs. The tissues become irritated, infected and weaker, leading to straining.

✓ Use of oestrogens to induce heat may cause vulval sucking/weakening which may last if other factors are involved, *i.e.* slight mycotoxin presence.

✓ Prolonged use of antibiotics to cure disorders in pregnancy.

✓ Bloat – too much whey when whey is cheap, also too much dried sugar beet pulp (both when conventional feed prices are high), causing straining to the connective tissue. DSBP swells about x5 when wetted.

✓ Waterworks problems, especially if urine is **bright** yellow. Infection can cause pain, thus straining and abnormal posture. See the veterinarian. In the meantime ensure ample water supply/check ease of drinking in a sow which is large and awkward just prior to farrowing. If needs be, bucket water into the feed trough for a while.

RECTAL PROLAPSE
A CHECKLIST – SOWS

Check management in particular. Some cases seem to be due to levels of hormones following oestrus, and the reasons for this are poorly understood. Check any factors which put strain on the connective 'curtain tissues' supporting the rectum such as:–

✓ Constipation *(see preceding checklist).*

✓ Excessive coughing.

✓ Overcrowding. Cold conditions for grouped sows can cause a herd flare-up.

✓ Internal abdominal pressure caused by gut fermentation. If gas production is suspected, refer the problem to a nutritionist. See preceding under bloat.

✓ Flooring *(see preceding checklist).*

✓ Water deprivation.

✓ Sow stalls with a horizontal back bar or chain. *Never* design a sow stall this way. The sow may sit on it causing distortion of the connective support tissues.

✓ Mycotoxins *(see preceding checklist and Mycotoxin section).*

✓ Penetration of the urethra or rectum during service, causing inflammation and straining.

RECTAL PROLAPSE
A CHECKLIST – GROWING PIGS

✓ Scouring. Excessive strain on the rectal tissues.

✓ Coughing / Respiratory disease *as above.*

✓ Colitis. Excess fermentation can occur.

✓ Overfeeding. Some food reaches the colon insufficiently digested and causes too much fermentation and distension/pressure on the rectal tissues. Try changing the food and/or restricting it.

✓ Water shortage *(see preceding checklist).*

✓ Mycotoxins *(see preceding checklist).*

✓ Changes. Sudden changes of diet, *c.f.* colitis. Change to a anti-colitis diet. Some outbreaks have occurred when accommodation is changed especially if small pigs previously in small pens are moved into large groups and seek warmth.

✓ Any pigs too cold can develop the problem. If there has been a cold snap then keep the pigs warmer and less huddled next time.

✓   Slippery floors.

✓   Tail docking (too short). Check your technique. (See Tailbiting section) Don't sever under 25 mm (1 in) at 2-3 days old.

✓   Constipation (*see preceding checklist*).

✓   Competition due to lack of trough space. Wet feed troughs too high for the age of pig. (> 150 mm, 6 inches, for 30 kg; >230 mm, 9 inches, for 70 kg pigs).

RECTAL STRICTURE

Progressive narrowing of the rectum, usually about 8mm (¼ in) inside, often caused by scar tissue after a rectal prolapse/suturing repair. This can occur about 3 weeks from the original prolapse. Alternatively infection at the site could be responsible. The condition is sometimes characterized by bloated bellies and a squirted diarrhoea. If many cases are apparent, always consult your veterinarian as medication (if infection is the cause) could save the pigs, otherwise there seems no treatment for strictures.

CHOOSING A GILT 7

Selecting a breeding female which will give a good first and second litter with minimum intervals between, and go on to produce at least six good litters thereafter.

TARGETS

Good choice of gilts can set the standard for a future replacement rate of under 35%. 70 offspring might be produced in a gilt's lifetime (20 of these in the first 2 parities) and 30 empty days per year, every year of her life achieved.

When I was young I worked for what was at that time England's largest pig farm. By today's standards it was small – 1200 sows, but in those days it was massive.

Our gilt replacement rate was about 8 per week and one of the jobs I had to do was select batches of 10 or 12 on Mondays.

After two years of this I got reasonably good at it – and developed a system which has stood me in good stead ever since. I made some pin money too, as local farmers paid me to do the same for them, or do a run-through of gilts which had been selected for them by the breeding company.

Last year I did my final job of this nature (I don't bend down so easily these days, and those teats/udder lines are so important!). So here's my own check-list for you. Generally I'm told that I rarely chose a bad one. This must have been helped by my need to use a guide to keep my attention on the points to look for, in sequence and in a deliberate, objective way.

CHOOSING A GOOD GILT

First rule

Don't try to carry everything in your head or in your mind's eye! You've got a superb computer between your ears, but it is still not good enough to compare a pen of gilts objectively. So write it down!

Second rule

Use a progress chart. I illustrate my own personal route map (Figure 1), but you can design your own. The diagram shows how my eye travels around the gilt, noting details as I go. A planned progression is important – or you'll miss something important. My own scoring system is 1 – 10 with ***anything*** below 6 not being selected.

That computer between your ears is a fickle instrument. It tends to pick up what you are looking for – what you are concerned about (narrow chest, poor hams, legs and movement may all be characteristics you know need to be improved in your herd's gene mix). So you'll very likely pick these up – but miss others! Those last two teat pairs that aren't going to develop; that crossed toe; that tendency to be nervous. And so on. A progress chart slows you down and makes it less likely that you'll miss things.

Third rule

Examine them one at a time. Do a pen of ten or so at a time but concentrate on one, get the stockmen to move her about a bit, then finish and select or reject her. Do not jump from one to another, *e.g.* "Is ***this one's*** ham better that than one's". It is a great temptation and you will get confused if you do this.

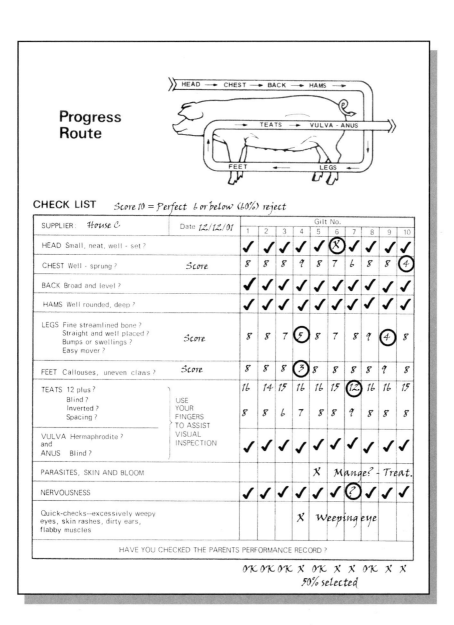

Figure 1 Chart for examining a gilt.

Fourth rule

Favour the docile gilt. This is my own personal rule and it is not scientific by any means, but in the UK with our wide range of sow temperament it has rarely let me down. Ignore it if you want to – I just find it useful. It is this:-

I try to *choose a gilt which is quiet and amiable* and quite unconcerned by me, a total stranger, giving her a visual check close-to and prodding her around gently, even clapping my hand close to her ear to startle her. *These girls almost invariably breed well!* Of course, if there's something wrong visually or in locomotion, of if she's low-rated on the index, out she goes. I wouldn't be at all surprised if, one day, some bright young PhD proved to us that hyperprolificacy is positively correlated to an easy-going temperament. I find it so after selecting thousands of gilts.

Fifth rule

Do not select from a group of less than 10 gilts (and 20 is better, if not always possible). This is because you need to select from as many replacements as you can so as to give yourself the best chance of improving your genetic traits over a period.

Sixth rule

Always get a gilt to trot. If she's an easy mover her legs will almost certainly stand the loads to come, fed properly, of course, and on the correct floor surface. Horse breeders know exactly what I mean by a 'fluid mover', and while pigs are not nearly so elegant you need to get the same feel about a gilt. Strangely, a well-known horse breeder and I both judged a group of 30 gilts on the move and selected the same dud to an animal – there were 5 of them and we compared notes and agreed exactly. Suspect gilts with short strides and a stiff gait. Conversely, long strides and a swaying back end (future back trouble?). Just drive them on a bit so that they do a fast walk – you need to see how they *move*. And do it last, on the way to the draft pen as it disturbs them.

Check for swollen joints, tendons and ligaments, weak pasterns front and rear, and extra straight and stiff hocks. Legs should have a good spring and cushion, but not to an extreme. The Legs Checklist Section deals with this in much more detail.

SOME OBSERVATIONS

Narrow chests

Countries like Britain can be cold and damp, so respiratory troubles are a problem to us. Unscientifically again, I associate a narrow chest (*ie* poor spring of rib) with more pneumonia, often in the offspring too, and this applies to boars as well as gilts. Also I *never* select an animal with a dip behind the shoulder blade, real trouble here: little stamina and maybe slow growth in the offspring. If the farm is a good warm one, maybe; if the index is good and maybe if animals are a bit scarce, I *might* select such a gilt – otherwise I reject them.

Teat troubles

You have to be very careful with blind, inverted and immature teats (button teats). I can't teach you this, it's experience of knowing when one cistern is not going to make it, now, or later. One tip is to feel the udder – if it is rough rather than silky, check for blind teats very carefully. Also eight teats forward of the navel is a good sign, but increasingly rare – breeders please note!

Vulva size

A small vulva coupled to a small pelvic spread is trouble ahead; they tend to make suspect breeders at farrowing.

Feet and legs

Extremely important, of course. Downgrade a limb structure which is extra vertically 'set' *i.e.* not sprung, when viewed from the side. This is very important in hocks and pasterns – if extra straight, score low on the chart. Durocs and Hampshires need careful examination in this respect. Feet and legs should have good springing to cushion the effect of hard floors, but some genotypes – often Landrace and Welsh – show extreme springing and thus hind leg weakness when older.

Downgrade gilts with swollen joints or inflamed tendons etc, and downgrade inverse toes if they differ by more than 1.2 cm (½ inch) on the same foot on a 70 kg + (155 lb) animal.

Further notes on legs and a Lameness Checklist will be found in the appropriate Section.

A CHECKLIST FOR BUYING GILTS

✓   **Check your prospective multiplier:** In most cases you will be receiving your stock from a pure or cross-bred multiplier, not the nucleus or nucleus multiplier the seedstock firms like to talk about. Check him out – ask for names of other breeders who have bought his stock and telephone them. More important, search out others who have not been recommended and ask *their* opinion. Once the seedstock house has been agreed as your supplier *go and see him* and learn how he prepares 'your' pigs for entry into your conditions.

✓   **Ask for the vendor's conditions-of sale document:**
These, like the multipliers, vary enormously. Read them carefully. If you don't like parts of them, say so and threaten to go elsewhere unless the condition is modified in writing or explained satisfactorily. Check closely what they say (or don't say) about animals which do not breed satisfactorily.

✓   **Check the general differences between the lines:**
Differences are appearing in the genetic strains coming from various seedstock houses *ie* conformation, appetite, type of finishing food advised, docility and mothering qualities, leg strengths and hyperprolificacy. There are also quite major differences within the breeding companies own line structure, so check that you are getting what your market or system of production needs, not what they think you want, or maybe is convenient for them to sell to you. For the progeny of outdoor sows, concentrate particularly on *proof* of fast, lean growth, as despite what is said, this still can be a weak area compared to (their) white indoor breeds.

✓   **Get to know the vendor's salesman/pig specialist:** Once you have established a relationship based on trust (this takes time) he or she will be a key factor in ensuring you get the right stock for your system of production *on time, checked personally by him/her and old/well grown enough!* Part of the process of getting to know the vendor and his salesperson is to know the questions to ask them.

CHECKLIST: QUESTIONS TO ASK A BREEDING COMPANY

Check carefully how they reply and what the answers are – it helps you decide when the choice is wide, as it is these days.

✓ **Genetics**
    1. a) What proof have you that the performance of your herd is normally better than average?

       b) What schemes to you test under?

    2. What selection intensity do you use on the farm I am buying from?

    3. How do you ensure that the variation you are measuring is due to genetic ability and not to variations in management?

    4. How quickly does the farm replace its females?

    5. How many years have you been making consistent genetic improvement? Also please *quantify* this in female and male traits.

    6. How large is the genetic pool from which you select?

    7. What proof can you give that the prospective breeding stock is free from genetic defects?

✓ **Health**
    1. What proof can you give that your stock is healthy?

    2. Can you provide independent evidence, such as from a veterinary surgeon's reports? Are you free from such diseases* such as:
       • Enzootic and Haemophilus pneumonias
       • Atrophic Rhinitis • Aujeszky's • Transmissible Gastro-Enteritis • Swine Dysentery
       • Swine Fever • SVD • Leptospirosis
       • Meningitis?
    State the current circovirus status of your herd(s).

3. Are any drug treatments given as a routine which would mask the symptoms of any undesirable condition, for example Atrophic Rhinitis, Swine Dysentery, Meningitis, etc?

4. Will you give permission for your veterinary surgeon to give me a full history of your herd? And for him to compare notes with my own vet?

* Note: This list refers to common European problem diseases – it may be different outside Europe.

✓ *Profitability*   1. What proof can you give that the prospective breeding stock are prolific, sturdy and will thrive under commercial farm conditions?

2. Will the progeny sell at an above-average price? Proof please.

3. Are your pigs likely to show a better than average profit? Customer records please.

4. Are your pigs good value for money? How would you compare them to your typical competition?

QUESTIONING SALESPEOPLE

This last question in the foregoing checklist is a very interesting one to ask. Salesmen may have been instructed not to denigrate the opposition. If so, accept it as part of selling ethics. But you never know - by judicious questioning I have obtained some useful and important cross-confirmed data on various genetic lines which diplomatically I will not publish here, and there is no reason why you cannot do the same.

For example, data from 7 or 8 sources have revealed important advantages between various commercial blood lines in the areas of:-

• Meat per Tonne of Food (MTF) on the same carcase yield.

• Leg strength and Appetite under hot conditions.

- Feed Protein needs in the last month before slaughter.

- Loading and haulage stress.

- Docility.

- Presence of marbling genes *e.g.* 0.7% marbling fat v 1.3%.

- Killing out percentage *e.g.* ± 1% under identical conditions.

This has enabled me to recommend certain breed lines which are more likely to be suited to the specific farm conditions I've encountered. I know this works because in most cases the follow-up resulted in comments like "Since we tried (or changed to) breed 'X' the problem has been **much** better." Remember, no one breeding company's pigs are necessarily 'the best'. The best one is the **right one** for your conditions and to compensate for/remove your commercial weaknesses.

GETTING TO THE TRUTH

Of course getting the relevant "classified" information out of people is difficult, and in a commercial situation most lay people regard it as impossible. Trade secrets are just that. Secret!

But an old journalist's trick is to 'float the negative'. You need to know the subject matter pretty well, and insert an assumption, statement or claim into the discussion which is just sufficiently and deliberately wrong for the victim to at once correct it with the right figure from his kindly or professional instinct to put you straight. There are a variety of conversational subterfuges like this, and the rest I'm keeping to myself, although if you go into a good bookshop and read up on interrogation methods you'll get the hang of it! Meanwhile – beware of journalists!

WEANER GILTS – A NEW TREND
(Junior gilts in North America)

This is an exciting new development. Many commercial breeders are now buying their replacement breeding females – not at 90 - 100 kg (198-220 lb) but at 25 - 30 kg (55-66 lb). I forecast that the majority of bought-in gilts will be purchased as early as this in Europe within the next 5 years – that is on the professional/efficient units.

Cheaper and better

The reasons are not hard to see. The economic and performance evidence is now coming through from the pioneers of the system who started about 8 years ago, as it is not until the fourth year beyond repopulation that all progeny are derived from sows bought-in as weaner gilts.

Cheaper cost

In Europe the cost of a selected maiden gilt at 100 kg (220 lb) bought from a breeding company is about £170 - £190 ($275-300). Of course the price of a 32 kg (70 lb) weaner gilt from the same source is not going to be as low as the value of a 32 kg home-reared female destined for meat, but prices have varied recently from £85 to £100 ($135-150) among European breeding companies. Table 1 gives a typical breakdown of comparative costs.

The Newsham breeding company, now merged with J.S.R. Genetics, has quoted savings of £20 - £25 ($30-38) at 95 kg (210 lb) (Brisby, 1998) which is a 12-15% saving on their average maiden gilt price.

Better performance

A comparison (PIC 1997) of 49 herds using standard gilts and 16 herds buying weaner gilts (called 'junior' gilts in the USA) showed a 5.9% advantage in farrowing rate, 0.07 more litters per sow per year, 17 fewer empty days per sow per year, 0.5 more pigs born alive/litter, 0.28 more pigs reared/litter and 1.39 more pigs weaned per sow per year on 60 kg (132 lb) less food required per sow per year.

Why is this? The rationale behind buying breeding stock replacements at an earlier age and lighter weight is to allow a longer and more effective acclimatisation period prior to full introduction to the breeding herd. At least six weeks (and with certain low level diseases present, 8 weeks) is now advised when buying in maiden gilts at 90 - 100 kg (198-220 lb). This delay is expensive in itself, and these extra costs alone would make a properly acclimatised maiden gilt kept longer before full introduction to the herd under the new recommendations, even more expensive. The extra costs are a further 5% per gilt to add to the 12 to 20% savings likely from buying 'junior' gilts at 30 kg (66 lb).

Table 1   TYPICAL COST OF WEANER GILTS IN THE UK

	(£)	
Weaner gilt at 32 kg (median price)	92.50	
Feed at £165/tonne 32-100 kg (FCR 3:1)	32.64	
Water, bedding, vet & vaccination	8.00	
Interest on cost of gilt and feed	3.35	
Combined purchase and production cost	136.49	($205)

Assuming 4 out of 5 gilts are selected at this stage the cost to 100 kg is ...

5 x £136.44	682.20	
Less sale of non selected gilt	64.00	
	618.20	
Net cost per gilt selected	£154.55	($232)

A saving of £15.45 ($23) per gilt or 9.1% on a £170 ($255) maiden gilt price

Source: Extrapolated from Beckett (1994)

Disease lower?

A much longer acclimatization period should result in less disturbance to the current health status of the herd. The weaner gilt herd owners interviewed felt that breeding herd health was better and there were fewer re-occurring health problems. We must wait for further evidence on overall disease incidence but sow mortality was lower, 4.0% compared to 4.3%. However mortality from born alives was higher on the junior gilt herds – 12.66 v 11.18 per litter. The absolute mortality figure per litter (A.M.F.) was 1.19 piglets (maidens) v 1.41 piglets (juniors), but the juniors piglets were weaned 2.5 days later.

Much more weaner weight produced per tonne of food

A very important difference hidden in the published figures was the amount of saleable weaner weight produced per tonne of sow and piglet feed. At 116.5 kg (maidens) against 142.2 kg (juniors) this is a 22% improvement. *Under European economics (for 2001) this is equivalent to a 9% reduction per tonne in the price of all breeding and piglet food.*

Comparisons to 36-38 kg

Did the considerable advantages of the weaner gilt system at weaning continue up the Acceleration phase of lean growth, which usually starts to ease off around 35 - 40 kg (77-88 lb)? Yes, and much more so!

Daily gain (7 to 37 kg/15-82 lb) was 585 g/day (juniors) as against 548 g/day (maidens) or 1.28 v 1.20 lb. There was a marked difference in FCR; 1.8 (juniors) to 2.23 (maidens). This in itself would suggest the junior-sourced pigs could cope better with disease challenges at this critical stage of growth. Because of this massive food conversion advantage, the liveweight produced per tonne of feed used through this stage was heavily in favor of the junior-gilt sourced herds – 698 kg v 559 kg, a difference of 139 kg (307 lb)or 25%! Even more dramatic – the figures on the PIC costings reveal a reduction of 70% on the cost/ kg gain to this weight.

If these results can be maintained by typical breeders it is no surprise to find that my forecast of big savings from buying junior gilts will be correct.

REFERENCES

PIC Easicare Yearbook (1997) pps 48-49
Beckett, M: Cost of Weaner Gilts. *Farmer's Weekly* Feb 1994.
Brisby, I: Personal communication

SEASONAL INFERTILITY 8

A decline in reproductive efficiency across the summer/autumn period seen as:–

- *Delayed puberty in gilts*
- *Problems with gilts cycling*
- *Extended weaning to oestrus interval*
- *Shorter oestrus periods*
- *Increase in number not in pig*
- *Increase in mummies and stillbirths*
- *Abortion storms.*

INCIDENCE

Muirhead reports that in the U.K. 70% of all abortions could well be due to this cause, while many farmers (seemingly mistakenly) believe abortions are largely infectious in origin. First noticed in the U.K. in 1970, it was particularly bad here in 1974 and seems to be getting worse, possibly due to more breeding being outdoors, better recording of farrowing rate and live births from season to season and a rise in summer temperatures.

• Abortions occur.	*Herd Target: 1 in 100 served sows, rises to 13 or 14 or more over short periods.* (Figure 1)
• Stillbirths increase.	*Herd Target: 3% true stillbirths, can rise to 8% or more.*
• Mummies increase.	*Herd Target: small 0.5%, large 1.0%, rises to 3% or more.*

Watch for the following in autumn & winter especially …
- Regular returns to service increase, *Herd Target: <10% rises* particularly among gilts. *to 20%+.* (Figure 2)
- Weaning – conception interval *Herd Target: <6-9 days rises to* lengthens. *10-12 days.*

Thus numbers born alive fall, and herd targets of 10.5 – 11.0 can decrease to 10.0 or less over a six month period, which improves again in late winter and spring. Bad cases can cost 150-250 pigs per 100 services, a severe drain on cash-flow, reducing annual gross margin/sow by 18% and more.

DEALING WITH SEASONAL INFERTILITY – A CHECKLIST OF LIKELY CAUSES

Seasonal Infertility cases are multi-factorial (many possible causes), complicated and not very well understood, although countries with distinct climatic seasonality – with bright springs and hot summers and rather dull, sometimes chilly winters (*e.g.* Australia and Central USA) – are more experienced in dealing with the problem. Difficulties have arisen because some good quality research has tended to confound some of the current advice and experience.

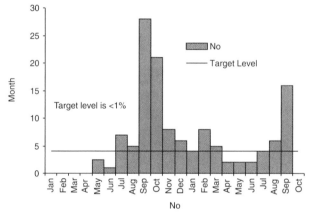

Figure 1 Abortions by time of year (500 sows). Northern Hemisphere. Extrapolated from Mackinnon 1994.

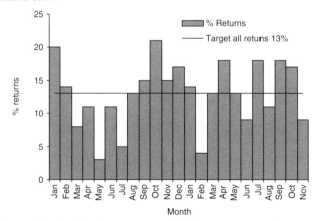

Figure 2 Seasonal infertility. Return rates allied to season. Northern Hemisphere.

A BASIC CHECKLIST FOR SEASONAL INFERTILITY

✓ High temperatures, especially boars.

✓ Too much bright light in spring; decreasing light patterns in autumn; too little indoor light at all times.

✓ Nutritional stress in hot weather.

✓ A variety of stress factors.

✓ Genetics – the remnants of the feral factor?

A COMPLEX SUBJECT – AND A DILEMMA

While experience, particularly in Australia, Mid West USA and Spain – and now recently in the UK, is contributing to knowledge on Seasonal Infertility all the time, there is a good deal yet to discover. In fact some scientists are quite vehement in their criticisms of some aspects of current thinking about the subject. So when experts disagree ….!

Take the embryonic mortality aspect of seasonal infertility. Dr Phil Dziuk, an eminent researcher, wrote to me setting out an apt and amusing analogy thus . . .

"Embryonic loss is an elephant. It is the same elephant as described by each of a group of blind men. One blind man who grasped the leg in his arms said it was like a tree, another felt the trunk and said it was a large fire hose, a third touched the tusk and described it as a spear and the fourth thought it was a rope as he held the tail. They were each correct but they were each wrong. Embryonic loss in the eyes of the nutritionist is a result of improper feeding practice. The veterinarian declares that subclinical endometritis or an infectious organism is the cause, while an injection of the proper combination of hormones will cure it, according to the endocrinologist. An undesirable set of genes that can be selected against is responsible proclaims the geneticist. Maternal-embryo histoincompatibility explains it, says the immunologist. The cytogeneticist finds chromosomal aberrations and deduces that these errors are at fault. Each may be correct but possibly each is also wrong. The elephant of embryonic mortality may be even more complex when viewed individually by the many research workers who have studied it over the years, or it may be many factors acting through a relatively common mechanism to produce one result, loss of some embryos."

Good stuff! Hopefully, those of us at the sharp end working on farm problems (while as blind as anyone) have felt all over the elephant for a period of many years before coming to a conclusion and while still puzzled – and still blind – have had a lot of elephants on which to practise so as to form an opinion of their shape!

This chapter therefore outlines my own experience, and is based on what other practical people – farmers and veterinarians in particular – have found helps.

EXPANDING THE CHECKLIST

Temperature

Boars : Temperatures over 27°C (81°F) may well affect boar libido and, after excessive heat of 5 to 14 days, damage sperm quality for 4 to 6 weeks thereafter.

Sows : Temperatures over 22°C (72°F) and especially towards 25°C (77°F) and over affect appetite, particularly in lactation. This can throw the sow into negative energy balance, when she has to use her body tissues to an extent which may affect reproductive efficiency, even to the state of abortion.

Gilts : Gilts are better able to withstand heat than sows but are particularly susceptible to increased stress especially if water is short or if they are overcrowded in groups. Allow at least 3 m² (32ft²)per animal. The typical effect of temperature on litter size is shown in Figure 3.

Light

Too much light : Very bright light seen on those clear late spring/early summer days especially if the animals have access to direct sunlight in outside runs or outdoor paddocks. Runs should be shaded with 'Galebreaker' mesh covers and paddocks have a shaded lying area which sows *are encouraged to use,* both in the cool, clear days of spring and the hot, muggy days of summer.

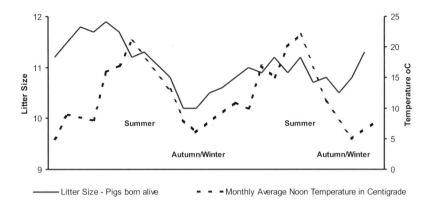

Figure 3 Seasonal infertility: Typical (actual) records showing effect on litter size of increased temperature

Decreasing light patterns : These are inevitable as mid-summer progresses through autumn, and result in a quite natural reduction in the hormones needed to maintain the pregnancies of sows and gilts mated during the peak summer day-length period. (This is known as the 'feral factor' achieved by sows in the wild who instinctively do not wish to farrow down and raise a litter in the depths of winter).

It is almost impossible to counteract this decreasing light pattern effect on outdoor breeding farms short of bringing the sows into a structure where day length can be controlled for 40-60 days after service. But this can be done with sows housed indoors, where 'autumn infertility' can also be a problem.

Some researchers have questioned the amount of time and trouble currently spent on this area of the whole subject in relation to the other factors involved.

Too little light : This is probably the main cause of 'Summer Infertility', where the problems occur much earlier than 'Autumn Infertility'. Here poor stimulation into oestrus and conception due to a combination and progressive build-up effect of poor light, stress from cold and damp quarters and nutritional catabolism (essential within body nutrient destruction as distinct from nutrient construction) occurring towards the end of winter or after a 'late-spring'.

The later arrival of 'springtime' weather seems to be a recent (1960's onwards) phenomenon, resulting in 'longer' (even if milder) winters, due to heavier cloud cover in NW Europe. Maybe this 'winter shift' is the result of global warming.

To combat this, lighting should be at least 350 lux (about as bright as a 100 watt strip-lit kitchen, and certainly light enough to read the small print in a newspaper) shining into the sow's eyes (not behind the head). Opinions differ on the daylight/darkness (<25 lux) mix but the author has found 16-18 hours 'on' and 6-8 hours 'off' to be satisfactory as the following quotation suggests:

"John Gadd, the British-based consultant who writes regularly for *Pork Journal*, was responsible for New Zealand producer Neil Managh "seeing the light". Gadd, who was in NZ on a speaking engagement, visited Neil's piggery at Feilding. "He wasn't in the operation one minute before he told us about what we should be doing in our mating area, and it helped us no end," Neil said. "He told us to light up our mating area, using a timing clock and good lights, so it is as bright as daylight but better. We played around and just set it. Now they come on at 6 am whether it is winter or summer and they go off at 9pm. Since we did that we rarely get a return. We just don't get seasonal infertility. We can wean 14 sows in a week or thereabouts. We usually do that on a Thursday or a Friday and by Wednesday it is very rare that the whole lot are not completely mated and finished – and we put it down to those lights. It is around the farrowing parts and the mating area that the lighting has really helped us. The funny thing is that it cost so little, about $200 or $300 and we were up and running."

Australian Pork Journal (1995)

On the other hand, we are being told by good researchers in the seasonal infertility area from Australia and S. Africa, that reducing the light to darkness ratio to 10 hours light, 14 hours darkness "restores good oestrus in sows and gilts, as well as more advanced puberty attainment in gilts across the summer". (Janyk, personal communication 2002). Is this the effect of light intensity or photoperiod (light to darkness ratio)? Does it mainly apply to strains where the 'feral factor' is still strong? We need to watch the research and resulting advice on this, and may eventually have two sets of guidelines for summer/winter or seasonal infertility/no-seasonal infertility circumstances.

Meanwhile, I still get plenty of comments similar to that in Table 1.

Table 1 EXTRA LIGHT BROUGHT FORWARD CONCEPTION

	Control	*Extra light*
Number of Sows	164	163
Average days to mating	5.9	5.5
Mated within 5 days (%)	68.5	83
Mated 6 to 10 days (%)	26.8	10.9

Source: Author's library data from another farm (UK 1993)

The ratio of positive comments from 111 farm visits on the subject since 1979 (90% of them followed up six months later or more) is 72 positive, 17 'don't really know' and 11 'no differences'.

STRESS

Stress comes in many forms, but in the case of Autumn Infertility, heat stress is paramount. Here are a list of things to consider to mitigate the effect of heat stress on breeding animals.

CHECKLIST : COMBATING NUTRITIONAL STRESS IN HOT WEATHER

✓ Is your food fresh enough?

✓ Is it stored in a cool place?

✓ Have you adequate water available? Troughs are better than drinkers. Water flow must be at least 1.5l/min. if so (0.33 Imp. galls).

✓ In the farrowing crate, a self-dispensing feeder is a good idea as the sow can eat when she feels like it during cooler periods evening, night, morning.

✓ Blow fresh outside air (even if it is hot) on to the food. The arrangement of a linking tube fed by a small 20 cm (8-9 in) fan gently ticking over delivering air to a down-pipe over each feed trough in the farrowing pen works well (Table 2) as the air is fresh over the food which seems to tempt eating in hot countries.

✓   Consult a nutritionist to alter dietary ingredients (*eg* more fat, less cereals). *See Hot Weather Nutrition checklists.*

✓   Feed pregnant sows and also boars earlier in the day – well before the stockperson's breakfast break.

✓   Use barley straw, not wheat, as bedding.

✓   Feed wet by pipeline. Especially consider it in hot climates.

✓   Do not overfeed thinnish pregnant sows 14 days or so before farrowing as many producers are tempted to do.

✓   The author has noted benefits from feeding a yeast-based additive – Yea Sacc (Alltech Inc.).

Table 2   BEFORE & AFTER RESULTS FROM SNOUT-COOLING LACTATING SOWS (PHILLIPPINES 1993)

	Before		*After*	
N°. of farrowed sows recorded	826		260	
Mean temperature*, °C (°F)	25°	(77°)	26.1°	(79°)
Airspeed m/sec*	0.3		0.35	
RH%	81		90	
Lactation feed intake, kg (lb)	3.8	(8.4)	4.5	(9.9)
Av. condition score at weaning	2.1		2.8	
Litter size (b/a)	9.1		9.9	
Sow weight loss in lactation, kg (lb)	n/d		10.1	(22)

* *Over the sow's back*   Source : Client's records

AI OUTDOORS?

Some may question the practicalities of using AI on an outdoor herd or less intensive situations, but the problems can be overcome. Feed can be used to entice the sows onto a high-sided, low-loading trailer where they are inseminated. It helps entry if it is the trailer used to transport their feed every day, but I worry about disease transmission. The AI materials are kept in an insulated box and the entire operation is carried out by one person. In the hottest weather when the sows were reluctant to move from the wallows or shade they were inseminated on the spot

CHECKLIST: COMBATTING MANAGEMENT STRESSORS IN HOT WEATHER

Particularly after service . . .

✓   Keep sow groups as small as possible and don't mix them until after implantation (14-22 days).

✓   Handle movements to and from boars gently, and confine to essential moves only.

✓   Check ventilation is adequate, especially fan capabilities. (*See Ventilation Section, Hot Weather.*)

✓   Drip cool farrowed sows, spray cool pregnant sows and boars (*ibid.*)

✓   Mate very early or very late in the day.

✓   Young boars seem to work better in extreme heat. If possible buy new boars in the spring.

✓   Use supplementary AI especially when it is hot.

Several results known to the author average out at over 2.0 more total-borns per litter from boar to AI service over boar only services *in hot weather* (Reed, 1990). At present AI prices, only 0.8 more total-borns (0.6 more b/a's) will easily pay for the AI plus labour, which must be considered as an extra cost in hot weather.

with no apparent problems (Sunderland, 1991) – but careful cleaning of the vulval area is essential! Insemination is often done in a central corral area, or by inseminating at feeding times, heat detection being done at the previous day's feeding.

Having said this, my experience is that outdoor AI is most cost-effective when:–

•   The herd suffers from infertility problems. Higher-performing clients who tried it showed little benefit over their own skilled conventional mating management, using boars.

•   During periods of hot weather, when it is very valuable.

- The outdoor stockmen attend the same hands-on training courses as their indoor colleagues.

  Here are some further tips gathered from producers across the world who are getting on top of seasonal infertility on their outside units.

WALLOWS AND SHADE

All pigs are prone to sunburn and this causes considerable stress leading to classic seasonal infertility. Mud wallows are an excellent sunburn protection and these must be kept topped up with water, but *not* diluted slurry effluent, when things are dry.

I have already mentioned shade as it affects too bright radiation in spring. Such days can catch the producer unawares, and I have driven past many paddocks near to roads and motorways where sows and gilts were out basking in the welcome warmth after a cold winter – and getting sunburnt in the process. It may take as little as 3 hours for this to affect productivity!

In general shades are put up too late as, in Europe, these days can occur in early April; shade is not so much a protection against heat, but against radiation.

To minimise trouble, posts and T-bars can be fixed to huts so as to erect plastic mesh-netting quickly in such cases. The placing of shade *between* huts' entrances encourages use but clear early spring days often come with cold, penetrating wind so a straw bale windbreak may be needed on exposed sites to encourage occupancy of an otherwise too cold area.

We have a similar problem with shaded areas outdoors in a heatwave – the sows must be encouraged to lie *in* the shaded area, so light bedding (permissible when it is dry) and some feed nuts in it may housetrain them at the start of a hot spell – they soon catch on. Again the between-hut mesh shading does the same job. Shading over the wallows is also used on some farms.

- In the USA wooden A-shaped outdoor huts are popular, with a window at the rear. This allows a through draft to cool the interior, which when closed in winter, makes it a snug enough refuge.

- Huts must be insulated.

A NUTRITIONAL ASPECT OF SEASONAL INFERTILITY: BOTH INSIDE AND OUT

Altering the nutrient allowance in pregnancy has been shown to lessen the problem.

Conventional advice on pregnancy is shown in Figure 4 where changing to a lactation diet for high performing sows or a special gilt pre-farrowing diet is fed about 14 days before farrowing to maximise litter weight.

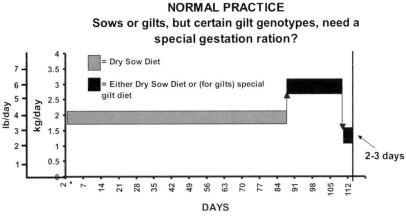

Figure 4 How specialised feeding for seasonal infertility in pregnancy compares with normal practice

However case histories of Summer Infertility trials, which I came across in 1993 (Love *et al.*) seem to be helpful (Figure 5)

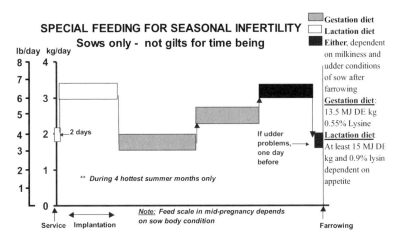

Figure 5 Specialised feeding in pregnancy under seasonal infertility cases

The Australian work suggests that relatively high post-service feed intakes (45 MJ DE/day compared to say 26-30 MJ/day) for group-housed dry sows minimised the adverse effects of summer infertility. In trials, groups of 22/23 sows were fed up to 25.8 MJ DE/day, compared to similar groups fed 43.5 MJ/DE day – both for the first 28 days after service. The higher intake sows had 80% positive preg-checks across the summer period as against 57% in the others.

After several years experience, the Australians have been advising 45 MJ/day *i.e.* 3.3 kg (7.25 lb) of a 13.6 MJ DE sow diet one month after service for all sows and gilts mated in Australia's 16 hottest weeks in Jan, Feb, Mar, Apr, which corresponds to our June, July, Aug, Sept. in the Northern Hemisphere. Summer infertility used to cost them about £3 million (sterling) annually covering the national herd of 290,000 sows, or £10/sow (US$15). I'm told now that this must have been halved – progress indeed!

MODIFICATIONS?

Some American contacts in the big conglomerates tell me that **farrowing rate** from increased post-service summer feeding is improved significantly in the autumn/early winter. Past research (at normal temperatures) has concentrated on the effect of post-service feeding on litter size and ignored what happens to farrowing rate, as to achieve a confidence limit – worth translating into advice – on a 5% hike in FR needs figures from at least 400 sows in the treatment group over a period. The US integrators can easily observe data from much larger samples than this on their 5000+ sow units, where each unit could provide 1000 dry sows as replicates, and some operators have 20 or more such farms to select from. Thus current American advice is also worthy of study.

As this aspect of summer sow nutrition is ongoing, you should refer to a pig nutritionist for the latest advice.

EMBRYO MORTALITY?
(ESPECIALLY IN GILTS)

These same sources advise a further modification, which is not to push the extra food in until perhaps **2 days after service** which should take care, they believe, of any adverse effect on embryo survival in multiparous sows, although whether this takes care of the gilt's problem of embryo reduction if fed more than 35 MJ/DE day immediately post service isn't commented on.

This probably helps to account for the increases found in litter size across several parities. The moral for me is – don't let the sow 'nosedive' in lactation, and consider adding organic chromium to the sows diet all the year round, assuming local regulatory approval is in place, of course. (New nutrient additives take some years to obtain full approval as being safe and effective).

MELATONIN – THE KEY TO THE DAY LENGTH PROBLEM?

Melatonin is a hormone released by the light receptor pineal gland *only during the hours of darkness*. Melatonin cuts down the production or release of gonadotrophin agents from the pituitary gland which can interfere with the development of the follicles and ovaries. Extending the day length would therefore reduce the inhibitory effect of melatonin very early on in the reproductive process. Conversely an increase in darkness in autumn/early winter reactivates melatonin's interference with early reproductive processes, hence the sows 'natural' disinclination to breed successfully as daylight shortens.

DO STRESSORS HAVE AN EFFECT?

Stress (anxiety, worry, fear, annoyance) is a poorly understood subject as it is so difficult to measure. Having been involved with pigs for so long convinces me that stress does interfere with many smooth metabolic processes and hormone pathways, and that in the case of seasonal infertility they are likely to involve, apart from disease…

Working in these hot conditions, as I find myself doing several times a year, suggests to me that what to do best for the gilt on farms with a seasonal infertility problem is a trade-off between summer infertility and embryo loss. A comparison between gilts' first farrowing performance in late autumn/early winter and how gilts do for the other 8 months could give a clue. We have a further problem in Britain in that we don't **know** if it going to be a really hot summer/early autumn, so perhaps it is best to err on the safe side and not feed our gilts too heavily for 3 weeks after service. The jury is still out on this one.

A STRESS CHECKLIST FOR SEASONAL INFERTILITY

✓   Heat stress, especially in the boar

✓   Stocking density, aggression, competitiveness and lack of adequate fleeing space.

✓   Putting small submissive sows (and gilts) next to dominant sows in the gestation stalls.

✓   Lack of access to shade, and where required, mud as a 'suncream'.

✓   Lack of access to water, both for wetting and drinking, including water/flow rate.

✓   Water temperature over 28°C (82°F).

✓   Indoor air heavily contaminated with gases.

✓   Rough floors, broken slats.

✓   Gestation crates too small.

✓   Diurnal temperature variation, especially causing night-time draughts.

✓   Noisy, rushed, non-empathic stockpeople.

✓   Noisy, unsettled gestation houses.

✓   Abrupt weaning.

✓   Hunger – especially lack of gutfill.

✓   Unpalatable food; mould/mycotoxin presence.

✓   Constipation.

All these stressors – you can probably think of others – can tip the herd over into a seasonal infertility syndrome.

TECHNICAL BACKGROUND TO SEASONAL INFERTILITY

While infertility problems in sows are rightly the province of the veterinarian, many management factors can be responsible on which the general adviser can help improve.

With global warming suspected to be on the increase, seasonal infertility is expected to become more of a problem as temperature stress and increased exposure to daylight are thought to be involved. In the UK, due to reduced capital costs and other benefits, outdoor breeding is increasing, hence the effect of season is likely to affect these problems too.

I find breeders adopt difficult remedial measures more rapidly if they understand a little of what is happening to their sows. First – as is often a good idea when dealing with animal problems – let's go back to nature, in this case the wild (feral) pig.

THE FERAL FACTOR

The wild sow always was, and still remains, a seasonal animal. But the wild boar will still mate a willing sow whatever the season. Thousands of years of conditioning has programmed the sow's biology to decide, if made pregnant and the season ahead seems adverse, to call it a day and dispose of the litter.

Shortening day length seems to be the key to this process – this is the advance-warning built into the sow's consciousness wherever she lives, so with a reducing light pattern, chemical reactions are set in train involving the hormones maintaining pregnancy to reduce to levels which will not sustain the pregnancy process, even if satisfactorily mated.

Shorter day lengths start to predominate from midsummer into early autumn and wild sows mated during this period may well conceive, but because the forthcoming litter will then be born when food is short with warmth and shelter difficult in the depths of winter, the hormone progesterone (remember it as being aptly named – ' the **pro-gest**ation-horm**one**'!) which is essential to maintain pregnancy, is dampened down by shorter day lengths.

Conversely as day length increases, the dampening effect is removed, and sows mated in mid to late winter are stimulated hormonally to farrow in the comfort of spring & early summer. The 'sensor' is the pineal gland which is a natural light receptor. Hidden away at the base of the skull, it is activated by light impulses from the eyes via the brain which then activates the adjacent pituary gland to set in process a chain of hormone events affecting reproduction positively or negatively.

This is why so much attention has been paid to 'fooling' the pituitary, by providing extra light to extend day length artificially in autumn, into maintaining progesterone levels. Trouble is, this is easier said than done. And, as we have seen, other factors may interfere.

These are nutritional stress, loss of body condition, social stress, extremes of temperature and disease. Trouble is, again, any one of these in combination with the others can tip the scales in favour of infertility. For example, a thinnish sow, slightly underfed, weaned into a pen of strangers, some of which are bullies, forced to lie out in a draught at night in a rather dirty resting area *may* not react – even to these insults – except in the late summer or early autumn.

So progesterone level may be threatened and nothing untoward results when such conditions apply, but the onset of shorter day length on top of too many of these adverse factors may precipitate abortions or reduced litter size. That is where management advice can help.

CAN WE INJECT PROGESTERONE?

This often doesn't work and would be expensive anyway – suggesting that the reaction may occur elsewhere in the hormonal pathway. Don't resort to hormones unless the veterinarian suggests it. A better solution – at least for the time being – is to do an overall audit of all the factors that are thought to affect the maintenance of progesterone by natural means and put them right if detected or suspected.

1. *High feed levels in early pregnancy of the gilt* may reduce embryo survival. This is because a high food level after service increases blood flow to the liver to deal with the arrival of these extra nutrients. This leads to a lower progesterone level. Meanwhile the newly formed embryos secrete a protein (RBP or Retinol Binding Protein) which helps them establish themselves in the uterus. If the progesterone level falls, the secretion of RPB is hindered and embryo mortality increased.

2. *The same high food levels in early pregnancy in the second litter sow onwards* do not seem to affect embryo survival or litter size to anything like the same extent, unless they are considerably overfed.

Why?

Sows seem to have, naturally, a higher blood progesterone level after lactating. But a sow which has nose-dived in condition may not show this. The hormone insulin is probably vital in likely feed energy metabolism and reproduction. If the sow is energy deficient in lactation due for example to underfeeding or a larger litter, insulin is reduced and very soon progesterone production is compromised. Chromium (*e.g.* organic chromium from yeast) is probably best and is an important precursor of insulin production. It is only recently (1999) being added to breeding foods (at the low level of 200-300 parts per billion), which probably helps to account for the increase found in litter size across several parities.

CHECKLIST: DEFENCES AGAINST SEASONAL INFERTILITY

Finally let's go back over some things you can do…

If outdoors…
- ✓ *Provide shades*, erected early enough in Spring.
- ✓ *Check for sunburn daily*; bring indoors.
- ✓ *Provide insulated huts*, facing down the wind direction.
- ✓ *Provide wallows*, sited not too far away from the social areas.
- ✓ *Ensure some pigmented genes* are in your females.

Indoors and in general…
- ✓ *Use AI 24 hours after each natural service*. A central service unit (or if outdoors an ecoshelter in the service paddock) is advisable to facilitate this, accentuate pheromone exchange and keep a check on services.
- ✓ *Serve in the cool* of the morning.
- ✓ *Keep boars present* for the first four to six weeks of pregnancy.

✓ *Up your service rate* by 10-15% if Seasonal Infertility is likely.

✓ *Do a mycotoxin audit* (*See Mycotoxin section*).

✓ *Check your lighting adequacy* and diurnal light pattern.

✓ *Discuss with your breeding stock supplier* if he has strains where selection for decreased seasonality are available, *e.g.* percentage of pigmented genes, and stress-tolerant (docile) strains.

✓ *Discuss with your boar supplier* about doing a boar sub-fertility test during hot weather. Some boars are more hot weather resistant than others.

✓ *Record the maximum temperature* in the service area every day.

✓ *If the outside temperature rises past 23°C (73°F)*, use AI top-up indoors or out in the summer/autumn.

✓ *Check temperature variation* within the service/breeding unit. In early spring and autumn this can cause abortions.

✓ *Provide lids* in winter or if the early autumn is unseasonably cold for yarded groups to even out temperature variation, but make sure the run outside is well lit, and that the lids can be raised in summer or air blown through in on hot days (not at night).

✓ *Check feed intakes* in summer.

✓ *Watch you don't cut down on space allowances*, especially in summer.

✓ *Ensure you have organic* chromium, selenium and iron in the breeding feed.

REFERENCES & FURTHER READING

Australasian Pig Service Assoc (1987-1999) Various annual volumes of 'Manipulating Pig Production' contain valuable research data.

Dzuik PJ (1987) Embryonic Loss in The Pig: An Enigma 'Manipulating Pig Prod' 1 28-39

Gardner JAA, Dunkin AC and Lloyd LC (eds) (1990) 'Pig Production in Australia' Butterworths (for Australian Pig Res. Council)

Love *et al* 'Summer Infertility Effect of feeding regime in early gestation on pregnancy rates in sows' cited in 'Animal Talk' (eds Close & Cole) Sept 2001

Mackinnon J (1996) Autumn Infertility – Notes (Pers Comm)

Reed H (1991) Summer Infertility, Pig Farming June 1991 30-35

Sunderland, C (1991) Summer Infertility NAC Newsletter Oct 1991 pps 6-8

BUSINESS AND
MANAGEMENT SECTION

CONTENTS

NEW TERMINOLOGY FOR THE FUTURE 9

Much has appeared in the technical and lay media concerning problems associated with industrial pig production – pollution, disease control, training adequate labour and the perception of food safety and pig welfare by the consumer. This is true, especially from large new units.

But largely unremarked on by the press is the rising trend for pig farmers and managers to become increasingly business-orientated – to measure their progress by *profit*, not necessarily by *physical performance*. The day of the 'art' of keeping pigs by dedicated enthusiasts rather than as vehicles for making money is over. Pig owner/producers of the latter half of the 20th century needed to see, touch and smell their charges daily; until recently they considered themselves as *pig* producers. Now after a storm of low pig prices they realise they might be in the business of producing *meat*, not pigs. Falling profits have accelerated this change in perception. In other words, they do indeed now see themselves as meat producers. With more businessmen pig farmers around, their need to use terminology based on profit, not necessarily performance, is also essential.

This chapter makes a case therefore for new terminology based on profit, but which also embodies most of the performance-orientated terms which have served us so well as farmers in the past, and which, of course, the academic will continue to use with the precise measuring facilities available to him.

At the sharp end the two will continue to run together for a few years more, but as more businessmen producers realise how simple profit-related terms help them make better business decisions, the use of the New Terminology will become increasingly popular.

The examples used to illustrate the use of the New Terminology draw heavily on matters considered in other sections of this book.

RESEARCH AND SUPPLY TRADES

Both academics and especially feed and seedstock company technical staff will need to familiarise themselves with the new concepts. This is particularly important to feed company nutritionists as the New Terminology makes it easier for the feed company to sell increasingly nutrient-dense and more expensive (per ton/tonne) feed in a pig production market place which is (sometimes to its own detriment) increasingly cost-reduction conscious.

The New Terminology also helps prioritise in the businessman farmer's mind how best to invest the 8 to 15% extra to his feed raw material costs which he can use to add value to his feed. Nutritional biotechnology products – like organic selenium, iron 'proteinate', the Bioplexes (especially zinc/methionine combination) and enzymes are all technical innovations which at first sight may seem to cost a substantial amount per kg in the bag or drum. Added enzymes especially are an exciting new area for pigs, too, especially in the future, and they are not cheap either.

Because of the relatively very low rate of use of all of these items, they can give quite dramatic *economic* returns of 10, 20 or even 60:1 across a year's use. *Such payback far outstrips their percentage improvements in physical performance terms*. This is why we need new measurement terms to highlight these hidden advantages, and keep pace with what they reveal in profit terms. In addition, because of this low rate of use both physical and economic space in the diet can be freed up for investment in other nutritional improvements – increasing amino-acid or energy intakes among increasingly appetite-challenged pigs, for example, and/or better/safer quality raw materials in the young pig, such as 'Ultimate Protein', rather than plasma protein.

SCOPE OF THIS SECTION

I am certainly no economist or mathematician, but from working at the sharp end recognize that we need to move on in our world of business pig production and have available a new set of easy to calculate measurements which will get us faster to making a satisfactory profit. On-farm consultants are successful when their advice is seen to generate more profit for their clients. Many years ago I began to realise that the physical measurements we all used were holding me back in persuading pig producers to take certain actions, and that an alternative set of terms was needed to reinforce my advice. To convince wary pig producers that a product or system which seemed expensive was in fact very cheap – far too cheap to ignore.

The complete New Terminology concept covers a wide range of disciplines *e.g.* EBV (Estimated Breeding Value) and WWSY (Weaner Weight/Sow/Year) in the case of breeding stock; PLR and ILR (Profit and Income to Life Ratios) in the financial, housing and equipment fields; and AMF (Absolute Mortality Figure) and others affecting disease costs in the veterinary area.

This chapter is mainly confined to those new terms useful in the economic assessment of the nutritional and feed supplement fields (examples are given in Table 1 but I also describe other terms mentioned in the book).

Table 1   THE NEW TERMINOLOGY (AS IT AFFECTS NUTRITION & NUTRITIONAL SUPPLEMENTATION)

Existing Terminology	*The New Terminology*
Food Conversion Ratio (FCR)	Return to Extra Outlay Ratio (REO)
Average Daily Gain (ADG)	(Saleable) Meat/Tonne of Feed (MTF)
Cost/kg gain	Price Per Tonne/Ton Equivalent (PPTE)
Return on Capital Investment	(Producing) More for the Same Cost
(ROC; ROI)	(MSC)
	(Producing the) Same at Less Cost
	(SLC)
	Annual Investment Value (AIV)

WHY THE NEED FOR NEW MEASUREMENTS ?

There is nothing radically wrong with the old terminology. After 70 years of use, it is certainly very familiar! Even so, it is not good enough for today's conditions. We can do better.

Problem 1   The existing terminology is based largely on *performance*. Today *profit* matters on pig farms far more than it ever did. You can have very good performance but still make less profit (Table 2).

Problem 2   The existing terminology also mainly covers *costs* of production (*e.g.* cost/kg gain). Again, a producer can have nice low costs but still suffer reduced *income* (Table 3) which can turn low costs on their head and result in less nett profit.

Table 2 PROFIT RATHER THAN PERFORMANCE

PROFIT   Here are some actual returns from people in each category. Sale liveweight av. 87 kg

(n) Pigs Weaned Sow/Yr (Pigs sold Sow/Yr)	Wt of saleable meat per sow/year		Wt of saleable meat/tonne feed		Relative nett profit per pig/sold
	(kg)	(lb)	(kg)	(lb)	(%)
2 29.7 (26.9)	1778	3921	229	505	109%
8 27.1 (25.7)	1642	3621	235	518	111%
14 25.0 (24.1)	1540	3396	240	529	115%
100's 22.0 (20.1)	1285	2833	209	461	100%

Comment :The producers at 25 pigs weaned/sow/year made most profit. Note how meat produced per tonne of feed is a more reliable guide to profit than meat produced per sow per year.
Source : Clients' records, and M.L.C. (UK) Annual Yearbooks (1996-99)

Table 3 REAL-LIFE EXAMPLE OF ADJACENT FARMS IN THE SAME FAMILY BUSINESS SHOWING RELATIVE COSTS AND SALES INCOME OVER A 3 YEAR PERIOD

	Low cost farm	Higher cost farm	Position of lower cost farm
Pigs produced/year	5320	5610	
Cost/kg gain (p)	36	40	4p/kg savings (2.6c/lb)
Deadweight meat sold year (1000 kg)	345	387	BUT ....
Income per pig sold (£)	66	80.7	
Income per liveweight kg sold (p)	112	117	5p/kg lower income (3.3c/lb)
Nett margin per pig sold (£)	8.02	8.68	66p/pig less margin

Comment : Lower cost/kg gain but much less profit per pig sold! Costs only give you half the story.
Source: Clients' records (1998)

Of course reducing costs is a good thing. But never to the extent that it affects income to a greater degree. The existing terminology doesn't necessarily identify or forewarn of such situations or trends, **which the new terminology does.**

Certainly, use both together. Scientists who need to assess physical performance accurately and use terms such as food conversion will prefer to stick with the old and familiar terms. But when dealing with pig farmers and the feed trade who supply them their advisers will find an increasing move towards the new terms will help the feed compounder, the veterinarian and the pig producer.

HOW THE NEW TERMINOLOGY HELPS THE FEED MANUFACTURER

There are two major sectors in the animal feed industry, the complete feed manufacturers and the supplement supply trade who service the on-farm mixer.

Both are bound by the same fact-of-life – **superior quality invariably costs more.**

So the salesperson's problem is to convince the customer to pay more for a good quality product.

The customer's problem lies in trying to distinguish between (possibly over-priced) good quality on offer and low-cost but potentially questionable quality.

The new terminology helps the feed and feed supplement manufacturer assess the cost-effectiveness of his products quicker and more easily and gives him confidence in passing on the same message to users of his feed.

HOW THE NEW TERMINOLOGY HELPS THE FARMER

The new terminology clarifies the issues in the customer's mind in two ways.

1.   It sets the quantity of his primary *input* (food) against the total *output*, meat (in our case, pork).

2.   Farmers are still very concerned with price per tonne (ton) of feed. While we can argue the validity of this attitude, salespeople can use it to help them sell successfully by using the new terminology to relate the value of the extra meat sold on a cost per tonne (ton) of food basis.

PPTE

This is of great help to the farmer in arriving at a decision because pig producers all know their current feed cost per tonne (ton) and are equally familiar with the current price they get for their pigs, or more accurately saleable (*i.e.* deadweight/dressed carcase) meat.

The new terminology, in this case PPTE (Price Per Tonne (Ton) Equivalent), presents performance data in a way which is very easy and swift for the farmer to do an econometric* calculation himself, because it shows the benefit if any, in terms of what it saves him on a cost per ton(ne) basis. If he does this calculation himself he immediately *convinces* himself.   As conversion to PPTE is a childishly simple calculation, he is very likely to do it.

*Econometric = the measurement of cost effectiveness.

THE NEW TERMINOLOGY HELPS THE SALESPERSON, TOO

He/she may be part of the feed supplement supply trade; alternatively many of their customers will buy complete feeds.

Virtually every product sold has between 2 and 20 competitive products also being offered to the prospective customer by other salespeople. That's bad enough, but in addition the customer only has a limited amount of capital to invest per tonne (ton) on feed supplements. And at the last count there are over 80 feed additives in Europe he *could* use – all different! If he used only a third of them his feed cost/tonne would rise eight-fold and the amount of remaining space for feed nutrients/ingredients fall to only 21%!

The producer has to prioritise what could be the best value for him and his pigs. The New Terminology, in this case REO (Return on Extra Outlay), MTF (Meat per Tonne of Feed), PPTE (Price Per Tonne Equivalent) and AIV(Annual Investment Value) are ways of demonstrating, *from published trial work*, the *Value for Money* of the products on offer. He can then see very quickly where he can get best value for the limited capital he can allocate – for example – to feed supplements.

SO WHAT'S WRONG WITH THE EXISTING TERMINOLOGY ?

FCR (Food Conversion Ratio)

FCR

A useful yardstick *if it is measured accurately*. This is difficult to do on a busy working farm. Careful tests suggest that even if they attempt to keep track of it (only 20% do) pig farmers still get it wrong by about 0.2:1, which is equivalent to a rise or fall in their price per tonne of feed of 15%! Most take a decision to buy or not on far less price difference than that.

While Table 4 is over 25 years old now, things don't seem to have improved much, as two investigations carried out last year revealed slightly greater discrepancies!

Table 4   FARMER'S CALCULATED AND ACTUAL FCR'S TAKEN FROM 5 COMPLAINTS OVER POOR PERFORMANCE (1975 - 1977)

Farm	Weight range		Farmer's estimated FCR	Actual FCR based on careful measurements on-farm	Likely reason for error
	(kg)	*(lb)*			
1	6 – 28	13 – 62	2.9	2.71	Input food not weighed
2	20 – 91	44 – 200	3.2	2.86	Poor recording
3	30 – 90	66 – 198	2.9	2.81	Mistake over input batches
4	25 – 86	55 – 190	2.6	2.92	Guesswork (!)
5	30 – 64	66 – 141	2.6	2.45	Poor recording.

Average error of 0.22 FCR over 48 kg (106 lb), an 8% error
Source : RHM Agriculture (Unpublished) 1977

Average Daily Gain (ADG)

ADG

This is a measure of growth. But how much of the growth is lean meat, or low value offals, bone, fat etc? It may be good growth, *but is it the wrong sort of growth*? ADG doesn't tell you. This situation often occurs when growing pigs 'catch up' with unaffected pigs after an illness. On recovery they can grow faster but it tends not to be *lean* growth so much as low value gut offals and fat, often with a poor FCR despite the fast growth. ADG doesn't tell us this. MTF does. Economically ADG is really only useful in determining the contribution that fewer days to market and thus reduced overheads make to profit – often not insubstantial, it is true.

Cost kg/gain (or cost/lb gain)

Only measures *cost*, it doesn't marry it to *income*. Any businessman knows the two together are needed if profit is to be made. Costs can be cut so low that income falls to a greater degree so that profit is reduced.

THE NEW TERMINOLOGY

REO

Return to Extra Outlay Ratio (REO)
Note: REO is not the same as ROI (Return on Investment)

REO is of great help to the feed trade, and should be to academics too. As we have seen, a farmer cannot possibly use all the additives / feed supplements on offer. Generally he is prepared to invest an increase in his cost per ton(ne) (usually about 8 to 10%) to include a protective, growth-enhancing or nutrient-sparing additive. The question is which ones will give the best value for money? REO helps considerably because it indicates, from published trial work (usually expressed in the old 'performance-based' terminology) which of them is the best value for money.

REO tells you, from each monetary unit invested per tonne of feed, how many monetary units are likely to be recouped also per tonne of feed. How much *return* you get for the *extra outlay*.

ROC

Is this the same as ROC (Return on Capital = Return on Investment, ROI) ? No; REO involves the *smaller* sums spent on boosting profit or performance – like feed additives, minor improvements in housing or alterations to diet density – while ROI is used for more comprehensive investments - like new buildings. Moreover REO is a measurement of the *extra* investment required, for example organic selenium costs more than Na selenite; non-GMO phytase more than dical, *etc.*, but the REO – the extra income likely from the extra investment needed – puts the requirement for extra working capital into perspective. In an urgent cost reduction situation, farmers object strongly to paying more for what they consider to be the "the same" additive or feed formula they have been used to. This is particularly true during times of marginal profits. REO helps to unblock this mindset, especially in the case of organic selenium which can cost at least 50 times more per unit of selenium compared with bimodal selenite popular in the last 10 years. REO shows that, despite this, it is still extremely cheap for what it can do.

Table 5 illustrates REO and ROI taking organic selenium as a replacement for sodium selenite.

Table 5 ECONOMIC PAYBACKS FROM INCLUDING SELENO YEAST (SELPLEX) AT THE ADVISED LEVEL IN PIG DIETS AS A REPLACEMENT FOR SODIUM SELENITE

Trial Source	Result	Physical Benefit*	ROI†	REO†
Janyck (1998)	Piglet growth rate	+4.7 %	7 : 1	17 : 1
Ibid.	Litter size	+6.7 %	4.4 : 1	11 : 1
Munoz (1996)	Drip loss, pork 72 hrs	- 12.0 %	3.8 : 1	4 : 1
Mahan (1998)	Sow litter performance	+ 0.5 pig/litter	20 : 1**	25 : 1

Key
* Over Na Selenite in feed
† ROI = return on total investment/tonne (of including Se)
† REO = return on *extra* investment/tonne (of Selplex compared to the cheaper cost of Na Selenite)
** From extra meat sold from one year's sow's progeny at slaughter (estimate, computer model)

	UK£ / tonne	US$/ton
Estimated cost of sodium selenite	<£ 0.05	$0.07
Cost of Sel-Plex	£ 1.54	$2.16

Conclusion

While the *extra* cost of selenium yeast inclusion provides a 1% rise in raw material cost/tonne the REO paybacks could vary from 4 to over 20:1. Thus REOs put the apparent massive rise in inclusion cost for the 'same' 0.3ppm Se into perspective. In simple language – something costing 30 times more can give a return on the extra investment of between 5 and 25:1.

REO

MSC and SLC

There are two main routes to making profit – produce More at the Same Cost (MSC) or produce the Same at Less Cost (SLC). Of course, there is a third way, which is to produce **More** at **Less** Cost, but this is only rarely achievable in practice – usually not the case in pig production! Historically livestock production has concentrated on increasing productivity and trying to hold costs down, but all too often this has resulted in *over*-production causing the pig price to drop. This fall in income swallows up any extra profit from producing more – and the producer is no better off! So if we all adopt MSC and produce more at the same cost, we are on shaky ground in profit terms.

What is more sensible, especially today when producers are becoming more proficient at many livestock tasks, is to hold production at an adequate level, but *concentrate on reducing the costs of doing so*. This way over-production does not occur and as a result the pig price keeps up, meanwhile the reduced costs contribute substantially to the profit. So SLC or producing the Same (assuming adequate performance) at Less Cost seems much better.

But How Good is 'Adequate Performance' ?

This is the key question. *How good does performance need to be to allow maximum attention to be given to reducing costs without damaging physical performance.*

Figure 1 gives two examples – the first involving sow productivity and the second slaughter pigs. The right hand area of each graph suggests the degree of importance the producer should devote to improving performance, and the left side to saving costs. As his performance improves, so the attention devoted to each sector moves from right to left.

Similar sigmoid 'S' shaped curves can be plotted from data provided by computerized recording schemes (preferably with more than 300 sow litters and their progeny) using bottom 10%, bottom third, top third and top 10% and the median averages as locator points. A wide variety of performance criteria can be used, such as empty days, farrowing index, piglet mortality, saleable meat per m^2, *etc*.

Remember these are *guides* to how much management effort should be allocated to MSC or SLC and not hard-and-fast criteria, but they do help to answer the question "How good does my productivity need to be before I can ease up on improving productivity and really spend time (and money) on cost-reduction?"

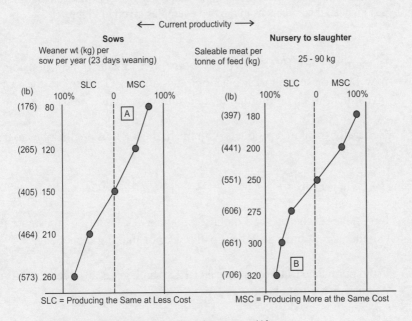

Figure 1 SLC or MSC - which gets priority?

This depends on current physical performance. I give above my estimate of the attention a producer should give to each dependent on his current physical performance (based on MLC Yearbook, 2002)

Instructions to farmers on how to use this figure

Calculate the amount of weaner weight produced per sow per year in kg (or lb); and the amount of saleable meat (kg/lb dressed carcase weight) produced per tonne/ton of feed from exit from the nursery to slaughter.

Read off on the graphs how much attention (in %) you need to devote to SLC or MSC in proportion. For example, if you are at (**A**) you need to devote about 60% of your time to improving sow/weaner performance and perhaps 40% to saving costs without destroying what breeding performance you already have. At (**B**) however, the physical performance (FCR:ADG) is so good that it is unlikely to improve much further without incurring high costs. So while trying to maintain this performance the producer should devote about 85% of his time towards *reducing the costs of doing this.*

The key (left hand) performance scales vary between national pig industries, of course. U.S.A. would have very different reference scales to Thailand, for example, which is why they should be constructed from local national figures. If 5 reference points are taken for each scale, the sigmoid shape will nearly always emerge, some more pronounced than others as in the two examples given in Figure 1.

ROI
ROC

Remember – ROI/ROC involves *major* investments in new housing, a new farm or section of a business, etc. and REO relates to the *extra* investment to be committed to a current programme or strategy.

REO

Generally speaking, products with the highest REOs are the ones to use first.

Using REOs in practice

Let's take an unspecified dietary enhancer. A variety of these are an excellent, if at present rather costly, way of making diets better.

From published trials they have an REO which gives an excellent return. About 7.5:1. First class!

But would it be better to use one or more alternative supplements **which have a much higher REO**? They cost less to add, take up less physical space in a tonne of feed, and while individually they don't approach the payback of the original additive, cumulatively they might do so for less cost, sometimes much less cost. And thus release more space in the feed, and more capital to be used elsewhere to improve the diet up to the 8% ceiling. Table 6 shows the concept.

Table 6   HOW REO CAN BE USED TO COMPARE POTENTIAL FEED ADDITIVES - total dietary cost £ 160/tonne. A good growth enhancer may cost up to 4% of the dietary cost (£6.40/tonne) and yields an REO of 7.5:1. £6.40 × 7.5 = £48/tonne)

	Rate of use & cost per tonne			*Expected return per tonne*		
Additive A	1%	=	£1.60	REO	8:1	=	£ 12.80
Additive B	2%	=	£ 3.20	REO	10:1	=	£ 32.00
Additive C	0.5%	=	£0.80	REO	5:1	=	£ 4.00
Total			£ 5.60				£48.80

These figures are based on current European costings, but the same principle can be used with US Dollar costs, where the REO will come out at 3:1 overall.

These three additives have given us virtually the same return as the more expensive one used previously. So . . . where's the benefit? Why bother to change?

(1) *Either* SLC is improved (Producing the *S*ame at *L*ess *C*ost). We now have virtually the same return *for four fifths of the original capital invested.* This reduces feed costs and improves cash flow because we've spent less.

SLC

(2) *Or* : MSC is improved (Produce *M*ore at the *S*ame *C*ost) – we can use the £0.80/tonne saved (12.5% of the cost of the major one-off additive used previously) for other dietary improvements *thus improving performance for no extra cost.*

MSC

REO

REO is an excellent way of comparing products to make limited capital go further

Please be assured, I am not against adding enzymes, growth enhancers, or increasing the amino-acid levels/energy, or the use of anti-mould protectants or any of the feed improvement products the producer can use cost-effectively these days. But using REO figures *obtained from reliable published trial* work used by most good suppliers to support their sales claims enables the pig producer:

• To assess which of these products are best value for money.

• To use his valuable investment capital in the most effective way, which means achieving the shortest route to obtaining the most profit.

Remember, REOs are worked out from the vendor's own claims. REO merely ranks these in order of cost-effectiveness.

REO removes the apparent disadvantage of concentrated 'expensive' products

Many new feed additives are concentrated, low usage rate but expensive *per kg* or *per lb* products in themselves before addition to the feed. Nutritional biotechnology products are typical examples. Using the easy-to-explain REO concept puts their value in a true econometric (value-for-money) light and helps the customer prioritise the options on offer, and dissuades him from choosing something just because it appears cheap or cheaper than most.

How come some REOs are over 20:1?

Any banker would sit up to attention if you proposed a scheme where he lends you a dollar and you benefit by as much as twenty dollars. Is this really possible in pig production, you may well ask!

When a good antibiotic growth enhancer like Carbadox or Tylan may have yielded what seems to be a modest 5:1 REO, how come some R.E.O's can achieve the high 20's or more? The example of Bioplex Iron in the sow's feed is interesting. For example, Table 7 shows trials to August 1999 which have suggested very high REOs.

REO **Table 7 REOS CALCULATED ON IRON PROTEINATE (BIOPLEX IRON)**

SOUTHERN HEMISPHERE (50,000 sows trialled)

Results :	Weaning to service interval reduced by 1 day		
	Weaned pig mortality reduced 1.5%.		
	Combined extra value sow/year	£ 15.23	
	Cost of Bioplex Iron per sow per year	£1.49	= REO 10 : 1

UNITED KINGDOM (Smyth, quoted by Close, 1999)

Result 1	26 day weaners 290 g heavier		
	Value per sow per year of these at slaughter	£ 15.23	
	Cost of Bioplex Iron per sow per year	£1.75	= REO 8.7 : 1
Result 2	Weaners were more even – 9% fewer 'smalls', 8% more 'larges'. This should give a 2 day quicker close-out at slaughter.		
	Worth per sow	£ 2.46	
	Cost of Bioplex Iron per sow per year	£ 1.75	= REO 1.4 : 1

UNITED KINGDOM (Brown, quoted by Close, 1999)

Result	0.52 fewer weaners lost per litter		
	This is worth £30 more income/ sow/year	£ 30.00	
	Cost of Bioplex Iron per sow per year	£1.39	= REO 22 : 1

The reduction in close-out time alone more than paid for all the Bioplex Iron needed.

AN REO CHECKLIST
FEEDS AND FEED ADDITIVE PRODUCTS

✔   Find out the *inclusion rate* of the product you are asked to buy.

✔   From its unit price (per tonne/per kg etc) work out its *inclusion cost*/tonne of feed. Include any mixing charges/extra labour costs if these are additional to your current procedure.

✔   Work out the *performance benefits claimed* for the product, being very careful to request evidence/proof of performance and from where/how the evidence was obtained. If in doubt refer the data to an independent consultant/scientist.

✔   Convert the physical performance benefits into likely *economic benefits* – for example a Meat per Tonne of Food figure is a very useful one to set against inclusion cost/tonne feed or extra inclusion cost/tonne feed if the product is an alternative to an existing one. There are several others like WWSY (Extra Weaner Weight per Sow Year) you can use.

✔   Calculate the REO.

✔   Use the REO to compare it with other REOs from alternative products or systems you can use.

✔   Generally speaking the product likely to provide the highest REO is the one to use (first).

✔   Finally, a refinement of the REO concept is to graft on to it an AIV (Annual Investment Value) which shows you how many times a year you would recoup the REO (AIV is discussed later on page 132).

MTF (saleable) Meat Per Tonne Of Feed　　　　　　　　　　　**MTF**

The new term MTF is as important as REO. This is because pig producers sell *meat* (pork and bacon) not pigs!

MTF relates the producer's total income against 58% to 66% of his total costs – food.

Using MTF as our primary yardstick means that we don't need to calculate FCR which as we have seen is difficult to collect and usually an inaccurate figure in the producer's hands. This is because meat is about 72% water and water is much cheaper than food! So the better the MTF figure the better *must* be the FCR – with no need to try to record FCR on the farm, as distinct from a research unit.

MTF   **MTF is easier to use**

The MTF figure is much easier for farmers to record. This is because pig producers are paid – or should be paid – on dressed carcase weight (dcw) at a variable price per kg or per lb. So each week or month they know accurately what their 'meat' income is per kg or per lb of the pigs they have shipped for slaughter. As for food, they know from their feed invoices how much food the pigs have consumed on a running basis for that week or month, and also know how much they will be paying for it per tonne/ton.

PPTE   **Price Per Tonne Equivalent (PPTE)**

MTF is also useful because it can quickly be converted into an equivalent price per tonne figure (PPTE). As we saw with REO (Return on Extra Outlay) pig farmers can often be overly concerned with price per tonne. While it is important, price per tonne can be a fickle friend, but if the preoccupation with it exists with the producer, let's use it, not fight it!

If the MTF reveals a higher or lower figure for the period in question – for example 20 kg more MTF – the producer knows what the current price is for saleable meat, say 90 pence/kg deadweight. 20 x 90 pence = £18, so in this case an improvement in MTF of 20 kg is equivalent to £18 per tonne cheaper food. Simple ! Work it out in your local prices to see how simple it is. PPTE has the advantage that it is easy and quick for the farmer to do his own calculations, thus he convinces himself the outcome is correct! If he is provided with the data and the method of working it out and he does the sums – the product is much easier to sell.

Table 8   HOW TO CALCULATE AN MTF FIGURE

(1)   Establish how many pigs are produced per tonne of feed
e.g. FOOD EATEN 250 kg

$$\frac{1000}{250} = 4 \text{ pigs/ tonne}$$

(2)   Calculate saleable meat produced / pig
e.g. 75 kg liveweight put on x 75% killing-out percent
= 56.25 kg (deadweight per pig)

(3)   MTF = 4 pigs x 56.25 kg   =   225 kg Meat per Tonne Feed

In the US we have heavier pigs, with a higher dcw and eating more food. Thus the figures may look like...

(1)   Food eaten (from 45-265 lb) 725 lb

$$\frac{2000}{725} = 2.76 \text{ pigs/ ton}$$

(2)   220 lb put on x 78% KO percent = 172 lb dwt/pig

(3)   MTF = 2.76 pigs x 172 lb = 475 lb Meat per Ton Feed

This is particularly valuable in feed trials, something which both academics and many feed firms so far have chosen to ignore. Everyone involved in pig production should use PPTE/MTF and benefit from the advantages.

What is a good MTF figure ?

MTF

Table 9 suggests what these could be on a world basis. Because of this note the modest FCRs quoted; even so they are fairly typical of average producers world-wide.

It seems that, at present, across the world a figure of 250-275 kg (550-606 lb) of saleable meat per tonne of feed is the one to achieve. However, in certain countries, and among certain producers, performances are higher than these world averages and 300 kg (661 lb) MTF is their current target with the top 10% producers easily achieving 325-340 (717-750 lb) MTF All these refer to the 25-100 kg weight range, with the Americans with their different production ranges, about 10-15% lower.

Table 9 PERFORMANCES, WORLD-WIDE 1998 (25-100 kg or 55-221 lb)

	Poor	*Typical*	*Good*	*Target*	*Exceptional*
ADG, grammes (lb)	518 (1.14)	617 (1.30)	681 (1.50)	763 (1.68)	1000 (2.20)
FCR	4.38	3.68	3.33	2.98	2.28

Extrapolated from these figures, at 84 kg (185 lb) of saleable meat per pig (73.5 K.O%) ...

Saleable meat per tonne of feed, kg (lb)	185 (408)	220 (485)	245 (540)	275 (606)	360 (794)

A word of caution

Just like the old term FCR, the value of MTF depends on the weight range taken. Pigs are much more efficient at turning food into meat early on in their lives than when close to slaughter. In Tables 7 & 8 the weight range taken is typically from 25-100 kg (end of nursery to slaughter).

Therefore a target MTF (world-wide) over this range is 275 kg.

However if the range is 7-70 kg (weaning to light slaughter) . . . the target MTF would be 313 kg.

And if 60 kg - 115 kg (the heavier grow-out period in U.S.A.) . . . the target MTF would be 230 kg, or about 500 lb in the US.

Does this mean that MTF is unwieldy / unusable ?

Not at all. We are using MTF to give a more profit-orientated method of assessing performance than FCR/ADG As with this old terminology we are using MTF to demonstrate the improvement a feed or feed additive product can make over the controls / competition in a *comparative* manner. So as long as the weight range is the same there is no problem. As was always the case with FCR (but rarely done) we need to record over what weight range the figure refers to, so as to compare like with like.

So when using MTF in a production target manner, please can we start correctly and always qualify it with the weight range cited, *i.e.* . . . "A target of 275 kg MTF across the 25 - 100 kg weight range".

Written as "275 kg MTF (25-100 kg)."

How the modern salesperson uses MTF & PPTE to sell products to farmers

At first MTF and PPTE may seem rather removed from the supplement-selling task as distinct from selling complete feeds. This is not true. At present, most trial results quoted in this area of meat production involve the company getting across better FCR's, ADG's or lower cost/kg gains of the products they sell. We've seen the disadvantages of . . .

FCR :   Yes, better, but at what cost ? Did it cost too much to get it lower?

ADG :   Yes, better, but how much of it was the right sort of growth (lean meat) ?

Cost kg/gain : Yes, lower, but did the income suffer as a result ?
(or cost/lb gain)

The intelligent salesperson takes the trial results given in the old performance-related terminology and grafts on to them the REO and MTF figures, because they mean more to the businessman-producer. And if it doesn't mean more to him, then the salesman is in the advantageous position of explaining the benefits of these new profit-orientated ways of looking at the 'old' measurements.

PPTE is a particularly effective weapon as it uses the prospect's weakness for cost per tonne. Benefits like 0.1 better FCR, 4 days quicker to market, 2p lower cost/kg gain certainly mean something in the customer's mind, but ... using PPTE to add. *"This is equivalent to a saving of £15 (or 10%) on every tonne of food you buy in future"* . . . is a far more impressive statement about the same improvements.

And using REO allows the continuation of . . .

REO

"And as our product with a likely REO of 4:1 only costs £3/tonne more (raising your cost/tonne figure by 2%), by using it you should be £12/tonne better off, equivalent to enjoying a 8% cheaper food in future if it is included."

So . . . both farmers and feed/feed additive salespeople should …

*   Look at the trial results supporting the product and the degree of expected significance.

*   Work out the REO, MTF and PPTE figures. It is not difficult and can be done quickly.

*   Use them to make a greater impact while others struggle with the old and narrow terminology based on performance rather than profit.

As feed salesmen, just becoming familiar with and using REO, MTF and PPTE is sufficient to help them sell more, sell it better (at a higher price) and encourage them to ask for the order more quickly. If feed and feed supplement companies do nothing else, get familiar with using these three "new terms" because the farmers and farm students reading this book will be doing so!

AIV   **AIV (Annual Investment Value)**

REO is a useful tool when comparing the price of feed additives and feeds based on trial significance. But a good high REO may take a long time to pay-back – and a lower one a much shorter time. So we need a further measurement of the return likely over, say, a year's investment. A year being the normal period of time over which a straightforward loan is granted.

For example, a product/additive giving a benefit in the creep feed may be turned over as many as 11 to 12 times a year; in the nursery 6 to 7 times a year; in the growing finishing stage down to 3 times a year, and in the sow only 2.4 times a year.

So the time it takes to recover the expected REO less the initial investment, less the cost of the interest on the capital – can be important.

A typical AIV based on an additive to growout diets has an REO of 10:1 by slaughter. Using dollar terms, turned over at 3 batches of pigs/

year at a $3 per batch the (extra) inclusion cost/ton would be, at 3 batches/year $3 x 10 = $30 *less* the capital investment in use at the time plus the 30 cents or so interest needed to repay the $3 borrowed. This is $26.70 extra nett income per tonne per year. In this case while the REO is 10:1, the AIV is $26.70 ÷ $3 or 8.9 : 1.

In contrast take a creep feed additive which might give a lower 2 : 1 REO at weaning for a higher $10/tonne extra inclusion cost. At first sight this looks a much poorer prospect in several ways compared to products with higher REOs. Indeed, this yields a modest $20/tonne return, but is spread over a much greater number of pigs eating one tonne of creep feed. Moreover $10 extra investment is turned over about 12 times a year with 24 day weaning (as is common in the U.K). Our apparently lowly 2:1 REO in AIV terms requires an extra annual investment of $10/tonne + 10% interest, but the head start of 2: 1 REO it gives each litter is magnified by a factor of 12 in terms of slaughter pig output over a year's borrowing costs, *i.e.* $20 x 12 = $240, an AIV of $240 ÷ $11 = 21.8 : 1.

The moral is that a little investment in the correct nutritional area of the very young pig can give encouraging benefits at slaughter. We've always known this, and the AIV principle helps quantify it across a year's fiscal 'trading'.

WWSY (Weaner Weight per Sow Per Year)

WWSY

1. *Output/sow/year* Pigs (usually weaners) reared per sow per year is the common current yardstick (*i.e.* 18 or 20 or 22 *etc.* 'Pigs per sow per year'). However, this takes no account of weaner weight per sow per year (Table 10) and can be misleading.

Table 10 TWO CLIENTS. FARM 2 WON NATIONAL PRIZE FOR PRODUCTIVITY IN HIS SECTION IN 1991, BUT FARM 1 MADE MORE MONEY!

	Farm 1 430 sows		Farm 2 380 sows	
Pigs weaned per sow and served gilt per year	23.7		24.9	+5%
Average 3 week weight, kg (lb)	6.7	(14.8)	5.4	(11.9)
Liveweight output at weaning, per sow and served gilt/year, kg/lb	158.8	(350)	134.5	(297) -15.3%
Potential Value at £1.20/kg	£189.60		£161.40	

Comment : 5% more performance but 15.3% less income !
(UK 1996 figures)

Weight at weaning has a significant effect on economic performance, thus the suggested standards (Table 11) give target performances at 12, 21, 28 and 35 day weaning as well as the conventional figures.

Table 11 WEIGHT OF WEANERS PRODUCED PER SOW PER YEAR, kg (lb)

	Weaning age (days)	Poor	Typical	Good	Target	Exceptional
S.E.W	10-12*	n/a	n/a	103	103	n/a
Conventional	21	81 (179)	89 (196)	133 (293)	147 (324)	202 (445)
Conventional later weaning	28	97 (214)	109 (240)	184 (406)	217† (479)	[—]
Swedish conditions	35	115(254)	133 (293)	261† (576)	299† (659)	[—]

* Insufficient data for SEW technique. [ ] Exceptional producers outside U.K. rarely wean over 24 days.

† These sorts of productivity place a considerable strain on the sow thus can be difficult to maintain consistently. All figures are corrected for farrowing index.

As Table 11 shows, **Weaner Weight per Sow per Year (WWSY)** is a better yardstick than the current Weaners Per Sow Per Year. Obviously the date/time of weaning will affect the weight produced per year, as it does the numbers of weaners terminology used currently.

So after the WWSY figure an indication of weaning time, in days, is needed. For example, weaning at an average of 24 days should be expressed as . . .

158.8 kg WWSY [24]

Secondly, the time of entry of the replacement gilt to the herd should be standardised. Opinions differ on whether recording should start as soon as the gilt arrives on the farm and needs feeding and looking after, or on completion of first service. However as this can vary by several weeks and affect the weaner weight sow year (and the weaners per sow per year) by some 7 kg (or one weaner pig/yr) it is better to standardize on a 'start-date' for the input gilt *at the date/time she is put to first service*. While this ignores the bulk of the run-in or acclimatization period, it does ensure all farms are comparing like with like, as a new female replacement has to be offered for service at some time or other and this moment provides an equable start date in performance terms.

Piglet mortality to weaning. Another vital figure, but with the advent of bigger litters in future due to genetic progress, misconceptions will increase. This is because the more piglets get born (alive) so the relative percentage mortality will increase. *Always relate percentage mortality to weaning to the born-alive figure.* Look at this, for example:-

Looks bad - **15%** **mortality** of 12 born-alives is 10.2 reared (12 - 1.8 = 10.2). While ...

Looks good - **5%** **mortality** of 10.75 born-alives is also 10.2 reared (10.75 - 0.54 = 10.2)!

Certainly it is better to only lose half a pig than nearly 2 pigs per litter but the wide **percentage** differences in mortalities quoted among producers can be misleading. A better measurement is A.M.F.

Absolute Mortality Figure (AMF)

AMF

Better than % Mortality is to give an Absolute Mortality Figure (AMF) – how many pigs died per litter of those born-alive. Not the % which died per litter, as we don't know the litter size when a percentage is quoted.

The clearest definition is to express it as........

AMF 1.2 of 12 BA (BA = Born alive).

Hopefully this could become familiar as AMF $^{1.2}/_{12}$

This would be a 10% mortality of those born alive.

However AMF $^{0.8}/_{12}$ is only 6.66% mortality and therefore good, while AMF $^{0.8}/_{8}$ is 10%; again, not so good at all.

As a rule of thumb, using the percentage method, a target for the typical producer to aim for today is 10% *i.e.* in A.M.F. terms this is : for 8 born alive = 0.8 piglets lost; and for 12 b/a = 1.2 piglets lost. When it is expressed like this we at once can see that the problem in the first case *is the born-alives, not the mortality*! So always ask what the b/a figure is, as it is too important not to know this when measuring mortality.

And with the likelihood of 12 born alives being normal in future – possibly even 14 with advanced DNA selection techniques – expressing mortalities in this new way becomes even more important. For example, losing 0.8 piglets litter in a litter of 14 b/a looks high, but it is still only 5.7% mortality – very good.

ILR and PLR (Income to Life Ratio. Profit to Life Ratio)

REO (Return to Extra Outlay ratio) enables the producer to compare paybacks from a wide variety of investment options all of which require extra capital. As we have seen in the REO section, using MTF or WWSY to provide a comparison of performance improvement, generally the option providing the highest REO is the one to adopt first.

But if a high REO takes a long time to payback – years instead of months as can happen with equipment and housing alterations – then perhaps the option with the highest REO may not be the best economic bargain. ILR or PLR helps refine REOs in terms of the *time* taken to get the 'Return' part of the REO figure.

Long term paybacks

So with ILRs and PLRs we are dealing more with equipment REOs, which should have a long life (5-15 years), as with feed additives the payback date is relatively short and finite – at the end of the 160 day growth period, for example, or one sow's reproductive cycle, or her output over a year, or a product used only in the 6-9 weeks of the nursery period.

REO – or indeed ROI/ROC – along with ILR/PLR can be used to prioritise the expected effects of big and expensive refurbishment jobs (like ventilation renewal) or a host of smaller ones like replacing troughs, putting in height-adjustable drinkers, or comparable piglet heating systems (pads, lamps, under floor pipes, *etc.*) or even the comparative value of equipment used only occasionally, like showers, or on a small percentage throughput, like a sophisticated hospital/get-better pen.

As with feed additive paybacks, the results/benefits must come from acceptable trial results. This extra income (ILR) or (Nett) Profit (PLR) is divided by the time it takes to achieve it against the expected life of the equipment. Table 12 gives one example.

What does Table 12 reveal?

A study of Table 12 shows that a wide PLR is encouraging if capital is scarce, as with REO the wide ratio indicates the projects which are likely to repay soonest and best in profit terms.

Table 12 HOW PLR (PROFIT TO LIFE RATIO) HELPS ASSESS CAPITAL IMPROVEMENTS

Situation	Client has the option of investing the same amount of money either to refurbish his ventilation, or install new low-waste feed hoppers. From data supplied by the manufacturers of previous before-and-after results, alterations to the ventilation provided 10% more nett profit/pig and replacing the feed hoppers saved 4% feed (25-90 kg or 55-198 lb).
Ventilation refurbishment	500 pigs. Cost £8/pig capital and interest over projected 8 year life before major renewal refurbishment is again needed.
	Nett profit improvement claimed as 10% (on £7/pig = 70p)
	Total benefit 70p x 3.5 pig places/year = £2.45 pig place / year.
	Payback therefore in 3.26 years leaving 4.74 years clear benefit.
	Expected net gain in profit, after payback, over life of product. £2.45 x 4.74 years x 500 pigs = £5758.
	PLR Extra profit per pig place over life of product £11.52. Investment cost £8/pig **PLR = 1.44 : 1**
Installing low-waste hoppers	500 pigs. Cost £8 pig capital and interest over projected 12 year life before replacement is needed.
	Nett profit improvement is 4% less food wasted 25-90 kg (on 150 kg)
	6 kg x 15p/kg = 90p/pig.
	Total benefit 90p x 3.5 pig places/year = £3.15 per pig place per year.
	Payback therefore 6½ months leaving 11.45 years clear benefit. **Expected net gain in profit, after payback, over life of product.**
	£3.15 x 11.45 years x 500 pigs = £18,034.
	PLR Extra profit/pig place of product £36.01 PLR = 4.5 : 1.
Comment	Both PLR or ILR (Income to Life Ratio) make you look at the *expected life* of the investment and the *expected financial benefits after payback*.
	Costs and benefits claimed came from the manufacturers in both cases.

The payback (from the manufacturer's evidence) and the expected life of the installation relate both ratios to each other. If you are satisfied with the evidence, choose the project with the widest PLR.

Both PLR and ILR further refine this comparison and demonstrate *how quickly the capital required can be recovered* – ventilation 3.26 years with hopper replacement only 6½ months.
The recovered investment can then be kept as profit, or reinvested in other performance-improving (MSC) or cost-reducing (SLC) projects. As in the example cited, this rapid redeployment of (scarce) capital *can be more important, in value-for-money terms, than a straight REO figure.*

Always, especially in projects requiring long term capital, use PLR/ILR *in conjunction with REO*.

ILR

PLR

Why ILR and not PLR ? (Income to Life Ratio v. Profit to Life Ratio)

Some businessmen pig producers prefer to use *income* rather than *profit* when calculating the effect of the life expected to achieve a payback on interest and capital invested.

This is because income is one, finite figure and is not subject to variation in interpretation as in 'profit'. Profit can be nett, gross, margin-over-food cost, profit per pig place, or per m² *etc*. Unless carefully defined this can cause confusion, and so some producers are more comfortable with ILR.

Again, the data from which the calculations are made can arrive at varying definitions of profit, or alternatively in straight, increased income per pig.

An ILR figure tends to be rather easier to calculate and compare between project options.

SUMMARY – POINTS ABOUT THE NEW TERMINOLOGY

The farmer

• To survive in the future pig producers have to become more business-orientated.

- Thus they need better measurement terminology to help them take the right financial decisions.

- *The new terminology* takes cost and income into account when measuring performance, thus is profit-oriented, not just performance-oriented as at present.

- Capital resources are finite. The New Terminology helps rank them in their most cost-effective manner.

- If the salesman is going to start using these new terms, whatever else he does, the farmer-customer needs to understand the logic.

- If the salesman isn't into using the concept, it is easy for the farmer to assess the *value of what he is told*, and *compare it to* the blandishments and claims of other salesmen interviewed.

The salesman

- Those customers who survive will be bigger and better businessmen to deal with and so the sales interview will increasingly be *financially*-based rather than *performance*-based.

- Salesmen must understand the *new terminology* so as to keep pace – and even think ahead of – their customers, because the new terminology involves *cost-effective* measurements.

- Good quality products are not cheap, and can never be the cheapest on the market.

- The *new terminology* makes it easier to sell good quality in a highly competitive sales situation.

- The *new terminology* makes selling more interesting as it opens up new/novel sales approaches.

- Understanding *the new terminology* and when to use it effectively to make a sale will impress customers that salespersons have their eventual profit rather than the physical performance of their pigs as the centrepiece of the sales interview.

The commercial company

- Has the advantage of highly professional sales negotiators fully conversant with business thinking and up-to-date terminology. Better-trained salespersons mean more sales for the same costs thus more profit to the company.

The competition (this includes other farmers technically 'in competition' with you, the farmer)

*   Other commercial companies will be immediately placed at a disadvantage if they are unaware of the new terminology and how it can be effectively used both to make a sale and retain repeat business. This itself will have a debilitating effect on the other company's morale in feeling 'left behind', especially if the new terminology is quoted at them by potential *and existing* businessmen farmer customers.

    This means you/your Company, not they, will get the potential customer's orders and you/your Company will take their existing customers off them more easily.

    In the farmer's case, he will be able to use his limited capital more cost-effectively than other farmers not using the concept, and so make more money, or in difficult times – survive!

The scientist

*   In the end the farmer pays his salary. He therefore needs to keep abreast of the changing *commercial* aspects of pig production. So he needs to immerse himself in the New Terminology as well as retaining, correctly, his use of precise performance measurements. In contrast to the busy farmer, he can measure and use performance accurately to help the farmer achieve the basic data so essential to his proper use of these new econometric-based terms.

    Also, very often, by taking on board the New Terminology I suggest, the scientist is in a better position to prioritise his own applied research from the farmer's viewpoint.

REFERENCES AND FURTHER READING

Campbell, R.G. Personal communication, unpublished. (Effect of Bioplex Iron on a large industrial sow complex.)

Close, W.M. "The Effect of Bioplex Iron on Sow & Piglet Performance: Preliminary Results" (1998). Internal Res. Ref. 44.034.

Close, W.M. "Organic Minerals for Pigs : An Update" Procs 15th Annual Symposium in Biotechnology 1999, pps 51-60.

Gadd, J. Monograph: "The New Terminology : Guidelines for The Feed Trade & Ancillary Industries Sales & Marketing Departments." Privately printed; available from the author.

Gadd, J. Monograph: "Pig Production Standards" (U.S. edition 1998, Metric Edition 1999.) Privately printed; available from the author.

Gadd, J. "Off With The Old, On With The New", Pig Farming 47, 1 (Jan. 1999)

Gadd, J. "New Terms, Better Yardsticks", Pig Farming 47, 2 (Feb. 1999)

Gadd, J. "Questions You Should Ask The Sales Rep", Pig Farming 47, 11 (Nov. 1999)

Gadd, J. "Man of Straw . . .and Iron", Pig Farming 47 7 (July 1999)

Janyk, Stanley W. Organic vs Inorganic Selenium Supplementation of Gestating Sows, Jan. 1998, Agricultural Research Council, Animal Improvement Institute, Irene, South Africa. (Presented at the World An. Prod. Conf. Seoul, Korea)

Mahan, D.C. Organic or Inorganic Se Fed to Lactation Sows – How Effective is Either Source and How Effective is the Combination? 1998. Poster presented at the 14th Annual Symposium on Biotechnology in the Feed Industry, Lexington, KY (1998).

Muñoz A *et al*. Effect of Se Yeast and Vitamins E & C on Pork Meat Exudation and Natural Antioxidants in Consumer Satisfaction and Meat Quality. Procs. 12th Annual Symposium on Biotechnology on the Feed Industry, Lexington, KY (1996).

U.K. Meat & Livestock Commission (M.L.C.) Yearbooks for 1995 – 2002. M.L.C., P.O. Box 44, Queensway House, Bletchley, MK2 2EF, U.K.

COMMON PROBLEMS WITH RECORDS 10
.... AND A LOOK INTO THE FUTURE

It may seem odd to need to define record-keeping on the pig farm, but perhaps the most arresting definitions are, for once, negative ones.

(1)   Without records the control of productivity is guesswork.
(2)   Without control of productivity, profit can never be maximised.

PROBLEMS WITH RECORD KEEPING

The first problem is that pig farmers are not good record-keepers. You still hear people saying about a stockman or owner – "he's good with the pigs but hopeless at records", although this criticism is not so common as it was. This is because pig production units are becoming specialised capital-intensive operations requiring full-time management. Record-keeping itself has moved much closer to being user-friendly with the arrival of the microchip and desk-top computer. Records are less of a drudge than they used to be – *if* you use the right record system.

I go on to many farms where record keeping is disliked by the staff. In most cases the records are too cumbersome, the staff are probably keeping too many records and they have insufficient time left to do them well. In many cases they are not using them properly – not getting the most out of them. A lot of hard work for a minimal result, which is wasteful. If the right record system is used properly then keeping the records becomes – well, not enjoyable perhaps, because compiling amounts of numbers and events is always going to be a repetitive task – but at least tolerable.

With a good system the stockman/manager actually looks forward to his weekly or monthly summary. Man is basically a hunter-gatherer and records satisfy the 'gathering' (or in the 21st century – collecting) instinct in us – it is satisfying to be able to compare current performance with previous achievements, and to see if past efforts have indeed borne fruit.

CAN DO BETTER

Whilst not denying the great strides forward made by computerised record-keeping systems over the past 20 years, with my on-farm problem-solving work – which uses a variety of such record systems – such experience suggests that the pig records industry has 'plateau-ed off' and much more can be achieved by them. Sorry – but I'm sure computer recording needs to move up a gear!

Table 1 lists the deficiencies I have run up against in consultancy work.

WHAT RECORDS ARE NEEDED

I have read many worthy but complex flow charts on how record keeping systems should progress. All this theory is fine, but what I hope you, dear reader, will use as a take home message in this section is much simpler:–

1. Correct and regular inputting is vital. (Cross-checks to pick up as many human errors as possible are needed.)

2. The software needs to collate only the essential information to be measured against the farm's preset targets. (Most systems provide a plethora of largely unused data.)

3. The correct interpretation of what the figures show is critical. (Any computerized matrix can assist here.)

    It can be seen from the above that both the stockperson at the start of the recording process and the manager (or section head) at the end of it are absolutely vital cogs in the productivity machine, and moreover have an equal responsibility in achieving profit!

Table 1 WHAT IS WRONG WITH RECORD-KEEPING TODAY?

Problem	Result	What needs to be done
Inputting is too onerous	Risk of error	Re-examination to make it as simple, for example, as the 'Pig Tales' method (pages 148 and 149)
Too cumbersome	Offputting	Re-examination to include only essential information
Not graphical	(1) Slows up recognition of a problem (2) Non motivational	Most columns of figures can be 'pictorialised' and the pictures updated frequently. Stockmen are better motivated/trained by monthly/weekly graphs in the rest-room than by columns of figures.
No statistical overlay	Wrong assumptions made	Simple statistical flags appended to changes to signify : ["*No problem*"] "*Problem may be ahead – amber alert*" "*Action required – red alert*"
Over-dependence on physical performance	Wrong conclusions may be reached	Include additional data on econometric as distinct from physical performance (*e.g.* The "New Terminology")
Underuse of "*what-if*" facility	Opportunities lost	Most recording systems are only retro-active in this important area, depending on the farmer or his adviser to call it up. Better is to insert **auto-periodic** "*what-if*" exercises when productivity falls away, thus stimulating thought, consultation and, if needs be, earlier action.
Weak in financial aspects	Overdependence on physical performance in determining strategies	To maximise returns, performance records **must be linked automatically** to financial records. Another opportunity here for more '*what-if*' promptings, at present rarely attempted, this time based on economic matrices.

A good record system makes their respective jobs of collecting and interpreting the data *easier*, not more difficult, as some systems do.

Targeting

A fourth area is equally important – that of targeting. You will find plenty of targets propounded, and indeed you will find them in this book, but at the end of the day it is *the business's own targets which are paramount*.

These targets are farm-specific and are tied to the farm's own capabilities and circumstances – no-one else's. Don't listen too hard to other people's targets; they are a useful sounding-board, but that is all.

Your targets will depend on what you / or you financial adviser / he who lends you capital / your immediate customer demand of the business.

Your targets will depend on the circumstances which affect you – your own required standard of living, your pig-housing status, your labour pool, your financial strengths or weaknesses, the stipulations of your immediate customers, the cost of money, credit facilities, disease status and so on. Targets have to be fixed for most of these – some flexible in a crisis, others immutably fixed whatever happens.

Having established your own targets, especially the output targets, a primary role of farm records is *to inform you of how you are doing against target.*

Records will also allow you to re-adjust your targets if the initial decision on that particular aim / forecast was wrong. *Every* record should have a target attainment figure alongside it, because by studying these individually, two things appear.

1.   The area of under or over achievement is pinpointed. (Yes, *over*-achievement is important too – you may not have enough future housing space to accommodate a surge in productivity, for example.)

2.   A general under-achievement of a final target can be quickly traced to the sector of the business where the deficiency lies.

SO ... WHY KEEP RECORDS ?
AN EXPANDED CHECKLIST

✓   *Daily activities* – Reminding and forewarning us of essential tasks. The most important reason is to identify basics which need to be completed on the farm (service dates, due return dates, farrowing dates, weaning dates) so as to maintain an efficient pig flow and work flow.

     Secondly – to record primary events so that your awareness of how things are progressing is accurate. Right decisions ride on accurate information.

✓   *Animal location* : It is important to know where every animal is on the premises. This is vital where breeding stock is concerned but the nursery/growout operator may only need to identify a sample of growing/finishing pigs to obtain growth and carcase quality efficiency ratings. On the other hand some producers carefully trace back market information to individual boars/AI and female lines, and recording the progress of individual growing pigs is again necessary in order to be able to do this.

✓   *Traceback:* Records are becoming increasingly important in terms of quality control on the farm and, from the farmer's point of view, in ensuring he gets his just financial rewards for investing in the means to provide high quality meat.

✓   *Monitoring Efficiency* : Records are used to identify both efficiencies – performance and economic – which influence possible profit. This is a huge subject dealt with later in this section.

✓   *Modelling* : Records provide data for the at present under-used practice of modelling. Modelling uses a computer program to establish a predictive "what-if" situation where different actions, target weights, production systems, market outlets, carcase grading goals, improvements to insulation or ventilation, feed scales or diets of different nutrient density, etc. are tested out against current performance. Modelling is a form of instantaneous forward trialling on paper and is becoming increasingly accurate in its predictions. Records therefore provide us with guidelines for future production targets.

✓   ***To help those who can help you***.   Two or more brains are better than one.   The vet, the environmentalist, the feed rep/ nutritionist, the accountant/financial adviser and the management consultant, like myself.   Every business needs advice.   Good advice rests on what good records reveal.

✓   ***Diagnosing Trouble*** : While modelling looks forward to predict the results of proposed action, good and accurate records are invaluable when seeking the past cause of poor profit performance.   At present the potential is barely explored and I can see a much greater part played by the computer in future. At present much of my job as a farm consultant is taken up with studying figures provided by the producer who is seeking help with a problem.   My experience is that about half of the likely solutions can be discovered lying hidden in the records – pointing to areas where attention is needed.   Sometimes, by analysing records sent in advance of a farm visit, a solution to the problem can be suggested so that the visit to confirm the diagnosis is unnecessary, which saves me much time and the farmer some money.   In my case this has happened in about one in five of my 'call-outs'; the rest do need a farm visit.   But this pre-studying of the records I find essential so that what I see on the farm tour confirms or denies my preconceptual study of the 'on-paper' situation.   Frankly I am surprised more consultants and veterinarians don't work this way.   Records first – *then* a look at the pigs, not the other way round, which is commonplace.

A future role for the computer

There is no reason why, when programmed in detail, a computer should not follow the same diagnostic processes used by the consultant when examining records.   I can see the computer monitoring not only the producers' management and stockmanship skills as it does now, but signalling warnings and *suggesting remedial action* automatically and at no extra cost.

Using the computer properly

The logic of this procedure is straightforward:–

- The pre-programmed computer will flag up below-target achievements in a wide range of performance subjects.

- As mentioned earlier a statistical overlay can be included so as to 'grade' the performance shortfall as, for example, something to watch out for or something to act on (Warning Level / Action Level).

- If action is needed, guidelines are pre-programmed in to the matrix in the form of checklists similar to the examples in this book. Thus the computerized recording system does much more of the diagnosis work than is achieved by present systems. *It can even go on to suggest what to do*, and/or suggest further checks.

  The producer is alerted to the problem, advised of how serious or important it is, and pointed towards areas/actions he should consider as likely remedies.

  Computer recording firms are failing us in not providing these built-in benefits, and I hope they read this section!

✓ *Monitoring Health :* This leads on to a very similar situation with regards to disease. Records aid the veterinarian to identify causes of disease and production problems and already computers are assisting the pig specialist vet, albeit in a limited way, to isolate the most likely cause of disease. At present the veterinarian works to a schedule of :

   ✓   monitoring the level of disease

   ✓   where and when it occurred

   ✓   what medicines and their dosage were used

   ✓   how the diseases that occurred might relate to the management, etc.

Computers will markedly assist and maybe even revolutionise this approach in the future.

✓ *Motivating Staff :* Records can be a great stimulus to improving the morale and efficiency of the stockman. This may seem to be a paradox, as I have already said that producers and stock people usually dislike compiling and processing records, but their attitude can be completely changed if:–

   •   compiling on-site information is made as simple as possible (Figure 1).

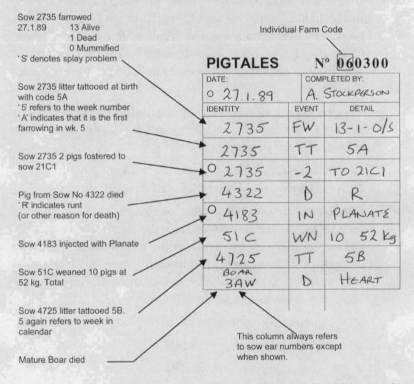

Example of daily diary entry from

The Farrowing House

Sent to database on same day each week

Sow 2735 farrowed
27.1.89   13 Alive
              1 Dead
              0 Mummified
'S' denotes splay problem

Individual Farm Code

Sow 2735 litter tattooed at birth
with code 5A
'5' refers to the week number
'A' indicates that it is the first
farrowing in wk. 5

Sow 2735 2 pigs fostered to
sow 21C1

Pig from Sow No 4322 died
'R' indicates runt
(or other reason for death)

Sow 4183 injected with Planate

Sow 51C weaned 10 pigs at
52 kg. Total

Sow 4725 litter tattooed 5B.
5 again refers to week in
calendar

Mature Boar died

PIGTALES   N° 060300

DATE:		COMPLETED BY:
○ 27.1.89		A. STOCKPERSON
IDENTITY	EVENT	DETAIL
2735	FW	13-1-0/S
2735	TT	5A
○ 2735	-2	TO 21C1
4322	D	R
○ 4183	IN	PLANATE
51C	WN	10 52 kg
4725	TT	5B
BOAR 3AW	D	HEART

This column always refers
to sow ear numbers except
when shown.

System devised by Pig Tales (copyright)

Figure 1 An example of a classic (and unsurpassed) inputting system from PigTales
– 20 years ahead of its time in the 1980s. Just a series of simple cards sent in to the
computer, using if required, the stockman's own abbreviations/code.

Continuing on with the motivational benefits ...

(b)   the computing aspect is done for them

(c)   the information is presented in an attractive and easily-
recognised form, which in the majority of cases means
3-colour graphics. (Figure 2)

(d)   This form of presentation is updated weekly with
performance against target demonstrated in an on-going
form.

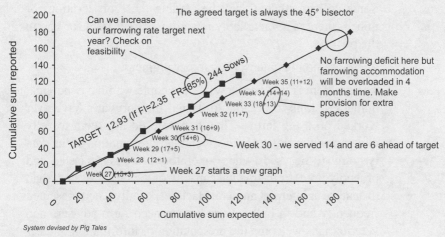

System devised by Pig Tales

Figure 2 Breeder – Services Cusum.

The value of graphics in getting information understood is not in question (Table 2).

Table 2 FARMERS UNDERSTAND PICTURES / GRAPHICS

Farmers were presented with a farm problem hidden in:
- a) columns of figures
- b) exactly the same data in graphical form

They were asked to provide a numerical answer to each question. Number of correct answers were:-

	(n)	Columns of Figures presentation	Graphical presentation
Class 1	30	42%	78%
Class 2	34	57%	84%
Class 3	37	36%	67%
Agricultural Students	20	80%	90%

Source: Gadd (1994)

Improvements needed

There is room for improvement in all three cases, but especially the first and last. There must be a hundred computer recording systems world-wide at the time of writing if those in-house variants used in veterinary practices or in the massive industrial units are included, yet I only know of two which meet all three criteria in every respect. The majority are pretty unimaginative, I guess.

Some college students will be reading these observations of mine – there is a remarkably rewarding career field for you in this respect.

There are no computer programmer pigmen!

The problem is that very few of the people who have the great numerate skills to be able to design the software, have ever worked on a pig farm. They therefore do not seem to fully understand the situations in which their presentation of the data is going to be used (where operators are busy, hassled and probably tired) or the mental capacity of the ordinary stockman to analyse or interpret numerate facts, especially when presented frequently and in bulk. Even experienced farm advisers and veterinarians welcome the presentation of information in clear pictorial form rather than in the columns and pages of printout with which we are presented!

✓ *Identifying the Pig :* The first step to establishing a record-keeping system is animal identification through ear tagging, tattooing, notching or, more recently by electronic implants – and way into the future even by a DNA 'barcode'. It is not well done, for example in Europe 10 to 15% of all pigs are wrongly identified due to lost tags or illegible tattoos. A good identification system has the following characteristics:–

AN IDENTIFICATION CHECKLIST

✓   It is unique to each animal. No two pigs have the same identification.

✓   It is readable from a distance.

✓   It is permanent over the animal's life.

✓   It is easily replaced if required.

✓   It is easily applied.

✓   It has a wide range of numbers and codes.

✓   It is humane.

✓   It is economical.

✓   It is tamper-proof.

TATTOOING

Tattooing, usually of the ear, but occasionally of the neck behind the ear, is the most popular method at the time of writing. Pigs may also be slap-marked before slaughter. (In some countries it is expected that ear *notching* will be banned on welfare grounds.) Because so many producers world-wide still use tattooing, I cover this rather outmoded technique here. (Tattooing can be used for growing pigs).

A TATTOOING CHECKLIST

Size of needle letters/numbers:

Sucklers 3 day to 10 kg (22 lb)   8 mm (0.31 in)

Pigs 10 to 100 kg (22-220 lb)   10 mm (0.35 in)

Breeding stock and slap marking   16 mm (0.63 in)

Paste : Use colour green for white-skinned pigs. A good brand such as Ketchum should last 10 weeks or more. Black is usually used for slap marking before slaughter. Grate polish is sometime used. Dark skinned areas will need an ear tag or tags, or notching.

Timing :   3 to 5 days old if possible but ear size may be a limitation. Before 21 days is advisable.

Positioning :   Figure 3 shows where to place the tattoo for pig identification purposes.

Figure 4 shows how to record birth date if required.

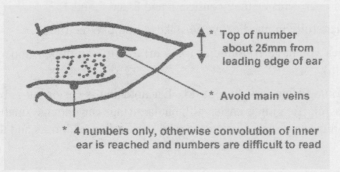

* Top of number about 25mm from leading edge of ear

* Avoid main veins

* 4 numbers only, otherwise convolution of inner ear is reached and numbers are difficult to read

Figure 3 Positioning of an ear tattoo

Lettering by day/week number is useful in
<u>**following performance of growing pigs**</u>.

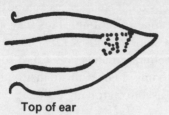

<u>Birthdate 3-17</u>

* 3 = 3rd day in the week = Tuesday
  17 = 17th week in the year = end of April

* As only 2 or 3 letters are used the ear
  tip can be used as it is easily read

Top of ear

Figure 4 How to record a Birthdate

Make sure you have . . .

✓   Applicator with at least four sets of numbers 0-9

✓   Die applicator/toothbrush

✓   Surgical spirit

✓   Record book

✓   Disposable tissues

✓   Disposable gloves

✓   Die soaking / cleaning tray

✓   Die cleaning detergent

✓   Mild die / applicator disinfectant

Observations :

Die pastes vary in persistence so it probably is worth tattooing both ears, one inside, one outside.

Avoid ear veins as they confuse 5 and 8 and 7 and 1.

Go carefully/methodically. Mistakes cannot be rectified.

Check you have 10 mm dies (0.35 in) if tattooing piglets under 6 kg (13 lb).

Tattooing is not too easy to do. Ear notching is easier but is brutal/ stressful. Tagging is easier still, but large tags can damage small ears; small tags are more difficult to read later in life, both ours and theirs!

EAR NOTCHING

The author declines to write about this subject as he feels it infringes on his own personal welfare beliefs. It cannot be humane.

If you are determined to ear notch – and ear-notch young pigs especially, you should consult your veterinarian so as to minimise shock, pain and the risk of infection. In any case, ear notches can grow out, get misread after fighting and are tedious to read. Please don't do it.

EAR TAGS

Plastic ear tags are a useful means of identifying breeding stock. Gilts and boars should be tagged when selected as replacements. A number of different tags are available, are relatively easy to read, but do have a tendency to pull out. The large tags placed centrally through the ear seem to be the most effective. Tags placed around the edge of the ear tend to pull out more easily. Care must be taken to allow for growth of the ear.

Permanence of ear tags is largely dependent on the conditions under which the pigs are kept. It is worthwhile trying different tags to find one which is most suited to a specific situation. It is definitely advisable to put a tag in each ear, and if possible have them numbered top and bottom.

CHECKLIST
HOW TO CHOOSE A GOOD EAR TAG

At the time of writing there are about 15 makes in the world market, so some guidance on choice may be helpful.

✓ **Flexibility** : The composition of the plastic must be correct to stand up to chewing, abrasion and weather conditions, including ultra-violet light in hot countries. A nice flexible but durable texture is best.

✓ **Hygiene** : Choose a tag where the ear piercing spear *is incorporated in the male half of the tag* and not in an ear-piercing portion of the applicator. Such an instrument has to pierce other ears and there is a distinct risk of infection, as it is impossible to swab-clean the ear surface well enough beforehand. The applicator should have a spring-back and surface-clearance action to avoid tearing the ear during penetration.

✓   **A sharp spear** : The brass spear point must be very sharp and have a gradual conical shape, *i.e.* a good piece of engineering turning. So when buying an applicator, a magnifying glass helps as it does when choosing a drinker valve, which also needs good precision engineering.

✓   **A strong shoulder** : Some tags wear and snap off at the point where the flap meets the spear (the 'shoulder'). Choose a tag which is strengthened at this point.

✓   **Permanence** : While it is possible to get the tags pre-lettered or numbered by a heat-sealing process, some farmers prefer to make their own codes. A special pen is available for this purpose. Preferably, the ink should actually infuse into the polythene surface which is made porous to allow this, thus the mark is permanent and will not fade or rub off with time. Alternatively, tags can be obtained with a covering disc to slip over a own-written number on the plastic disc. These must be cleaned with methylated spirit first, as if the cap is put on greasy it will slip off.

✓   **Problems** : Are usually abscesses/infection. Did you use a clean applicator? Wipe the site first with surgical spirit? Are the lying areas dirty? Or too wet? Are the tags stored clean and dry? A few cases of allergy are reported – change the brand of tag?

✓   **Legibility** : If possible choose tags numbered top and bottom, especially in group-housed sows and gilts. As colour coding is useful in following parent lines and crosses, legibility depends on size and colour, bearing in mind the larger the tag the easier it is to lose. Research by Lecurier suggests the following colour contrasts as being the most legible from a distance in poor light, or under a light layer of dirt.

Table 3   ORDER OF PREFERENCE OF COLOUR CONTRASTS ON EAR TAGS

1-12 Order of Preference	Decoration	Background
1	Black	Yellow
2	Green	White
3	Red	White
4	Black	White
5	White	Black
6	Yellow	Black
7	White	Red
8	White	Green
9	White	Black
10	Red	Yellow
11	Red	Green
12	Black	Blue

Do not use blue on ear tags unless a sixth background line colour is essential.

Losses

Welded iron rod partitions, sow stalls, scratching and rubbing (mange), fighting, outdoor fencing and huts and poor insertion are all common causes of loss. Use both ears and immediately replace when one tag goes missing. It is better not to use tags in growing/finishing pigs – use a tattoo. Generally other markers and sprays don't last, but a concentrated solution of crystal violet dye crystals used at 1 kg per 4.5 litres water (about 2 lbs/Imperial gallon) and applied with a paintbrush can last 5 weeks on fattening pigs when aerosol sprays, though considerably more convenient, seem to last much less.

SLAP MARKING

Producers selling pigs directly to processors may wish to use the packer's carcase classification information as part of their herd improvement program, by linking carcase weights, lean meat percentage estimations and backfat measurements to sires and dams, or as will be increasingly common, on different feeding strategies under farm test. This requires an individual tattoo for each carcase in addition to any registered brand. This number will be recorded against the weight and measurement and passed back to the producer. The individual number is applied using a rotary tattoo.

A SLAP MARKING CHECKLIST

The slap mark should be applied to the shoulder of the pig so that the brand will be clearly readable on the carcase at the place of slaughter.

To achieve this:

✓   Use only approved branding inks, that is, carbon-based fluids or pastes. Non-approved substances such as aerosol stock markers, boot polish and even sump oil should not be used!

✓   Apply ink to the needles each time the pig is struck with the tattoo brander.

✓   Use a shallow tin with a sponge pad about 20 mm (0.70 in) thick to coat the tattoo needles with the fluid or paste.

✓   Make sure that all numbers and letters are placed correctly in the brander.

✓   Make sure that tattoo needles are clean, sharp and in good condition. Bent, broken or blunt needles will not work properly.

✓   Ensure that the slapper strikes the pig evenly, and when the pig is still.

✓   Ask the buyers of your pigs whether your tattoo brands are readable. Better still, regularly follow-up your carcases on the hook.

Make sure you have . . .

✓   Slapper and sufficient sets of numbers.

✓   Code/record book.

✓   Ink pad.

✓   Some restraint device is advisable as the pig should remain still to get a good mark.

ELECTRONIC IDENTIFICATION (EID)

Safe and cheap electronic identification will eventually replace conventional methods of labelling all livestock in the near future, although some work has yet to be done on the concept, particularly on international standards and agreements.

Experts believe a more comprehensive system than conventional tattooing or tagging is needed, perhaps including an electronic identity tag carrying information which starts with the animal's birth date and the farm of origin.

Electronic tags are based on the passive or active transponder (more expensive) principle. At the stage of writing the experts seem to prefer the former because of cost. Required information is "read" on to the tag and then extracted by passing the detached tag over or through a reading machine or instrument. An active transponder transmits its own signal.

What is a transponder?

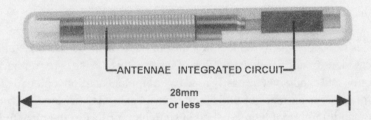

An injectable transponder, also known as a subcutaneous semiconductor chip, consists of a small transmitter and antennae. When a readout unit is brought close to an animal, the transponder becomes charged with energy and transmits its identification number. The number appears on the display of the readout unit. The transponder is shaped in such a way that it can be easily injected using a syringe. It is important that the encapsulation protecting the transponder be accepted by the body and that it does not migrate within the body.

But existing ID tags are not perfect. The ideal system would allow relevant information to be read on to individual animal tags starting at birth on the farm, and whenever stock changed hands at store or the final slaughter stages.

MICROCHIP IMPLANTS OR TAGS (BUTTONS)

Implants must not migrate through the body or break (early models were glass-cased) and of course must be fully recoverable at slaughter. However, the carcase must still retain its identification through to final processing and dissection.

Solving the processor's problem ..

The same transponder technology can be embedded in an ear tag of button which can be removed from the head/ear at slaughter and re-attached manually to the carcase with another pin, so that identity is registered both at the weighing and grading points. Thereafter individual primary cuts can be identified with the same number from similar bar coded pins.

... But not the farmer's

The problem still remains that the relatively expensive transponder button/ tag may be lost on-farm, as with any tag – especially in the hurly-burly of the finishing pen and that metal rails etc can disturb the read-out in the case of breeding stock.

It is now possible to reduce the transponder to a very small size (as dog/ cat identification) but at the time of writing this device is for numeric ID only, and cannot yet be easily read by equipment at a convenient distance for a larger animal not under restraint. Again, its very small size may mean it might not be recovered at slaughter, or lost in the busy turmoil of the slaughter plant, and these small microchips are more expensive too, though mass-production will help with the cost angle, should every pig be so labelled in future.

As well as ongoing technical research to solve these headaches, therefore, governments need to co-ordinate their efforts to use the concept of an individual electronic animal passport in the form of a microchip so that traceability is made easier, especially as existing tags cannot be interfered with. The concept also makes much simpler and quicker the tracing of pig movements after an outbreak of Notifiable Disease.

The use of such microchips to incorporate weighing, electronic feeding, environmental controls and biological body monitoring as animal records seems relatively simple compared to reconciling the continued but basic headaches of cost, loss of the device on farm and recovery/registration from slaughter onwards.

REFERENCE

Gadd, J. (1991) 'Getting the Message Across' Zen-Noh (Japan) Conf Proc
on Pig Production, Tokyo.

RISK MANAGEMENT 11

A business management technique which anticipates possible difficulties and then plans to reduce their consequences.

At the time of writing the subject of Risk Management is poorly understood by pig producers world-wide. Broadly it concerns two areas:

Avoiding Risk   Anticipating that a difficult situation could occur and putting in place actions to avoid it.

Minimising Risk   If the difficult situation does arise, taking action to reduce its impact.

Only the best ordered, most tightly-disciplined pig industries in the world are minimally affected by pig price and pigfeed price volatility – two of the most common risks affecting most pig producers.

The main risk areas are:–

- *Production risks*, including that of disease and a rise in the input costs, of feed and pig purchase price. These are the three main risks, affecting pig production almost on a daily basis.

- *Marketing risks* such as a fall in the pig sale price. This is the most influential risk of all and is likely to become even more volatile in future.

- *Financial risks* such as unfavourable cash-flow and inability to meet debt repayments.

- *Human risks* to your health or that of your employees.

- *Legal risks* through being caught out by failing to meet mandatory legislation/regulations.

The problem for all businesses – and especially the volatility-prone pig industry – is that these main areas overlap and can be difficult to disentangle to form a rational business plan. (Figure 1)

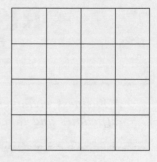

How many squares are there
in this diagram?

Figure 1   An example of how risks overlap (Martins, 2000)

At first sight it is easy to count 17. Then on perusal, 4 more appear.
And so on. After detailed study however, you will discover 30!

To help you disentangle the many risks you face, a Risk Management
Plan is essential.

Here is a checklist of the factors which involve your business risk.

RISK MANAGEMENT
A BASIC CHECKLIST

To do a Risk Assessment Audit you should check:–

✓   Your information.

✓   How you know about and respond to the market.

✓   Your business flexibility.

✓   Records (performance monitoring).

✓   Records (economic monitoring).

✓   Your technical knowledge and willingness to adopt new
     technology.

✓   How much you talk to others (co-ordination).

✓   Membership of key bodies – buying group, internet web , pig
     discussion group, NPA.

✓ How co-operatively minded you are (networking).

✓ How much of a gambler you are (see Table 1).

✓ How flexible your mind is.

✓ How and how much you delegate responsibility.

✓ How much you have examined / thought about diversification.

✓ Your staffing / workload.

✓ Leasing not buying.

✓ Using industry credit.

✓ Use of forward contracts.

✓ And forward buying of feed.

✓ Use of fixed rate loans.

✓ Hedging / use of options.

✓ Ensuring an adequate capital base.

✓ Keeping your bank informed.

✓ Your use of performance consultants.

✓ Your use of financial consultants – tax, etc.

✓ Your use of the vet as a regular adviser.

✓ Your use of the New Terminology.

✓ Compliance with Regulations.

✓ Insurance (disease, key men, etc).

✓ Carrying out a biosecurity assessment.

✓ Alarms and back-up systems.

✓ Key equipment maintenance / monitoring.

Do you see what I mean about 30 squares ? There are 31 here, and many interlock and overlap.

COMPILING A RISK INVENTORY

But set out baldly in a list like this is not all that helpful. Let me now group these individual, disparate squares into larger multi-square groups, what is called a Risk Inventory, and put some brief action lists to them.

You are the most important part of your business. Your brain, however much others might deprecate it, is light years ahead of the most advanced computer yet constructed. So know yourself.

RISK INVENTORY – YOURSELF, AND PRECAUTIONARY ACTIONS

Risk factors	Possible actions
Health	Insurance back-up, delegation, training, medical check-ups. Know your staff's emotional issues : What happens if … audit on staff shortage.
Are you a gambler ?	See Table 1
Accidents	Insurance. Key man insurance. Compliance with COSHH, Health & Safety Regulations, *etc*. Unexpected shocks… Emergency Telephone numbers. First Aid Training. Fire/fire drill. Flood / wind damage. Pollution and its financial consequences. Staff negligence. (Do a "What happens if…" audit with the staff concerned.) Theft and theft protection.
Deaths	Do a "What happens if…" audit

How much of a gambler are you ?

Risk-taking is, of course, a gamble. So how much of a natural gambler are you ? Table 1 (Harris, 2000) will give you an idea. Fill it in and turn the page to get a psychologist's view of where you may be in this respect.

A MORE DETAILED ACTION CHECKLIST

In compiling this series of Risk Inventories I give merely those items I've found useful in my own consultancy work with pig farmers. There are bound to be others which I've not had need to consider or which will be specific to your business. But among the variety of suggestions offered there may be one or two that you haven't yet thought about.

If you have a high gambling score you will need to review your protective actions very carefully – you'll need them!

Table 1 ATTITUDES TO RISK   From Harris (2000)

Mark an 'X' in the box that provides the best indication of how you rate yourself on the scales below in relation to the statements given.

I would not sign a fixed price contract because it could limit my income

I would sign a fixed price contract if it provided me with a basic income.

I like to try out new ideas

I prefer to stick to what I know

I will invest in new systems ASAP and develop them to suit my business

I like to see how others get on first with new systems

I am happy to borrow to invest in my business

I am not happy farming on borrowed money

I like to have lots of variety in what I do

I tend to use tried and tested methods.

If thieves want something, there is not much you can do to stop them.

I like to keep things locked and safe.

I think staff should be trained in more than just technical issues.

As long as staff do what they are told, I am happy.

My ideas are not always practical.

I am a practical-minded sort of person.

People see me as an 'ideas' person

I like to deal with factual information.

I often lose things.

I can be something of a perfectionist.

I take each day as it comes.

I never leave things to the last minute.

I find it easy to put myself in someone else's position

I don't back down in an argument.

I assess the likelihood of success of a project on individual merit

Past events always influence the likelihood of success of a new project.

Score – total of each column

Answer to Table 1

The higher the total score in the two left-hand columns the more of a risk-taker you are.

The higher the total score in the two right-hand columns, the less of a risk-taker you are.

If towards the low (cautious) end, you will need to consider actions which may improve your chances of progressing your business without involving too much risk – in your case I'm sure it will involve, nevertheless, not spending too much money!

If you are of a cautious nature - consider:–

- Getting better information on which to base decisions.

- Joining a web-site buying group.

- Keeping your bank better-informed.

- Examining what you can do to adopt as much as possible of a *full* biosecurity system (see Biosecurity section).

- Leasing, not necessarily buying.

- Spending more time on continually comparing sales contracts. The small print often makes them very different.

RISK INVENTORY CHECKLIST -
THE PHYSICAL PRODUCTIVITY OF YOUR BUSINESS

Risk Factors	**Possible Actions Checklist**
✓ Disease	Biosecurity. Correct use of the vet. AIAO/ Segregation. Batch production. Delivery and collection. Vermin and birds. Drug records. Awareness of local disease situation.

✓ Weather — Information. Frost and heat precautions. Weak structures re heavy rainfall and wind.

✓ Services — (*e.g.*) electrical, water, carcase disposal: Back-up systems.

✓ Staffing/labour — Work flow. Training. Motivation. Sickness. Part-time help. Your *own* contribution *i.e.* time spent on management rather than helping out manually: measuring, checking, forecasting, quotations, record analysis, your own housing, updating yourself, talking to others.

✓ Food quality — Home-mix / complete feeds? Wet *v.* d r y ? Knowledge of and control of raw materials. Dialogue with nutritionist and vet (*e.g.* current immune levels). Wastage, spoilage (mycotoxins). Feed and raw material analysis. Farm trials. Availability and correct use of by-products. Water adequacy.

✓ Housing — Adequacy, internal monitoring (measurements), use of consultant. Repairs and renovation. Provision of cheap overflow accommodation. Understanding basic technology – thermo-dynamics/air movement/pig behaviour.

✓ Records — Avoiding non-superfluous data. Action on data available (action lists). Graphics not numbers. Staff motivation. Outside inputting help *e.g.* accounts and tax help, and secretarial assistance.

✓ Contracts (termination or altered) — Good, up-to-date knowledge of alternatives available including altered transport costs. Alternative options – like the US share-cropping idea. "Think beyond the pig price" (Thornton, 2000), *i.e.* market security could be as attractive. If others haven't considered it – propose your own contract.

✓ Carcase quality — Good feedback from processor, then close liaison with nutritionists. Close liaison with geneticist. Concentrate more on male genes?

Keep your vet in the picture. 'Profit-box' recording system on each batch. Liaison *i.e.* (checking on) slaughterhouse/grader, especially if charges occur. Campaign for payment on lean. Stress audit. Check on transport competence/lairage time/ cleanliness.

✓ Productivity

Diagnostic records, use of consultant in interpretation. Understanding and using the New Terminology.

✓ Levies and impositions

Join your local organization to assist strength of industry voice. Help organise local agitation in cases of illogicality/unfairness, but get facts right first.

✓ Reducing exploitation by suppliers

Diversification, *i.e.* contract processing, farm shop (with other foods), local brands, exploit brunch/sandwich market, joint ventures, home mixing/wet feeding, more AI use, weaner gilts, commodity buying, internet buying (buying 'clubs'), Farmers' Markets (in future, Farmers' Supermarkets, *i.e.* open always)

RISK INVENTORY CHECKLIST - REGULATIONS

Risk factors

Possible action

✓ Non-compliance

Information on COSHH, HACCP, Drug Use, Health & Safety, Home Mixing, Fire, Stock Movement, *etc.*

Do an audit on each. Discuss with staff. Secure hazardous/restricted substances (insurance claim).

✓ Hidden shocks

Employ specialist consultant/share him with a group. Tap in to E.U. internet service (advance warning of possible unfair/restrictive legislation).

✓ Consumer pressure   Seek information. Listen. Be quick to exploit fears if public opinion seems justified as the big battalions and governments are often reluctant, disinterested, uncomprehending, thus slow to respond. Keep an open mind; some past apparent disadvantages have turned out to be benefits, farmed correctly *e.g.* outdoors, sows in groups, bedding, swing-away farrowing crates.

RISK INVENTORY CHECKLIST - ASSETS

Risk factors	Possible action
✓ Fire, theft, unexplained losses	Do an audit on each. Farmwatch. Rapid-contact phone numbers. Insurance. Guard dogs/kennel & chain. Alarms/Farmwatch/Notices/Electronic farm road gates.
✓ Livestock	Perimeter fence; pigs away from public access. Leasing/'share-cropping'.

A RISK MANAGEMENT TOOLBOX

Here are some tools to deal with likely risk. While directed at the British pig producer, and reflecting his weaknesses at the start of the 21st century, there are lessons here for any pig farmer across the world.

Your own attitude to risk - what is crowding in on you

• The number one threat to your business is the increasing amount of price risk in pig production.

• Price volatility is worsening, doubling every 5 years.

• Not only are costs (input prices) rising, but returns (output prices) are more volatile and trending downwards, or at best only keeping pace with inflation.

*   Production constraints (welfare, drugs, pollution, the effects of BSE and salmonella control) are increasingly affecting intensive livestock production in NW Europe.

*   It is essential that you keep abreast of what is happening not just locally or even nationally, but for your long-term planning purposes (5 years plus) in critical sectors of the global pig industry like Europe, Russia and Eastern Europe, North America, mid- and mid-west USA, mid & Western Canada, Brazil, Mexico and China. This is especially true of longer-term planning (10 years+) as it could well be your best option to take your production expertise overseas from now on and/or link up with overseas processing marketing opportunities which could well be supplying a share of your home market by then.

So the first tool in your toolbox is:

Information

You must put by more time in future to collection information on:

Markets : local, national, European, global.

Herd size and growth : British pig census data here has slipped badly over the past 20 years and reinvestment and reorganization is overdue. World market information is much better and, at the time of writing, various data sources are available, such as *'Whole Hog' Monthly Newsletter* and the UK Meat & Livestock Commission bulletins.

Output price trends : There is always a reason for a pig-price movement and you need to keep abreast of what causes them. Informed retrospective comment is good in parts of the UK agricultural press and such weekly newspaper's pig sections *e.g. Farmers Weekly* and *Farmers Guardian* should be read regularly, eventually to build up a good market knowledge and a 'feel' for making the right strategic decisions.

Competitors' behaviour : One competitor, well recognized but often overlooked by pig producers nevertheless, is poultry meat. Keep an eye on what they do and the way they think, from production to marketing. You will learn much and could be forewarned of any new impact on pig production.

Input price trends – retrospective monitoring : Feed contracts and forward buying are valuable cushions against unpleasant price rises at little cost to the producer. Also, a record of feed price movements, while less cyclical than 30 years ago, is still a valuable graph to have on your office wall as it can suggest reasonable predictive success up to 6 months ahead. *I see almost no such evidence in the pig farm offices I visit*; when I was buying I monitored raw material price graphs assiduously, and it was well worth it.

Pro-active buying : One or two clients have become so experienced in this technique that they have entered the feed raw material market themselves/on behalf of others/ their buying group, normally on 25 tonne minimum parcels. This has been worth, after admin costs are deducted between 3% to 6% reduction in final feed costs, and while this could be below a wholesale or brokerage discount, it is nevertheless a reduction in purchasing risk if managed well. The main risk remaining is to be landed with a substandard parcel, hence membership of a buying group or farm partnership is valuable. Experienced operators know the right questions to ask of the right brokers – on origin; length on the market; declared critical analyses; dry matter (if applicable, *i.e.* on by-products); storage/warehousing conditions and reason-for-sale. Again, taking time and patience to acquire information before action is taken is critical, and itself is an insurance against risk.

Action :   Work much harder in knowing *your* market, your immediate purchaser's market, *his* buyer's market and the world market. *You are much less likely to be exploited*, even if many events are outside your control. You narrow the exploitation-gap, or see how it might be narrowed.

The second tool in the risk management toolbox is:

Co-operation/co-ordination/integration

This needs a degree of selflessness which, it must be admitted, is not a strong feature of the global pig producer and especially so in Britain. Are you a sharer? Of information, of risks, of profit? You need to be. **Isolation, euphemistically called 'independence' is a non-starter these days.**

Several parts of the world are well advanced in co-operation where pigs are concerned, such as Denmark and the USA, with Holland, Canada and even Mexico showing progress in the intention and then determination by farmers to band together to share risks and secure less volatile markets, or at least have a calming influence on future local instability.

Modern pig production needs a co-operative mind-set. Choose sharers as partners. A sharer is prepared to go more than halfway in a partnership, those sort of partners markedly lessen risk. Again, I speak from experience!

The third tool in the toolbox is:

Relationships, not transactions

Following on from this, relationships are more important than transactions. Stop and think about this. An auction is a transaction. While it can be a satisfactory method of buying or selling it is still a temporal, ephemeral act, not one which is necessarily committed forward to a long term arrangement. To investigate business relationships you need to talk. Talk and propose. Talking has always been a prerequisite of negotiation.

Negotiation in its turn has always been a preliminary to co-operation. Proposition – making suggestions – is also a positive part of co-operation, too. So think positively and anticipate in advance the likely partnership problems which are always there, his and yours.

Pig farmers are not good at co-operation because they don't talk enough – to each other or to their primary customers, the processors/retailers. Too often any talking there is just revolves around problems. If the possibility of problems are too prominent in the discussions the issues become soured and defensive attitudes occur.

Other people have problems too. Understanding them and providing positive suggestions will get to a solution more quickly than highlighting (justified) grumbles or erecting barriers. Get round to them later but only after the advantages have been well-aired and understood.

So talk and think positively. Then once the relationship is established, the transaction often falls into place.

CONTRACTS

Contracts for pigs minimise risk. They are, if you like, more of a transaction in farmers' minds, but a good relationship helps materially, like this:

Develop a *relationship* with the buyer by ...

• Listening to his needs/problems. Ask what they are/may be – this often starts you both off in a positive mood.

• Selling a product based on his requirements (*He* is the customer, not you).

• Setting up a linkage which helps him defray costs and for you to obtain a guaranteed outlet/obtain a premium.

• In this way both his and your risks are reduced.

• Suggesting opportunities to jointly develop products so that his customers are pleased.

• Thus all parties benefit – you, him, his customers.

So *think beyond the price!* Sure, price is of great importance, but demonstrating a positive attitude to a relationship as well is crucial. UK pig farmers complain to me that the processors won't talk to them. This is partly because they suspect that pig farmers, beset as they have been over the recent years with economic impositions, will be negative and 'difficult'. Remember a good partnership tries to go the extra mile and a positive approach to a negotiation is one way of doing so. Positive talking does not cost very much, and usually nothing at all.

BEING POSITIVE

Here are some examples from some of my clients across the world.

1. Written evidence from a local butcher that a certain small number of your carcases are taken each week because his customers pay a premium (especially over supermarket prices) for taste and appearance.

2. Evidence from your farm's AI/breeding policy that carcase quality is uppermost in your mind and the essential genes to achieve it have already been invested in your business – and how much/kg this has cost you.

3. Evidence that you try to produce meat '*cleanly*'. A demonstrable biosecurity protocol can convince the processor of this in your case.

4. Use organic selenium in your food as it can cut the retailer's drip-loss by up to 10%, worth 2p/kg to the supermarket in more pork *sold* in relation to that *purchased* by them. The research and econometrics

have been published – so use it as another positive contribution you can make with your pigs to *his* business, *and* how much it costs you, about 0.5p/kg or whatever, **on his behalf**.

5.   Evidence of how your pig flow is planned to allow for evenness of supply both in number and weight. Evenness is crucial to factory processing. Several of my clients did this very well, but never mentioned it to the buyer!

"JUST SIGN THE CONTRACT!"

Producers often claim that similar approaches to those given above are all very well but all the buyer is interested in is your getting on and signing the contract to help fill his procurement targets. This may be so in some cases, but more and more contractors are responding to the longer term relationship aspect of contracting – they don't want their procurement base constantly changing, which is inefficient, stressful, adds to their costs and puts them into a risk situation themselves. Remember, talk to these people and provide positive suggestions. You are the seller, they are the customers and you need to be pro-active in the selling sense. Being pro-active in any negotiation puts you in an advantageous position anyway.

A PIGMEAT CONTRACT CHECKLIST
(UK   2002)

Deliveries	✓ Number per week.
	✓ Delivery time window.
	✓ Variability in numbers?
	✓ Confirmation of time and date needed?
	✓ Late delivery penalty?
	✓ Transportation requirements.
	✓ Dirty pig charge?
	✓ Growth promoter ban?
	✓ Antibiotic residues?
	✓ Haulage costs.
	✓ Biosecurity clauses?
	✓ Estimated time in lairage.

✓ Estimated time of weighing-in lairage.

✓ Situation re liability for strikes, government restrictions, flood, fire, etc. Situation in holiday weeks?

Price

✓ Base price? How is it calculated/fixed?

✓ Location of base price in relation to weight band penalties and % outside agreed weight band.

✓ Valuation of underweights?

✓ Valuation of overweights/uncleans.

✓ Non-contract price.

✓ Deductions – MLC classification
   – Standard levy and/or
   – Promotional levy (called 'check-offs' elsewhere)
   – Meat Inspection Charges
   – Residue Testing Charge
   – Waste Offal disposal charge

✓ Feed price rise/fall allowances/deductions (+ VAT situation called 'sales tax' elsewhere) Pricing alteration situation.

Other

✓ Length of contract.

✓ Length of notice of alteration.

✓ Length of notice of termination.

✓ Length of payment delay.

✓ Insurance situation.

THINGS TO GET STRAIGHT IN YOUR MIND ABOUT A CONTRACT

Comparing contracts

Pig farmers are too haphazard in comparing contracts, which must be done continuously with all the outlets within a feasible haulage range, not just when approaching contract termination.

A manual – or much better a computerised – spreadsheet should be compiled for all these possible outlets, on to which the farm's current actual and the past year's average performance can be overlain. A nett cash flow benefit is then obtained based on the actual performance of the pigs sent in.

The differences between current UK contracts can be substantial. An exercise I did recently on a contract involving 130 pigs/week compared to three others showed differences of between –£1.47p/pig to +£1.93/pig nett income over each 6 month sample across a 110 mile transport range compared to the farmer's chosen contract. In gross return terms for readers outside the UK this was –2.2% to +2.9% – a difference of 5.1% between the best and worst contract – over 33% of nett profit on the farm concerned. About 87% of these differences come from the positioning of the base price and the implemented or decreased price/kg in relation to backfat measurements. Only about 13% of the price differentials obtained came from other deductions/costs.

However when differences in transport distance were taken into account, and building-in accepted shrinkage losses, the nett profit difference overall in theory might have narrowed to 20% overall and to +9% in the most favourable contract. Even so, the exercise revealed surprises in the financial differences between the contracts and the importance of carefully comparing all costs and benefits together with transport distances when examining contracts.

Negotiation

On presenting our calculations to the existing contract supplier, price adjustments were made by them to bring it in line with the most favourable competitors, in mid-term too. On another occasion the contractor looked at the farm's track-record and also increased the premium allowed. Experiences like this confirm that talking, negotiation and presenting positive information is important, as evidence, even within the very competitive industry of pigmeat procurement/processing, can change things in your favour. So work at it.

CONTRACTS – A SUMMARY

- Be positive, look forward, not back. "Everyone has a horror story over a contract." (Strak, 2000).

- In your business, what areas of your production costs *could* be contracted? Pigmeat, feed, veterinarian, cleaning?

- Study various contract price formulae. Work at it. Become knowledgeable about other people's contracts.

- Start a contract comparison spreadsheet. Update it at least monthly.

- Consider production costs (are yours realistic?) and long-term margin shares. Don't be afraid to talk about margins, his and yours.

- A contract is give and take, both sides. Be wary of contracts where the other side is cagey – they could let you down. Redouble your efforts in probing, checking, getting confirmation in writing.

- Look beyond just the price.

   Action :   Talk, talk, talk. Work at it. Spend more time researching and negotiating, thus start early. Once you've contracted for a period, start looking at other options to use as future ammunition or reasons to change/re-negotiation next period.

LIFTING THE LOAD - CONTRACTING OUT

You should examine employing others to grow out post-nursery stock. Under these agreements the person finishing your pigs provides labour, power, buildings, straw, water and insurance. The owner provides pigs, feed, vet and med. and management input if needed. This allows you to dispose of perhaps 25-30% of your labour costs and/or concentrate more fully on the – let's face it – more labour-intensive and skilled aspects of breeding and nursery work as distinct from the not so onerous and less risky skills of the grow-out process. Some opportunities also occur for contracting out the nursery rearing aspect alone, but only if the contractee has the correct housing and labour expertise and time available for looking at every weaner in his care at least twice daily. Nursery rearers have taken pigs from 7-35 kg at a £2.50/head fee (15-77 lb at $3.50/head), while finishing fees seem to be around £4.50/head ($6.30). Overall savings of £5 to £6/head (7 to 8.5p kg/dwt but more typically 4p/kg - 2.6c/lb) have been made on fixed costs and you still have the pigs to sell at the end of the period.

The system does have all the disadvantages of a transport move at a critical stage of the pig's growth curve, but this is usually more than recouped by the advantages of segregation for the finishers, and the benefits of batch production to yourself as the breeder.

CONTRACTING OUT CHECKLIST

✓   Insist on a ***proven*** track record. Ask for records, especially sale dockets if available, and evidence of performance.

✓   By judicious questioning, get an idea of how much time the contractee is able to spend on stockmanship. This is absolutely vital in the case of nursery contracting-out. For example what other tasks/jobs is he/she likely to be engaged in?

✓   Check the buildings to be used very carefully, especially ventilation adequacy.

✓   Tour the premises with your vet and listen to his advice.

✓   Stipulate a proven biosecurity protocol; you to supply the products used.

✓   Draw-up a written, signed contract.

✓   Sale dockets go to both of you, but pig cheques go to you the owner, with agreed payment dates signed in advance by both of you.

✓   Ensure insurance cover is in the contract and who pays it.

✓   Keep your market outlet in the picture.

✓   Visit regularly. (Unannounced visits however can cause friction.) Have clear ***written*** guidelines on the agreed management protocols.

✓   Check on weighing and loading-out facilities and how smoothly it is likely to be done.

✓   Never contract with those who have other pigs onthe premises or adjacent to them. Consult your vet if in doubt.

✓   Renew the contract annually, not longer, by including a satisfaction or target-clause in the contract.

✓   To ameliorate this requirement, you can make provision for a modest performance bonus. Incentives are often worthwhile in a contracting agreement, and could be essential if you have found a conscientious partner and need to retain him.

Partnerships succeed when the terms are clearly written down and signed by both sides. They can always be modified next time.

Partnerships prosper when both parties go 51% towards each other.

FUTURES (FORWARD PRICING)

After introduction in 1984 and compared to vibrant pig futures markets in Chicago and Amsterdam, this has not been a success in the UK, the scheme barely lasting a decade. The demise was due to insufficient trading seemingly caused by disinterest in risk-spreading from both processors and retailers who, in the UK market, have largely been able to control their own input price structure so as to minimise their own risk in this area.

However, increasing globalization of pig trading could involve large UK pig producers in an EU Futures market, and a watching brief on developments in this direction is advisable if a UK Pigmeat Futures Market based on a better pricing structure than AAPP can be devised. There are good sources of advice and information on the situation from various sources *e.g.* MLC and Euro PA at the time of writing.

While International Futures markets may seem remote, there are benefits in learning how these markets operate.

Input price risk spreading on feed ingredients (especially, for pig producers, grains and soya) presents a happier picture. The UK feed trade already has forward buying contracts on complete feeds and straights, some of them even providing capped forward options where, for a form of insurance premium, pig producers could protect themselves against upward price movements of grain and soya and yet could still benefit from a price drop. In 2000, only 5% of all farmers use this facility.

Strak (2000) reports as an example: "It would only cost a fraction of a penny/kg pigmeat not to be caught out by the price of soya doubling." In such cases there is surely no need for farmers to engage too deeply in futures or options for feed ingredients when the UK feed trade can be approached for the same risk reduction at what seems to be a reasonable cost.

SUGGESTED ACTION

* Larger producers should subscribe to the relevant information services/ updates available.

* All producers should explore the costs and attractiveness of feed ingredient forward option schemes within the EU feed trade rather than attempting to beat the market by speculating on their own. At the same time a watching brief on spot price movements on raw materials not covered by such schemes can be rewarding in making your own forward buying decisions.

INSURANCE

Insurance cover has always been a popular way of reducing risk, and some companies have specific risk insurance policies. However, disaster-recovery policies don't cover price risk.

Farmers with fields close to public roads are not natural insurers – for example, I found few outdoor breeders have Public Liability Insurance and many field vehicles used very occasionally on roads are not covered. Recent floods and gale damage caught many people out.

All the time reputable insurance companies are designing new policies and it is worth keeping an eye on them by means of a once-yearly visit from their salesperson. You don't have to bite, as the cost may be still too high, but a counter-proposition may yet be accepted by them. Test the market. Explore. Propose. Talk to them.

REDUCING RISK BY MANAGING YOUR BUSINESS BETTER

MINIMISING CAPITAL RISK CHECKLIST

✓   Work out a **Debt Management Plan** before you borrow money. Prepare a range of likelihoods where pigmeat prices, production costs and interest/exchange rates[1] could move.

[1]   Exchange rates have a direct impact on prices. For example ± 10% in the Euro/sterling rate produced ± 3% change in the U.K. pig price from 1993-2000.

✓   Prepare contingency plans for each scenario.

✓   Review your borrowing intentions on each of, say, three situations. Optimistic, Forecasted Reality, Pessimistic.

✓   In the light of this preparatory work, use your accountant or financial advisor to work out *Gearing Ratios* for your business, *viz.* Debt-to-Equity, Debt-to-Convertible Assets; Debt-to-Disposable Income. Monitor these ratios frequently and discuss significant changes with your financial advisor.

✓   Assess your *True Asset Position*.   Identify liquid assets which can be converted to cash quickly.

✓   *Debt Management Capability* : Establish what funds you have available to meet operational costs on equipment, housing, breeding stock and food, all of which will determine the payback period. Review loan arrangements in relation to expected sales and cash-flow.

✓   *Research Borrowing Options*.   Your present deal may not be the best available.   Discuss with your financial advisor how they compare.   For example : *Venture Capital* – used particularly for major expansion or diversification, this can provide an alternative to conventional borrowing or complement it. Explore.

MANAGING INCOME CHECKLIST

✓   Keep accurate and up-to-date[2] *financial records*   *viz. :*

   •   *Timing* : Projected and actual cash flows.

   •   *Net Equity Position* : Balance sheet of assets and liabilities.

   •   *Net Profit or Loss Position* : Profit and Loss statement (earnings and expenses)

✓   Be careful to separate *living costs* from *business costs*. (The author has two separate bank accounts.)

✓   Be careful to separate *diversification income* and off-farm income from the on-farm income.

✓   Consult your accountant to have *tax management and planning* strategies in place.   Taxation is a significant risk in

---

[2]   On the past 100 farms the author visited and questioned globally, these were on average 9.7 months in arrears.

cash flow management, and risk can be mitigated by tax-smoothing arrangements. If necessary seek advice from a taxation specialist.

MANAGING YOUR ASSETS CHECKLIST

✓   Review *insurance cover,* not only on disaster protection, but also health and accident to all personnel, including yourself/your family. Also review areas of possible loss not covered or poorly covered by government/EU compensation.

✓   Draw up a list of *improbable/rare, possible and likely losses* and review the need for cover and its cost.

✓   Set up an annual *Risk Assessment and Insurance Review.* Both the farm, political and the insurance industry circumstances can alter radically in 12 months.

MANAGING LEGAL RISKS CHECKLIST

✓   Discuss your *business description* with your financial or legal advisor, *e.g.* sole owner/trader, partner, limited company, trustee.

✓   Check your *taxation, accounting and financial reporting*, as well as any auditing obligations and succession arrangements. Also what mandatory government returns (annual census, *etc.*) are required.

✓   Check that *COSHH, HACCP, IPPC, Welfare and Farm Safety requirements* are in place and being followed.

✓   Check that your buyers' *Codes of Practice/Quality Assurance Rules* are being followed.

✓   Check all contractual arrangements. Make sure you have the *original documents safe and quickly accessible*.

✓   If applicable (*e.g.* farm shop) check that all *trading practices*, including fair trading and product claims, are in order.

✓   Footpaths, outdoor sows, farm shops *etc.* Have you *Public Liability Insurance ?*

DIVERSIFICATION

'Having another string to your bow' is an attractive option for pig producers who have suffered the pain of reduced cash flow due to price volatility. This is especially true if the alternative source of income can be linked to the farm itself (on-farm income) such as bed & breakfast, gites/chalets, excursion centre, horse-riding, B&B plus stabling, farm shop, organising producer marketing, e-business / IT centre for other pig and livestock producers, AI service, vermin control, golf driving and shooting ranges. Areas like these are not far from the expertise or facilities a pig farmer already has.

Outside the farm (off-farm income) can seem to be limited in choice and also daunting due to the unpredictability of the novelty involved, apart from merely offering oneself as contract labour or doing another job. But there maybe more ambitious opportunities in the wider world of business where franchising something completely different is a route through the minefield of a stand-alone on or off-farm business.

Generally speaking farmers are not good at marketing. This is just what most accredited franchisors with a track record are very good at, so a well-chosen franchisee will have had the marketing of a product/ service done for him, with the likelihood of a trade name already established. Nevertheless the franchisee has to have operating organizational skills and, particularly if a pig producer has been a successful breeder he will already have had the natural ability in organizing and forward planning to contribute to the success of the franchise.

Later on we will look at franchising in more detail - it could be just what you or your family need to diversify.

REFERENCES

Martins, A. (2000) Risk Management Initiative Course, ADAS Wolverhampton. National Pig Association.
Thornton, K. (2000) Personal communication.
Strak, J. (2000) *The UK Pig Industry - Can we Compete*? JSR Healthbred Technical Conference

ANOTHER STRING TO YOUR BOW? 12
STARTING A QUITE DIFFERENT BUSINESS TO PIGS

AN INITIAL CHECKLIST

✓ *Is your new venture viable?* Are there enough people within your planned catchment area to make it so? How many competitors are working in your catchment area? What about suppliers? Local, national or overseas?

✓ *Research the market you are entering.* Take time to do this. Too many new business start-ups fail within the first two years because the owners are so enthusiastic about their new idea that they are over-hasty in getting started. It probably takes about a year of intensive investigation in addition to your existing workload to prepare a sound business plan, which will not only reassure you and give you the necessary confidence and determination to surmount the initial hurdles, but also convince others to join with you or provide funding.

In Britain the government's Small Business Service has a website – www.sbs.gov.uk – which is useful starting point for newcomers to this complex field.

✓ *Add up the money* you will need to organise a company outside your pig business – to buy equipment, build, alter or rent an office and carry the business in the first few months.

✓ *Locate funding.* Most pig businesses are already under-capitalised, so will not be able to fund another venture. You will need to approach some banks and also locate other sources of borrowings. Interview several potential sources. This is where a professional's advice is helpful. Start with your accountant and work outwards from initial contacts he should be able to suggest. This is why you need to put the time – and some travelling, into this important stage. You need to know what to ask for, and how much, and why you need it. So...

✓   Have a business plan ready for any potential lender.

A shrewd lender will want to know: -

- What the opportunities are/seem to be in the chosen market.

- Exactly what the project's activities will be. Be clear and concise on this. You know clearly in your own mind,   but the lender won't. Get across to him how it will operate.

- What you've considered could be its weaknesses. This impresses lenders.

- What your predicted profits will be, and how you have arrived at them in a realistic manner.

If he bites, then you can compare his terms with others.

✓   *How do you want to trade?*

Do you want to operate on your own (as you possibly are now in your pig business), or as a partner with someone else (who may be more experienced than you in the new project or product), or as a limited company?

Each of these has different tax implications, and once you have set up the business, you must register it with the Inland Revenue in the UK. And if the company's annual *income is likely to be* over the current threshold, you need to register for UK VAT *in advance* to avoid a subsequent retrospective penalty – I speak from experience here!

✓   *The legal side*

If you decide on a limited company (done to protect yourself should things go badly wrong) you need, after researching the pros, cons and costs of doing this, to get professional advice, read up on the UK Data Protection Act (or consider a patent if you have an innovative product) and get advice on insurance. Some insurance is a legal obligation anyway. Again, get professional advice on this.

✓   *People working within the new venture*

If you've been a family-based farm and now plan to employ people (including yourself if you decide on a limited company) you will need to set up a PAYE scheme and explore employee pension provision as well as read up on employment law.

A YEAR'S RESEARCH AND CONSULTATION

This is why, in my experience, you need a year to get familiar with the whole pattern of a new business outside the farm. You will definitely need a variety of professional advice to tell you what needs to be done and thought about in advance. New businesses which fail all remark on the 'nasty surprises' which caught them out in the first year. With advance help, none of these need have contributed to their downfall

FRANCHISING 13

---

*A business operated and part-financed by you (the franchisee)
which is originated, promoted and part-funded by the franchisor.*

Franchising offers the chance of becoming part of an already successful business at a significantly reduced risk.   As pressures on pig farming profits grow or become more volatile, franchising can offer an easier and safer route into the outside business world. At the time of writing "the chances of success from a franchisee are likely to be up to five times greater than starting up your own independent business". (Barclays, 2000).

CHECKLIST - FRANCHISING

✓   ***Avoid the hard sell.***   Preferably choose a member of the British Franchise Association, or who fulfills their criteria.

✓   ***Ask for a track record,*** especially the number of successes to failures, and length of experience.

✓   ***Enquire about the exact level of training*** and assistance provided both at the outset and ongoing. Don't be fobbed off.

✓   ***Choose a franchise where your rights and obligations*** are clearly set out. Present them to your bank for comment, then choose a specialist franchise lawyer to go into the fine detail. In my experience – and I'm not one for lawyers – it has ***always*** been worth the cost. Lawyers are more useful – and cost-effective – in a risk management situation than in expensively picking up the pieces afterwards!

✓   ***Choose a product/service area*** likely to grow in demand. (An example at the time of writing and applicable to both town and country areas, ***and for someone with a farming background*** is sandwich/snack provision).

✓   ***Carefully research the existing competition.*** Franchisors seem quite prepared to take on new franchisees even if the market is near saturation. If so can you add value to the local competitors' package?

✓   ***Talk to your bank*** at an early stage.

✓   ***Question the franchisor closely*** on his future growth and development plans. Does he seem to be coasting?

✓   ***Can you afford*** the (£10,000 to £30,000 at the time of writing - about $13,000 - $44,000) entrance/start-up fee charged by the franchisor? Are there hidden costs?

✓   ***Have you/your family the same degree of personal commitment*** to the franchise as you have for your present pig enterprise? Have you a part replacement pig manager to give you the necessary time and training needed, at least for the first 2 years? This is because ....

✓   ***You will need a high degree of personal commitment.*** The requirements of a good franchisor and his methods of operation/control can be onerous, but not impossible. He has a name/trademark and market stance to maintain and he will not let you damage it.

✓   ***Look at the close-out clause*** in the contract. Can you keep to its requirements? Get the franchise lawyer to explain it carefully to you.

STAND-ALONE DIVERSIFICATION

A few notes on some of the author's clients who have already diversified may be helpful to those pig producers who are thinking of hatching rather more eggs in their future income basket.

Some disadvantages to consider

*   **Land-based diversification** (golf driving or shooting ranges, specialist horticulture/aquaculture, quarrying, *etc*) usually have substantial costs, but the banks look more favourably on them. Most specialist pig farmers are not in the area of land-based diversification however.

*   **Capital costs** of both forms of diversification – especially land-based – are high in the UK compared to other European countries and especially those further afield.

- Your specialisation skills will be diluted so your core business may suffer. With franchising you may need to have delegation in place.

- Remember that it is difficult to be good at everything you take on so ...

- Do not underestimate the mental energy needed to run two very different businesses successfully.

- Be sure to run two separate sets of accounts. Diversification gets confused and fails if income 'fudging' occurs – when income from one source dulls the appreciation of loss from the other. Cross-borrowing of finance is common, but when temporary troughs occur in both ventures at the same time, without careful monitoring and control the end result is unfortunate; moreover your accountant or tax inspector won't be impressed!

- Other farmers diversifying into pigs tend to come in and out to suit their cropping activities and/or their end product prices, and so destabilise your market – and possibly you theirs. Thus the overall price expectancy of both markets may suffer. Also the financial costs and overheads needed to enter a market may lower, not raise productivity.

Having said all this, diversification can be an attractive option to core-business pig producers. In future, due to the spreading of risk, it could and probably will, in my opinion*, be a normal part of your business strategy*. Stop and consider this. Farming will be far more diversified in future. Have you the right mindset to accept this major change?

Some successful clients of mine have already addressed the checklist of positive actions given below, and one of them even commenced a diversification consultancy for other pig producers, which soon expanded into all livestock farmers and is now across into crop husbandry.

Some tips on diversification

- 'Information is not a luxury' (Strak, 2000). Assiduously collect data on other on- or off-farm ventures.

- The internet can be a good source of ideas. There are plenty of non-land based options both on or off-farm. Surf and search.

- If sufficient capital is available, franchising can be an immediate three steps up the diversification ladder. Successful franchisees tell me it put them at least 3 years 'ahead of the pack'.

- So with capital available, examine franchising as well as other diversification options demanding a financial contribution. It could get you to a supplementary profit much more quickly.

- Diversification can often beneficially involve family members. One problem here can be to convince sons and daughters to realign their own career prospects with a family venture. Another is to place too much reliance on the wife (to run bed & breakfast for example) especially when young children are around or another is likely.

- Be sure to keep the book-keeping separate and employ part-time secretarial assistance if this is considered prudent.

- Having said all this, I have seen more successful part-time diverse businesses succeed on pig farms *where the whole family was involved than where the producer himself went it alone*. Often the successful principal had looked at what members of his family could contribute rather than what he could.

- However the successful family head is more likely to possess a definite skill or enthusiasm he can develop – restoration of machinery, craft skills, shooting expertise *etc*. But where delegation is not in place then both ventures suffer, especially dangerous where travelling is involved, even local travelling involving regular periods away from the core business. I speak from experience!

- Mental attitude is a vital part of diversification. Diversification works when you *want* to diversify, often not, or not so successfully, when you *have* to diversify, which has unfortunately been all too commonplace in the difficult years 1998-2001. Again it hinges on having the right, positive mindset.

- Producers who have to diversify have a higher mountain to climb, that's all, and need even more determination and encouragement. I have seen some remarkable transformations, as farmers – and the British – are good when the going gets tough, and they are shaken out of their complacency.

- These days there is help and advice available and in Britain ADAS is a good start. Go out and see what they can suggest. Talk to your bank, to your friends. Just *talk* to as many people as possible and you'll be pleased with what ideas come to you.

MOSTLY ABOUT PEOPLE 14
YOU, YOUR STOCKPEOPLE AND THE PEOPLE WE DEAL WITH COMMERCIALLY

ERRORS COMMON TO MOST PIG PRODUCERS

I've been fortunate to visit about 4000 pig farms in my life and talk to their owners and managers. It is presumptuous and maybe impertinent of me to list their faults as I see them as in the list below, but perhaps it is worth you running your eye down them all the same. Pig farmers are surprisingly similar in any country.

On the positive side, pig producers are dedicated, hard-working, courageous, resilient, good-humoured, they do want to care for their animals within the bounds of convenience and cost, and they are tolerant, often to a fault.

Now for the bad news!

ARE YOU HERE ... ? SOME COMMON FAILINGS AMONG PIG PRODUCERS

✓ Ignorance of what is possible/how lamentably we fail to achieve current genetic ceilings.

✓ Not measuring things well enough on paper (recording) or in the piggery (monitoring devices/controls).

✓ Thinking you know best. 'Experience' often holds you back!

✓ Not being observant enough.

✓ 'Tailchasing' by too much time spent on daily chores.

✓ Doing too much hard work themselves thus ...

✓ Not investing in automation to remove hassle and drudgery.

✓   Not understanding ventilation/the way air moves.

✓   Overstocking.

✓   Not using cheap, temporary – even 'throwaway' – housing to 'defuse' production bulges and isolate sick pigs.

✓   Wastage, especially food.

✓   Underestimating the importance of the post-weaning phase to finished pig profits.

✓   Not training labour to modern demands.

✓   Not being present at farrowing.

✓   Pushing gilts into production too impatiently.

✓   Not using the vet properly.

✓   Not using AI sufficiently.

✓   Slow to espouse business partnerships/linkages/collaboration.

✓   Falling behind in biosecurity requirements (the poultry industry is way ahead of us in cleaning and disinfection, for example).

✓   Not treating pig production sufficiently as a business.

✓   Not spotting where or when to invest, *i.e.* poor prioritisation.

✓   Tending to delay spending completely rather than spending an affordable amount in the right place, and so building on that extra income to reinvest elsewhere.

Quite a long list. Even so, I find many people are in 50% of it.

On the other hand …

During these 4000 or so farm visits I have also been privileged to sit at the feet of, marvel at and learn from, a couple of hundred top-class owners and managers.

Maybe this is what they have in common – and it is not so much an opposite list to the foregoing as you may think.

PORTRAIT OF A PROFESSIONAL PIG PRODUCTION MANAGER – A CHECKLIST

Their one object is to **maximise profit**. Not necessarily physical performance or even income. Their methods can be divided up into short term and long term goals as follows:–

Short term
- ✓ Set production targets to achieve projected income.
- ✓ Maintain sufficient replacement stock.
- ✓ Breed to a pre-calculated production target.
- ✓ Reduce mortality and stillbirths.
- ✓ Wean a quality pig.
- ✓ Wean a sow suitable for rebreeding.
- ✓ Maximise rebreeding effect.
- ✓ Minimise rebreeding time.
- ✓ Minimise disease/maximise health.
- ✓ Reduce costs/identify waste & inefficiency.
- ✓ Improve animal and staff welfare.
- ✓ Motivate staff.

Long term
- ✓ Talk to various market outlets all the time.
- ✓ Select the correct genotype of stock for the market outlet.
- ✓ Maintain a recording system to maximise pig flow.
- ✓ Maintain a recording system to identify problems, especially to forewarn of potential fall-off against targets.
- ✓ Select, monitor and train staff.
- ✓ Pay staff adequately.
- ✓ Purchase feed correctly.

✓ Sell pigs effectively.

✓ Plan maintenance and repairs at minimal disturbance to the manhours available.

✓ Make cost-effective alterations.

✓ Train himself.

These are all key tasks. Most top managers employ them.

EFFICIENT USE OF LABOUR – YOURSELF

Examples from my own philosophy ...

Know yourself and your job. Mine is writing and giving correct advice, so in my case ...

Writing	• The idea is more important than the writing of it.
	• An idea needs to be written about four to six times to make the research and fact-checking involved cost-effective. There are many ways of writing up the same information.
	• I don't type a word. Why should I when I can employ top experts to do it so much quicker with far less effort? The time saved is put into being creative and acquiring and storing information.
Getting and storing information	A third of my working hours is spent on this. Obtaining, cross checking and cross-referencing information so that I am pretty confident it is supportable.
Computers	• I haven't a business computer in the house. Yet I consult 4 computer models, have long had one of the first e-mail addresses issued, can call up

the web at will, *etc*.   This is because I employ others to do the legwork more quickly and with less effort than I ever could.   So many of my peers are hooked on the computer drug – I am sure it is actually restricting their creativity, tying them down.   Beware!

Advertising
- Initially as a consultant I advertised my services, in those days frowned on by my peers. But it got me established and as far as I could see paid back 30 times over anyway. Any business needs (judicious) advertising until it is up to speed/depends how far you want it to develop.

- Advertise a farm? Why not? If you are producing something you are proud of, tell people, even (especially) if it is "only" *your processor.*

Travelling
- Is a wonderful source of ideas on what to do (and what not to do) in pig production. If you don't go-see, you are working with one hand tied/missing opportunities/getting out of date. Travelling improves your sense of judgement and self-knowledge.

Training yourself
- I've always spent 15% of my annual income and 5% of my time being trained by others. Even today, after 40 years' work in pigs, I am never too old to learn – neither are you! I learn something new about pigs every day; something important once a month; something of quite earth-shattering impact once a year! Pig production is rocketing ahead – we all need to keep up with it.

DOING A GOOD JOB – A CHECKLIST

Think like this about your own job. The approach to what you do won't be so very different from my own experience – even if your job is.

✓   Question everything you do.

✓   Can someone or something do (some of) it better or cheaper?

✓   Just because everyone does it one way, is it really right for you? Think laterally.

✓   Be a people-person. Talk to everyone. Never be afraid/too proud/too embarrassed to ask if you don't know. For each rebuttal/refusal you'll acquire a whole crop of useful advice – and it will be free.

✓   Silently question everything you are told; there is a lot of misinformation/half-truths about.

✓   Keep people in the picture. Your staff, your bank-manager, your vet, your nutritionist, the accountant, your family.

✓   Write things down. Busy people cannot remember everything. Then review your notebook regularly.

✓   Keep quiet about good ideas – or write a book about them!

Think more : stop chasing your tail.   Then you'll work less but work more efficiently.

WHAT GOOD PIG MANAGERS TELL ME

Selling his output effectively : The manager has to keep a very sharp eye on selling his pigs to the outlet which can influence his income in the most positive way – which sometimes may not mean the buyer who is currently offering the best short-term price.

The good manager realises his job is not pig production, *but meat production*. His job is to *make it easy for the processor to buy his pigs* which means maximising the output of exactly the quality of meat which the buyer needs, on time, in level deliveries (which is what a contract means, otherwise why bother to contract?) at the minimal cost to the farm enterprise.

Cost control

In exactly the same way that the most efficient pig farm manager is the one who manipulates productivity correctly so as to maximise profit, so it is not necessarily the farm which spends the most money which makes the most profit. Control of costs is a vital management area in any business, and with a modern pig farm likely to invest more money per head of staff working it than most industrial businesses in Europe (surprising but true; output per man on 'the larger' pig units in Britain is over £230,000/year – US$350,000), spending the capital wisely is a vital area of eventual profit. The ways in which money can be saved, or better, redirected into areas which give a more promising return can be counted in hundreds, not tens on any pig farm, big or small. But here are the 'big three' in my book as they affect *cost savings* …
I attack, on my clients' farms, when spending on **things** is involved, waste of *food*, waste of *warmth* and waste of *space*. Attention to all these areas rarely fails to provide the client with 20% (sometimes 35%) savings in costs which is a huge benefit to cash flow – and pays my fees 20 times over!

As to spending *capital*, the big three here – in my opinion – are on *precision feeding*, reducing *disease* and accurate *environment control*. An investment of one dollar on facilities which improve each one of these will yield never less than $3 return on *each*, and that's a huge hike in income.

Staff motivation

The instinctive manager knows that people – individuals – make or break businesses, and that a good proportion of his time must be made over to keeping people well-briefed as well as training, monitoring and encouraging them. Not forgetting himself – keeping up to date and in-the-know is vital in this fast-changing industry of ours.

The 'working' manager – how much work ?

I know of very few pig farm managers who don't do some manual work – only on the very biggest farms, perhaps.

Trouble is, most pig farm managers *don't allocate enough time to the key areas above*, and get caught up in an increasing spiral of manual work. Sometimes it is not directly their fault – the owner expects it and they may be strapped into a job specification which is out-of-date for the 21st century. Thus I hope as many owners will read these notes as managers!

Secondly, *managers are still very much output-minded* to the exclusion of all else. When I managed a pig farm I was offered (or asked for) a very low wage but a 10% share of the profits. That taught me very effectively how to manipulate output so as to maximise profit. If I got it right I was well-rewarded, but so was the owner – 9 times more! Thus he should have been much happier than I – but of course he wasn't, as owners are never satisfied, are they? We employees have to live with this. If you work for somebody, it usually goes with the job.

Third, too many pig farms I visit are understaffed. *This means the unit is always chasing its tail.* Repairs and maintenance tasks in particular are one long round of emergency action and the manager often has to assist to ensure the work-flow comes back somewhere near on-schedule again. This happens nearly every day!

This is often as bad on the larger 800 to 2000 sow units than the smaller farms. Such units should have a full time maintenance-man – electrician, plumber, builder, welder, either contracting out the work, or if large enough an in-house employee. Managers should sit down and work out the cost-effectiveness of such a policy. Pigmen should be trained as professionals in *pigmanship* – it is too important a job to leave to half-knowledge about peripheral tasks because he or she is busy at something else which also won't wait.

I often hear it said with pride that the modern pigman does indeed have to be stockman, electrician, computer operator, plumber, carpenter, welder, midwife, nurse – and so on. This is absolute nonsense! It is out of date and muddle-headed thinking. No modern production industry could ever survive on such a disorientated set of half-skills. The good manager isolates blocks of these skills and employs the correct professional for each one. Only on the smallest, non-industrial unit does the manager have to be knowledgeable about all these.

HOW MUCH 'WORK'; HOW MUCH 'THINKING'?

This depends on the size of farm, but I'd have thought managing 400-500 sows properly, with a just-adequate labour force, a manager would need to spend at least 22-24 hours a week on non-manual management tasks, between 3½-4 hours a day. Too much, you think? Not at all, because in this time the good manager must achieve a variety of 'thinking' tasks.

The manager must fit in to his daily schedule…

1. Look at every pig at least once a day (very small pigs and weaners twice a day). This cannot be done in under 1 hour/day, can it?

2. Plan pig flow. This depends on the visual monitoring and mental scrutiny I describe above and below. With discussion and issuing of guideline pig movements to the staff this, I found, took up 2 hours/week at least.

3. Monitor performance. With a computerised recording system it can be got down to ½ hr/day. Non-computerised it will take twice as long or much more. Managers must resist the temptation to 'play with' the computer and only stick to examining routine sets of marker figures against predetermined targets (or better, graphs), only using the programme's diagnostic facilities to probe a shortfall. A graphical computer programme is a superb tool for a manager to have, especially if it has a predictive ('what if') facility for input costs.

4. Staff control. I always put by at least ½ hr/day to give clear instructions, talk to, train and motivate people. Managing the work flow is as important as planning the pig flow.

5. Buying and selling. Vital. Regular telephone and personal contact with suppliers' reps. and prospective buyers must take up ½ hr/day at least, often more in emergencies, and time to visit feed mill suppliers and processing factory outlets/supermarkets purely on a watching-brief/personal-relationship basis must be allowed for.

    Again, the use of computer software to examine 'what-if' programs - both for buying feed and assessing selling price negotiations or offers - are underused by pig farm managers. The best managers I know spend 2 hrs/week on this sort of homework. It must also be remembered that 'tuning' these models accurately to your own set of current circumstances *takes up 75% of the time* spent on these worthwhile exercises in shaping your decisions. If this is not done (especially seen in hurried exercises done with visiting sales specialists with their lap-tops and prepared programs – beware of these) then the results will be misleadingly dangerous.

Understanding what the computer can do

Those managers who are using a 'what-if' computer program on feed nutrition intake, for example, have already discovered that the way the feed compounder specifies and formulates feed for him may tend to be in their interest and not necessarily his. A few progressive feed firms have seen the light on this. But the present system (to make manufacturing easier and cheaper) is, as you know, to fix nutrient specifications first and hope that by advising daily feed allowances target *performance* is achieved (not necessarily profit). A better way is to assess likely target *profit first*, then input the farm's specific details on pigs, housing and management – and *last, specify nutrient requirements to meet the target profit*. Think about it. This is a totally different approach, which will mean (as farms get larger) that these farms will have their own feed formulation which may change in nutrient specs as the profit target shifts due to the effect of market forces and as their pig's immune status changes. A good manager will pick-up technicalities like this from using computers and understanding what they can do for him – but he will never do so if he is shovelling manure or weighing pigs or mending things.

Half-a-day at least

All this, plus time to take the vet round once a month, prepare budget plans and reassess production targets with experts (accountant/bank manager/adviser-consultant) will take, I suggest, not less than 4 to 5 hours/day – a long morning or an afternoon – leaving working with people on preplanned tasks for the rest of the day if you want to do this.

The most difficult job of all

Is checking upon what is going on – which means what people are doing – without raising antagonism and being fair and equitable in pulling things back satisfactorily. Very few managers (in any business; not just pigs) find this easy. I certainly found it very hard. It helps if the manager has been professionally-trained in handling and motivating people, because the first thing a course of this nature does is to teach you how to handle yourself! Often I find where there is a poor working owner/manager relationship it is the *owner* who needs training, not the manager!

As a consultant, I find that is a tricky message to get across.

One of the secrets of a good people manager is to be in the right place at the wrong time – and make it look accidental!

THE TEAM SPIRIT / 'ESPRIT DE CORPS'

Pig farmers in my own country (Britain) have a lot to learn here, compared to the larger German, North American and Japanese farms. British farmers would argue that an average individual stockmanship is better in Britain than in any of these countries, which is true, but they miss the point of those nations who are naturally more group-orientated. This is that team-wise, they take a great collective pride in the unit. As a result the whole farm is neater, tidier, and better-organised; it is *good organisation* on the larger pig farm which is a vital prerequisite of *good stockmanship*. The larger pig farm in the future needs both. The first-class manager can improve both the organisation and the stockmanship, but of the two his main role lies in organisational skills. The key task where stockmanship is concerned is for the manager to recruit good section heads and get the labour load right, then the general stockpeople will improve by leaps and bounds, as it is a self-fuelling process.

PARTNERSHIPS 15

All pig producers are being exhorted to become 'businessmen-farmers' these days. To keep abreast of this movement I now have quite a few books on business management on my shelves, and have put myself on one or two specialised business courses recently.

This doesn't make me an expert in the subject – rather the opposite, as it happens, as they revealed my own shortcomings only too clearly – but I was surprised that no-one seems to have covered the subject of choosing a business partner either on the courses I attended or in the textbooks I purchased.

CO-OPERATION

Important? Yes, very, and not an odd choice of subject at all, because co-operation is now the buzz-word in many pig industries. Either co-operation up the marketing pyramid or alternatively (if you are are wary of the big battalions taking control) then forming localised mini-businesses where small groups of 3 to 6 producers co-operate both to produce pigs and then market their own brand of pig products. This can also be very successful – and, more satisfying!

THE FAMILY FARM HAS A FUTURE

All over the world I encounter family farmers apprehensive – even defeatist – about their chances of survival. OK, take that attitude if you must, but doing so tips you even further over the cliff. I've seen it happen with past clients of mine, now no longer in pigs or even in agriculture. Very sad.

In uplifting contrast, some clients have taken one look at that cliff edge and said "*No way* am I going to fall over it." Faced with the impossibility of making sufficient profit from, say, 150-300 sows on their own, they have gone one of three ways.

1. Mixed farmers have tended to close the pig side and concentrated on other sectors of the farm, and/ or engaged in diversification off-farm (*see item 3*)

2. Specialist pig producers have formed mini-businesses, usually 800-1500 sows or more, with other producers, sharing the profits. Increasingly, while no-one actually likes profit-sharing, the *improved performance from business restructuring has more than made up for the need to distribute what profits there are inter alia.*

    (This is where co-operation/partnerships come in and why I am writing this section).

3. Both specialist and mixed farms have diversified. On-farm if they can, but more often off-farm, even into becoming franchisors.

DISAGREE

I totally disagree that 'there is no future in the family farm'. With the right attitude of mind and good guidance, family farms can have a future. There is no shortage of conventional or venture capital, interest rates are affordable; but you must create your own local market, or supply into a national or global market who can sell what you produce. So what you produce must be seen to be wholesome, tasty, safe, attractively packaged (do your own) welfare-friendly and available at a perceived value-for-money price. Price is still crucial.

The latter, vital factor of securing a competitive price comes from co-operation and partnership. So let's talk now about partnerships.

EXPERIENCE

I've been involved in two personal business partnerships, one was in pigs and the other wasn't. Both were successful, but are now very amicably dissolved as my partners wanted to move onwards and upwards, which they have.

Lately I've become involved as a background consultant to other multisite production ventures started by pig farmers. So – from their and my experience here is a list of what I think are basic principles for a successful partnership – something the textbooks never cover!

A PARTNERSHIP CHECKLIST

✓ Both parties should be prepared to go more than halfway towards the other.

✓ Are you a 'sharer' and your partner likewise? If not, go no further.

✓ Like and respect your partner. If not, go no further.

✓ Ensure the partner's family are supportive. If not, go no further.

✓ Clearly define, *in writing*, each partner's horizon of responsibility and agreed commitment.

✓ Do not interfere. Discuss the partner's decisions, but be positive not negative.

✓ Talk daily; meet weekly.

✓ Do not have an open-ended agreement. Agree and define a time-scale, with a 'what-if' closure/ withdrawal clause built in. Each partner should have a 'silent' alternative partner or strategy in mind for their part of the business. This need not be talked about, but prepare yourself on what to do if a partner wants out.

✓ Mutually agree performance standards for each production section/unit and agree a 'what-if' action plan *supported by all parties* if one section is in trouble. One advantage of co-operation is that you can help each other, providing *each knows in advance* what this might entail.

✓ Accounts must be independently compiled and audited, and not involve any of the partner's family in their construction – so as to save money, for example.

✓ Do not have too many partners. Group businesses over 6 tend to fail in my experience, and 10 or more always have.

✓ One of the partners should be a butcher. There are plenty of these leaving their trade (sadly) so getting a good experienced man should not be difficult.

✓ Agree on a team leader, with (only if needed) a 3 year rotation.

✓ Use the big battalions, but keep them at arm's length. Never enrol them as a partner.

✓   Consider venture capital, but present a good case with an enthusiastic, united front. The venture capitalist will be a 'townie', so think your case into his shoes – *i.e.* his likes and dislikes about raising meat/food – from an urbanite's viewpoint.

✓   Plan well ahead on where the market may lead. Do your research; seek out the information – it is out there.

✓   If you can produce a good product, sell it to any outlet who (a) will listen (b) who clears the precautionary business hurdles (part of the research process) (c) never restrict your options or give exclusives (d) seek out concerns that need to steal a march on their competition.

DEALING WITH A CASH CRISIS 16

The author hopes that most readers won't have need of the advice in the following pages. However, since 1997 plenty of pig producers have needed help due to recessions in several pig industries. This sadly has been especially so in the UK pork and bacon industry where a combination of overproduction, rising costs, added financial burdens due to mandatory welfare, safety and meat quality Regulations and recently virus diseases – all served to produce a two-year period of losses quite unprecedented in the history of *any* pig industry, let alone our own in Britain.

During this time approaching half of my farm consultancy work involved perusal of clients' businesses, first, to see how costs could be reduced, then to defend cash flows, and finally to sit-in with farmers either called to account by their bank managers or who have to request an extension of loan facilities to help in maintaining confidence in the business.

These years, which involved a succession of tricky interviews, were very worthwhile in learning valuable lessons when producers' backs were against the wall. We both learned much about the attitude of lenders and I learned much about how unprepared some borrowers were over what was required of them. The following pages are a distillation of this experience, gained from the hardest school of all in weathering the greatest economic storm ever to hit a pig industry under modern conditions.

DEALING WITH YOUR BANK MANAGER / LENDER

First, know thy lender! It is important to understand the process which your bank will go through when considering your request to extend your overdraft. The following checklist gives a thumbnail list of what banks want.

WHAT BANKS WANT FROM YOU - A BASIC CHECKLIST

- ✓   What amount you need to borrow?
- ✓   Why? (What your business does).
- ✓   Your projected cashflow for the next 3 years.
- ✓   Your projected sales & profits for the same period.
- ✓   What you intend to do / what is likely to happen if sales income, or profits, fall by 10% and 25%?
- ✓   Who takes the business decisions – are you totally in command, or exposed to decisions by others (feed & breeding stock firms, etc)?
- ✓   What/who is the competition and how are they doing?
- ✓   What makes your business better than its rivals?

Let's now explore this in a little more detail so as to put you 'ahead of the herd' in the lender's eyes.

1.   Your bank manager may already have some idea of how your business is performing and the current outside circumstances likely to affect it. If so…

 Advice: Don't overdo the explanations in the run-in as to why you are here.

2.   Your business track record is important so you should have kept him in the picture in the past.

Advice: Even if you think you have no need to do so, keep your lender updated on your progress and conditions maybe once or twice a year.

3.   Unlike a long-term loan, an overdraft is repayable on demand so the bank will be checking that your business is capable of keeping up the interest payments where they fall due on the extra finance you require.

Be clear on what you need the overdraft for, how long you are likely to need it and the amount you want to borrow. Be positive – it gives both of you confidence.

4. Provide a business plan.

 The fact that you need more money to defend, sustain or progress the business means circumstances have changed, so even if you have submitted one before, you need a new one.

SOME TIPS ON PRESENTING A BUSINESS PLAN

There are two types of bank manager – those who know about pig production and those who don't. In either case you should be aware of what they need to know right at the start.

Both will enquire what your current rent and finance charge is per sow (or per 100 pigs kept for finishing only). This figure is the total of rent and all finance charges (bank, agricultural mortgage, loans, leasing and HP) for last year divided by the sows you own. A high figure will caution him, a low one will please him.

The knowledgeable bank manager knows this figure but I find tends to ignore aspects of output, so have ready answers to *"What's your rent and finance load as a percentage of gross output?"* Or a better question from his point-of-view, *"What is it as a percentage of gross margin?"* as this at least relates rent and finance charges to output and other major costs.

Bank managers seem to be obsessed with one single overhead cost – finance charges, and while it is wrong to use any method of business analysis which relies so much on just one area, you'd better be aware of it as it is so deeply engrained into them. Occasionally you will meet one who delves deeper, or sends off your figures to a livestock expert at Head Office who certainly will do so. So prepare the following:

PREPARING FOR THE INTERVIEW - A CHECKLIST

✓   **Your level of technical efficiency:**

You will need to have adequate records and be able to present your number of pigs reared per sow and served gilt per year; your growth rates and feed conversions - or better your MTF (see New Terminology section); your variable costs and gross margins. The analyst must know this as your technical efficiency influences the levels of other costs and after-profit expenses which are acceptable to the business.

Caution: Use of New Terminology. Few, if any, bank managers have the slightest idea of this modern concept, so don't overplay it. Just mention MTF in relation to FCR if the latter arises, as an indication that your sights are fixed on profit not just performance; this helps gain confidence in you.

✓   **Your fixed cost burden:**

The analyst will have details of local or national averages and will need to compare yours with them. Have yours ready including all funding costs - remember lenders are dead keen on these!

✓   **The balance between gross margin and fixed costs:**

Recent experience in helping producers who have rebelled against the high-output/high-cost method of pig production (most American hog producers farm exactly the opposite to high-cost production) has shown me that many good businesses are based on technically modest (but just adequate technical) performance, low gross margins, but extremely low fixed costs. You don't hear much about such farmers as pig performance is not impressive, but they can be secure and profitable businesses nevertheless.

A gross margin/fixed cost balance which is wide is looked on favourably by financiers providing your own output is not below local average.

✓ **Your level of profit:**

A very fickle jade! The profit figure (per sow, per pig sold, per year) on its own means very little as it must be looked at with the costs which it has to support. Sluggish cash flows threaten survival, as costs have to be met before the profit is there to pay them. So you need to prepare…

✓ **After-profit expenses:**

A viable business creates profits which are sufficient to cover the substantial costs that have to be met from that profit in the near future. Main items are: Capital expenditure, loan repayments, income tax (and due dates), private drawings, investment in new enterprises, tests of recent ideas/products, etc.

So, because after-profit expenses are just as important for securing a long-term viable pig enterprise, as are the more usually-discussed gross margins and fixed costs, you need to have a list of what and how much they are.

Sometimes a client and I have presented a good performance/profit case and have still been turned down eventually. Amazed, I phoned the bank managers to ask why – and it was due to high after-profit expenses in every case!

✓ **Income tax with due dates**

This last factor brings in a third adviser. If needs be, the income tax specialist. Find out what tax offsetting rules operate through your accountant. There could be substantial deferrals available dependent on each country's taxation system. Tax authorities don't exactly rush to tell you what these may be!

✓ **Confidence**

The object of this exercise is to give the lender confidence in you and your business. The underlying objective is to raise your pig business into the category where the lender marks your loan / mortgage / proposition – whatever it is – in the 'favourable' category. In other words, to put you ahead of the others that he will also be interviewing that week.

✓   **Seek help**

Don't be afraid to seek expert help. While an accountant (whom the bank manager realises knows your business) need not necessarily be present at the interview, a phone call from him to the bank manager or a letter-in-advance from him will materially help your case. Bank managers are reassured that you have taken such precautions. But take him along if you feel it is essential; usually I find it is not vital.

Someone else who is useful to have with you is a consultant or farm adviser, for two main reasons. In a crisis it is difficult to think clearly / objectively. So get the unbiased view of someone who knows your business to interpret your business from the outside, and give you technical support when it matters during the interview. It helps if the adviser himself has a national or local track-record *i.e.* is known by reputation to the manager or to his Head Office livestock specialist.

✓   **Advance information**

It is a good idea to send to send on your business plan before your meeting. Ask your bank manager how much notice he needs to peruse it – this builds his confidence in you as a 'forward-thinker' and reduces any preamble to the interview itself.

CASH MANAGEMENT

Modern pig production, like any business these days, is all about managing money. It is not a lack of profits which finishes many pig businesses, but a lack of cash-flow to sustain them.

Forward cash flow essential

You must work out a forward cash-flow. A cash-flow plan is nothing more than a farm budget on a cash basis for a period of time, and details monthly outgoings (cash outflow) and returns (cash inflow). In a crisis it is essential to have a cash-flow projection, but you may need help from an extensionist or accountant. But be honest with yourself and don't be afraid of seeking expert help. I find from dealing with farmers in financial trouble that there are three phases:

1. "It can't be happening to me",

2. "Let's try a few things" and

3. "Realisation".

    Farmers are too slow in getting to stage 3 and this is the stage to get a financial counsellor (rather than a technical adviser) who can calm you down. Cash-flow projections are as accurate as the records on which they are based. If you haven't got records, or have got poor records, it is essential to get this rectified at once. Plenty of good recording systems exist. Projections depend on the manager's realism which is why it is useful to do two projections when a recession seems likely, *i.e.* Cash-flow (A): "What do I really think will happen?" and Cash-flow (B): "What if it gets as bad as this?"

    If (A) starts to descend towards (B) in reality, then you should have a contingency plan.

Common mistakes in cash-flow preparation:

*   Initial bank balance incorrect.

*   Physical and financial targets unrealistic

*   Financial data inadequately supported by physical data

*   Cash income and payment items omitted, especially creditors and debtors.

*   In a crisis a weekly cash-flow is better than a monthly or 3-monthly one.

CASH-FLOW CHECKLIST

Everything to do with the pigs must be included, or your initial platform will be wrong.

✓   **Priority rating on bills:** When things are tight it can come down to "them or me". First, make a list of all bills owing, then realistically assess those who can probably be held off. Establish a priority list, and if you have to, use that in your realistic cash-flow (A).

✓ **Look for temporary credit**: Some seedstock and feed firm deals are very good. They fight each other with incentive credit deals to capture scarce trade. Yet farmers "don't want to be tied" or "don't want to lose my independence".

Rubbish! Be tied for a while if it helps generate better cash-flow.

✓ **Check the cost of credit**: If a producer is taking credit from his feed or livestock supplier, he should make sure he knows exactly what it is costing.

✓ **Overborrowing**: Don't overfund for longer than 5 years unless interest rates are fixed at a reasonable rate. Even very good production performance cannot match high interest rates and low cash inflow. High fixed costs, housing costs (such as in our industry) eat up profits but make gross margins look deceptively good because gross margin is the difference between output and variable costs – they don't include funding costs. Beware of gross margins!

✓ **Check whether personal savings can be used.** At least temporarily or to gain valuable major creditor confidence. But do check costs and likely income carefully otherwise overall financial security can be compromised.

✓ **A warning.** Selling pigs at lighter weights will boost cash-flow and lower total valuations for a while, but cause major headaches and risk future business collapse. This is not the same as selling faster-growing pigs to outlets which pay well for heavier weights and selling the slower-growing finishers to an outlet favouring lighter weights.

✓ **Can you sell anything?** Unused machinery and vehicles, unused land, timber, horses?

✓ **Re-evaluate your culling policy.** So as to remove any likely poor-doers which add to costs and may be a drain on sales income. Please read the section on culling carefully.

✓ **Look at your grading policy/achievements.** Again, read the section on grading returns. As MTF calculations repeatedly show, selling as much lean meat as possible is a major boost to widening the food cost/meat income gap and therefore easing cash-flow.

✓ **Review labour costs realistically.** Labour costs are growing world-wide.

Increasing productive man-hours reduces labour cost and using family labour can dramatically reduce it too. But first, look at yourself. Hard-workers rarely work for a lazy boss! Can you take on more work in a crisis?

And what about your sons, daughters, wife? Hold a family conference; it can be a temporary way through as it keeps the money in the business and so eases cash-flow.

Do you think a worker is lazy? One reason for laziness is that nobody set out a plan with achievement objectives for the day. And that costs nothing but a lot of thought and forward thinking, with the discipline to follow it through.

REDUCING COSTS IN A PRICE CRISIS

Having got as sound a factual financial base and forward forecast as you can, creditors are impressed by what you are going to do to reduce the costs on which these calculations are based.

You are asking them to defer money owed to them. They will be more likely to agree if they can see that you are strenuously attempting to reduce your costs and to minimise your short-term losses.

Feed costs

Feed costs anywhere in the world are from 55% to 68% of costs, so start here – savings will have the most impact on total costs.

1. *Wastage of food* averages 6% on most farms. This can be cut to 2%. Do a waste feed audit.

2. *Are you feeding the right diet ?* 18% of producers are not feeding what they think they are, or feeding the wrong diet for the genotype or environment. Check with your nutritionist to re-check your feed formulae and feed scales. Make the serious financial position clear to him, provide him with your present performance records and ask him if there are any alterations he can make even temporarily. Areas to discuss with him are:

Emergency diet alterations to reduce feed costs

> *Warning:* These are emergency measures only and must be discussed with a nutritionist and preferably an adviser who knows your farm.

i)   There is evidence that reducing *protein* (not energy) in bought-in gilt diets during the acclimatisation period will not harm production. And the use of some organic chromium (cheap to add) can improve litter size the longer it is used, so start now with all sow feeds.

ii)   Reducing protein (not energy) in *older* gestating sow diets to 14% or even lower is permissible in the short term. This may reduce litter size a little and cause some re-breeding problems but the latter is irrelevant where sows are approaching their culling date.

iii) A small reduction in protein content of the lactating diet is permissible in high milking sows / genotypes. Seek advice.

iv)   Revert temporarily to lower feed scales for sows from weaning to slaughter if litter size and returns to service are reasonable. Seek advice.

v)   If you are not already phase-feeding or split-sex feeding, do so.

vi)   Search for cheaper alternative by-product feeds, but *nutritional advice and analysis profiles are essential* to use these cheaper foods effectively.

vii) If you are weaning at 3 weeks or less and your sows are milking well, consider cutting out creep feed. If weaning later, on no account do this.

Do not reduce specs on weaner starter, nursery or growing / finishing feeds. This is always too costly in reduced short-term income (Meat per Tonne Feed). A small saving may be possible in reducing or omitting mineral and vitamin supplements in the last 3 weeks of the finishing period. Consult your feed supplier.

A 5-7% drop in grower feed specs allowing a 10% reduction usually causes a 16% drop in performance. Translated into economics the damage is much greater… If protein is reduced by only 1% (20 kg-100 kg; 44-220 lb) plus 0.05% less total lysine, DLWG is reduced by 10g day (over ¼oz), FCR by 0.05 and backfat increased by 0.12 mm. Thus a £3/tonne cost saving will result in a £9/tonne reduced income by slaughter at current European costs and prices.

Figure 1 explains pictorially what will happen. It helps with the profit situation short-term, only to fling you headlong into a potentially unsustainable cash-flow problem if the price crisis doesn't evaporate quickly, like within three to five months.

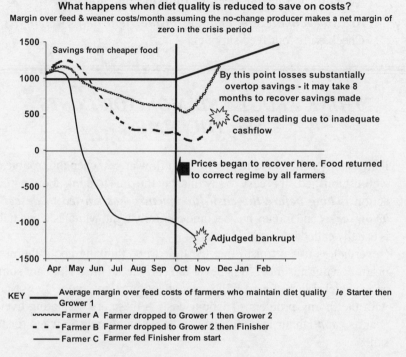

Figure 1 What happens when diet quality is reduced to save on costs?

═══════════════════════════════════════

CHECKLIST: WHAT YOU <u>CAN</u> DO
IN A CASH CRISIS

═══════════════════════════════════════

Do a feeding audit, or better, get someone impartial to do it for you: Here are some key points.

✓　**Double check feed costs.** Check the quote you have is the best you can get. But be careful, he who plays off one supplier against the others too hard and too long risks getting a food which looks OK on declaration but is down on critical specs (energy and digestible lysine). There must, at some point, be a bottom line for both of you, and a hard-pressed formulator can easily camouflage things.

✓   ***Are you feeding the right diet?*** A survey in Britain showed that 11% of producers (over one in ten) were feeding the wrong diet through misunderstandings with the rep, or misreading the leaflet (or not having one), or **thinking they knew best!**

✓   ***Are you making diet changes at the right times/weights?*** Check with your feed supplier.

THINK ABOUT AND MONITOR COSTS ALL THE TIME

I am usually called in to help with cash flow crises when the trouble is well-established. It is easy to say this, but the time to think about taking action is *long before the cash-flow picture has started to descend below target* and money has become painfully tight, which is when help is usually called in.

Nevertheless it is a truism that the time to be thinking about the next profit downturn is *now*. It is never too early to do this. Here are some areas which I have found will repay attention at any time. Let's start with the thorny problem of labour costs – these are escalating every year as *good* labour is rightly demanding a more realistic wage for its skills.

Know your labour cost per pig sold

Here are some differences between labour costs on typical pig farms, taken from UK Signet records (Table 1).

Table 1   LABOUR COSTS PER PIG SOLD

	Good	*Average*	*Savings per kg/dwt*
Breeder feeder (96 kg)	£7.05	£11.00	5.6p
Weaner producer (25 kg)	£4.70	£7.20	6.6p
Feeder only (96 kg)	£2.70	£4.20	2.1p

If our target in Europe is to reduce production costs by 20-25p/kg dwt (13-16c/lb) to withstand global assaults, better use of labour can contribute 10-20% of this leeway.

CONTRACTING OUT

Examine employing others to grow out post-nursery stock. Under these agreements the person finishing your pigs provides labour, buildings, straw, power, water and insurance. The owner provides pigs, feed, vet & med, and management input if needed. This allows you to dispose of perhaps 25-30% of your labour costs and/or concentrate more fully on the – let's face it – more labour-intensive and skilled aspects of breeding and nursery work, as distinct from the less onerous and lower-risk skills of the grow-out process.

Some opportunities occur for contracting out the nursery rearing aspect alone but only if the contractee has the correct housing and labour expertise and time available for scrutinising every weaner, in this case at least twice daily.

Nursery rearers have taken pigs from 7-35 kg at a £2.50/head fee, while finishing costs seem to be around £4.50/head. Overall savings of £5-£6/head (7-8.5p/kg dwt but more typically 4p/kg) have been made on fixed costs and you still have the pigs to sell at the end of the period. In the USA this can translate out at about 5c/lb savings.

The system has all the disadvantages of a transport move at a critical stage of the pigs' growth curve, but this is usually more than recouped by the advantages of segregation for the finishers, and batch production to yourself as the breeder.

CONTROL ENERGY COSTS

While food contributes between 55% to 68% of variable costs, electrical energy, on a typical pig farm only represents 2% or 3%. Thus a £1 ($1.45) saved on your electricity bill *will cost considerably more on the extra food required to maintain thermogenesis in the pig*. While keen to sell you electricity, most private and state-run electricity concerns will readily help you with an electricity-use audit to help you use their power more effectively.

Most of these energy-saving exercises I have recommended in Europe have yielded as much 0.5p/kg deadweight savings in costs, on typical electricity costs of 2.3p/kg dwt (1.5c/lb), a massive 22% saving. Methods used to do so are covered in the environment section of this book.

In the UK, since deregulation, you can ask different companies to tender for supply. Average savings have been in the region of 0.1p kg/dwt. In the author's experience much of the high remaining costs in the weaner accommodation is due to poor control of heat loss. For example, if the minimum ventilation rate in units is 20% of maximum instead of 10% or less, (as it should be), this can double your heating costs to achieve a correct LCT threshold and so imposes a further 1p/kg dwt on the final power costs. A good integrated control system (eg Farmex) can repay itself in a year on this basis.

REDUCING PIGLET DEATHS TO WEANING

A persistent and serious problem which has hardly improved over 20 years. Pre-weaning mortality from born-alives has 'stuck' at 11 to 12% since 1980, yet many clients have halved that to 6-7%. How they've done it is described in the Post-Weaning check section.

Reducing pre-weaning mortality of born-alives by 5% produces 1.27 more pigs sold per sow year, and reduces breakeven cost by 3.1p/kg (about 2c/lb) (Richardson, 1999).

PARTIAL OR LARGELY TOTAL REPLACEMENT OF NATURAL SERVICE WITH AI

Surprisingly this is, at the time of writing, still too slow to be adopted by certain countries, despite the facilities being in place and practicable on most of their farms.

AI can be practised by on-farm collection/extension of own-boar semen, and/or by using an AI centre.

The cost of service from naturally-mated indoor herds in the UK is about 2.8p/pig dwt, and outdoor herds (herding a 30% closer sow:boar ratio) 4p/kg dwt. This should be compared to AI, via an AI centre or via on-farm collection of 1.4p/kg dwt and 0.84p/kg dwt respectively. Richardson sardonically but correctly states:

Assuming that AI boars are within the top 5% of boars performance tested then the calculated breeding value of their progeny relative to those sired by natural mating would be 2.5p/kg deadweight.

*The overall cost saving and performance benefit being 3.9 –
4.5 p/kg deadweight when using semen from either an AI centre
or on-farm collection relative to natural mating is a high price
to pay for voyeurism!*

REDUCE DIRECT FEED WASTAGE

This can be confidently stated as typically 6%, when a reduction to 2%
is achievable on any farm. Wasted food is a direct charge on costs and
a 4% reduction is worth:-

**Savings, from attending to direct food wastage, p/kg deadweight on a 96 kg
(212 lb) finisher**

Breeding farm	0.45
Nursery unit	0.43
Grow out unit	0.49
Total 1.37p/kg deadweight (4.4c/lb)	

Consider batch production

While the highly desirable system of segregation by all-in/all-out
production is difficult to achieve on the smaller unit (<200-300 sows)
due to insufficient throughput numbers justifying the extra expenditure
on space needed, by farrowing every 3 weeks the problem of continuous
occupation can be resolved at less disturbance and cost. The pigs are
produced in discrete age groups and are therefore segregated by age.
The success of this system has been reported by Kingston (1998) (Table
2).

Table 2   WEEKLY VERSUS 3 WEEKLY BATCH PRODUCTION

	Weaning interval			
	Weekly	3-weekly	% Gain	p/kg dwt saved
Growth rate g/day (lb/day)	490 (1.07)	547 (1.20)	12	2.8 (overheads) (1.84c/lb)
FCR	2.36	2.26	4	1.0 (food) (0.65c/lb)
Medication £/pig (US$)	2.19 (3.17)	1.31 (1.89)	40	1.3 (vet & med) (0.85c/lb)

Extrapolated from Kingston (1998)
Calculated on overhead costs of 14.3p/day (20.7c), plus food cost of 14p/kg (20c)
and 70 kg (154 lb) deadweight

SOME COST SAVING MEASURES FROM JUDICIOUS INVESTMENT

Can you sink your own borehole?

Water costs in many parts of the world are rising steeply, *e.g.* in the UK some water companies have loaded 34% on to their charges, especially for those farmers in remoter areas, which is where, perhaps, pig farms should be located!

Two pig producers report the transfer to borehole water has been paid back in two years even though costs may seem high at £14,000 (the sale value of about 200 finishers). Subsequently savings have been 0.8p/kg dwt (0.52c/lb). Water quality and hardness needs to be checked out – borehole water can be hard on heater coils for hot pressure washing, needing more frequent replacement.

Tent type housing

This tends to be controversial – not because it doesn't work – it does. For example the Japanese using the Ishigami 'pipe-house' system grow pigs at 850g/day (1.87 lb) from 35-108 kg (77-238 lb) on double-skinned custom-built tent type structures on sawdust over earth floors, with relatively unskilled grow-out labour. The cost per house is under £30/pig which is $44 (even at Japanese prices) and the structures last at least 5 years when only the skins are renewed.

The controversy arises first because the American hog industry, now surging ahead with its professional large units, is going exactly the opposite way in eschewing temporary structures for permanent custom-built tunnel-ventilated housing on a grand scale.

But other countries don't enjoy the American's cheap food, low interest rates and economy of scale, thus the tent idea is an attractive if very different option for quite different conditions.

Secondly, as with all revolutionary innovations, the design and operating rules governing tent housing are strict. Failures have been common where these are ignored or short cuts taken and the author, experienced in what the Japanese do and from many visits to Japan over 18 years, has been called in to sort it out.

Costs of tent housing, assuming equal performance to slaughter as has happened has varied from 3 to 4p/kg (about 2.3c/lb), while conventional rearing/finishing housing has cost 6.5p – 7p/kg (about 4.4c/lb), a saving of about 3p/kg (2c/lb).

The tent idea has spread now to breeding tents for outdoor housing systems. The improvements to farrowing rate and liveborns may be worth 0.4 to 0.5p/kg.

Tent housing is also being applied to the 'streaming' concept, where previously sick pigs once recovered, are not returned to their previous accommodation but streamed in cheap structures and reared separately to slaughter. This has yielded benefits of 1p/kg (0.65c/lb) in better performance both in the original healthy pigs and in the affected-but-cured pigs, presumably from less immune demand in both groups when they are not housed together in the same structure.

Lidding open-plan accommodation

Many open plan structures, especially deep litter straw yards are cheap, costing slightly more than half the price at £50 per pig place ($73) when built by farm labour and contracting-out the steelwork, compared to a commercial structure at £112 per m^2 costing £90 pig/place ($131).

However these manually operated buildings can chill smaller pigs – a point seized on by package-deal suppliers, and can soon erode the capital cost savings, which is true enough.

Installing sloping kennel lids and either straw bale fronts or kennel flaps increase the initial cost by only £3/pig ($4.35) but has clawed back over £2 pig ($2.90) in improved performance due to less pneumonia and better warmth at night. This improvement is worth 3p/kg dwt for an extra cost of only 0.005p/kg dwt on every 1000 pigs reared! All such houses should be lidded in this way.

In the UK a well-designed deep straw kennel saves about 1.2p/kg dwt (0.78c/lb) after straw costs are taken into account and assuming equal performance with a more sophisticated and expensive totally enclosed structure.

The author is firmly convinced that in many pig industries growing/ finishing housing costs must come down – and come down markedly.

However farrowing house and nursery accommodation needs *more* investment per square metre/pig produced. Nursery accommodation in particular needs to last longer; some structures need urgent repairs after only two years so as to maintain the environmental status required.

REFERENCES

Kingston, N (1997) 'Strategic Options for Endemic Disease Control in Farrow-to-Finish Units'. Proc British Pig Association Spring Pig Conference April 1997.

Richardson, J 'Low Cost Pig Meat Production'. Proc British Pig Association Conf. Oct 20th 1999.

GROWTH RATE 17

Growth is the progressive increase in size of a living thing. Growth rate on the other hand is the rate of increase in body weight for a unit in time, e.g. grammes per day (g/day) or pounds per day (lb/day).

It is better to measure growth rate metrically (g/day) rather than in oz/day or lbs/week as the Imperial system is too imprecise, especially for small pigs.

In practice Average Daily Gain (ADG) or Daily Liveweight Gain (DLWG) are the preferred terms but in nutritional and genetic research papers you will also encounter Lean Tissue Growth Rate (LTGR)

TARGETS

Table 1 gives an idea of the growth rates possible on a farm where good lean-gain genetics are used (especially in the male traits). In this case, the AIAO housing is well-designed and operated (with a move at 10 weeks) and the disease levels low. The pigs are fed ad-lib on a multiphase system.

Table 1 may surprise producers, but they are achievable targets for the next 5 years.

If Table 1 gives the top end of the growth spectrum Table 3 illustrates the bottom end. If a producer approaches this level of DLWG an investigation is needed.

Table 1   TARGET GROWTH RATES/DAY, TOPSTOCK AND CONDITIONS - METRIC

Age Days	Age Weeks	Liveweight (kg)	Pigs growing at (g/day)	Weight put on per week (kg)	Growth Rate from 21 days (g/day)	(Out of) Nursery to Slaughter Targets (g/day) From 25 kg	From 10 weeks (31 kg)
21	3	6	–	–	–	–	–
28	4	7.20	171	1.2	171	–	–
35	5	9.80	357	2.6	271	–	–
42	6	12.85	435	4.25	326	–	–
49	7	16.50	521	3.65	375	–	–
56	8	21.25	679	4.75	436	–	–
63	9	26.10	710	4.97	479	692	–
70	10	31.35	750	5.25	517	739	–
77	11	36.75	771	5.40	549	754	771
84	12	42.49	820	5.74	579	784	814
91	13	48.65	880	6.16	609	799	824
98	14	55.16	930	6.51	638	824	850
105	15	62.09	990	6.93	668	851	878
112	16	69.16	1010	7.07	694	873	900
119	17	76.65	1070	7.49	721	897	924
126	18	84.86	1115	7.81	751	927	956
133	19	93.09	1175	8.23	784	951	980
140	20	101.49	1200	8.40	828	973	1002

Notice how (as with FCE and MTF) it is important to have start and finish reference points when examining DLWG, and especially when comparing them with other quoted performances.

Table 1a TARGET GROWTH RATES/DAY, TOPSTOCK AND CONDITIONS - IMPERIAL

Age		Liveweight (lb)	Pigs growing at (oz/day)*	Weight put on per week (lb)	Growth Rate From 21 days (oz/day*)	(Out of) Nursery to Slaughter Targets (oz/day*)	
Days	Weeks					From 55 lb	From 10 weeks (69 lb)
21	3	13.2	–	–	–	–	–
28	4	15.8	6.0	2.64	6.0	–	–
35	5	21.6	12.6	5.72	9.6	–	–
42	6	28.3	15.3	9.35	11.5	–	–
49	7	36.3	18.4	8.03	13.2	–	–
56	8	46.8	24.0	10.45	15.4	–	–
63	9	57.4	25.0	10.9	16.9	24.4	–
70	10	69.0	26.5	11.6	18.2	26.0	–
77	11	80.9	27.2	11.9	19.3	26.6	27.1
84	12	93.5	28.9	12.6	20.4	27.6	28.7
91	13	107.0	31.0	13.6	21.4	28.1	29.0
98	14	121.4	32.8	14.3	22.5	29.0	29.9
105	15	136.6	34.9	15.2	23.5	30.0	30.9
112	16	152.2	35.6	15.6	24.4	30.7	31.7
119	17	168.6	37.7	16.5	25.4	31.6	32.5
126	18	186.9	39.3	17.2	26.4	32.5	33.7
133	19	204.8	41.4	18.1	27.6	33.5	34.5
140	20	223.3	42.3	18.5	28.8	34.3	35.3
147	21	242.2	43.2	18.9	29.1	35.4	36.1
154	22	261.5	44.1	19.3	29.9	36.6	37.2
161	23	281.2	45.0	19.7	30.6	37.7	38.1

Notice how (as with FCE and MTF) it is important to have start and finish reference points when examining DLWG, and especially when comparing them with other quoted performances.   * For lb/day divide by 16.

SOME THOUGHTS ABOUT ADG

- An important measurement, as fast growth to slaughter generally needs less food. A slaughter pig finishing a week faster saves 7 days food, at maximum food intake, from its total feed requirement.

- Generally speaking you cannot grow a pig too fast to 30-35 kg (66-77 lb) because of the young pig's superior food converting ability. Beyond that growth rate has to be balanced with food conversion and grading (sufficient but not excessive fat deposition) to maximise income and keep costs to a minimum.

- It is advisable at least to keep a *weekly* check on growth rate of groups of pigs in each separate environment. In my experience fewer than one in five producers do so. But a proper representational weighing system of both food eaten and growth rate is much more important to profit than most producers are as yet prepared to accept. The reasons will be discussed later.

- When balancing the advantages of reduced days to slaughter, food conversion and grading to achieve maximum return, many trials rightly emphasise the importance of the reduced food effect on improved FCR but omit mention of overheads saved. These can be between 35 to 50% of any food savings from faster growth (Table 2) which is a significant improvement to profit.

Table 2   OVERHEADS ARE OFTEN OVERLOOKED WHEN EXAMINING FASTER GROWTH

Assumptions:	Pigs 30-100 kg (66-221 lb). Dressing % 73%. Av. food cost £140 (US$196)/tonne. Av. overhead costs/day (including capital depreciation).

Days to slaughter	Food eaten (kg)	(lb)	Overall FCR	Overall ADG (g)	(lb)	Food Cost/pig		Overheads Cost/pig	
100	210	462	3.0	700	1.54	£29.40	$42.63	£14.00	$20.30
90	189	416	2.7	777	1.71	£26.46	$38.37	£12.60	$18.27
80	168	370	2.4	875	1.93	£23.52	$34.10	£11.20	£16.24

Savings per pig from improved F.C.R. *vis-à-vis* lower overheads:–

Days to slaughter	Savings in food/pig		Savings in overheads/pig		% of overheads compared to F.C.R.
100	–	–	–	–	
90	£2.94	$4.26	£1.40	$2.03	47.6%
80	£5.88	$8.53	£2.80	$4.06	

WHAT CAUSES THIS
POOR GROWTH RATE?

The causes of these poor action level growth rates will be found to be:

1. A larger than acceptable ***post weaning check*** (see the *Post-Weaning* check section).
2. Lack-lustre growth once the post weaning check is surmounted. This is usually due to ***disease***, with ***respiratory infection*** the most common cause in my experience. As well as veterinary consultation, an audit of the ventilation system especially in winter is advisable. (see *Ventilation*, section).
3. ***The 12-16 week check.*** The reasons for this phenomenon are unclear but the following checks are advised. Those producers recording weekly growth rates do see it on their graphs while it may go unnoticed otherwise.
4. After 16 weeks to slaughter the pig's immune status, peck-order, appetite and thermoregulatory system should be well-established. The areas to check here are disease, again respiratory disease, ileitis/colitis and overcrowding. Too stuffy air rather than a too cold environment is usually the culprit, both in summer and winter. Occasionally nutrition is at fault, in my experience poor amino acid balance can be responsible, but also overfeeding protein in the last month before market weight is rather too common. This protein would have been better fed in the first third of the pig's life.
5. Temperature is often at fault (See Food Conversion, Ventilation and Hot Weather sections)

CHECKLIST

SLOW-UP IN GROWTH OR 'STOP-START' GROWTH BETWEEN 12-16 WEEKS OF AGE (USUALLY BETWEEN 30-45 kg (66-99 lb))

Consult the relevant sections in this book for remedial advice.

✓ Check stocking density.

✓ Assess ease of access to food and particularly water in warm weather.

Table 3   ACTION LEVEL GROWTH RATES/DAY - METRIC
These can be considered as borderline performance levels

Age		Liveweight (kg)	Pigs growing at (g/day)	Weight put on per week (kg)	Growth Rate from 21 days (g/day)	(Out of) Nursery to Slaughter (g/day)	
Days	Weeks					After 8 weeks in nursery	After 10 weeks in nursery
21	3	5.5	–	–	–	–	–
28	4	6.6	157	1.1	157	–	–
35	5	7.8	171	1.2	164	–	–
42	6	9.5	243	1.7	190	–	–
49	7	11.5	286	2.0	214	–	–
56	8	14.5	429	3.0	257	–	–
63	9	18.0	500	3.5	298	–	–
70	10	21.75	536	3.75	332	–	–
77	11	25.75	571	4.0	362	571	–
84	12	30.25	643	4.5	393	607	–
91	13	35.0	679	4.75	421	659	679
98	14	39.75	750	5.25	451	696	714
105	15	45.25	786	5.5	479	707	738
112	16	51.0	821	5.75	503	720	768
119	17	57.0	857	6.0	526	735	786
126	18	63.0	865	6.0	548	750	798

Table 3 CONTD.

Age		Liveweight (kg)	Pigs growing at (g/day)	Weight put on per week (kg)	Growth Rate from 21 days (g/day)	(Out of) Nursery to Slaughter (g/day)	
Days	Weeks					After 8 weeks in nursery	After 10 weeks in nursery
133	19	69.0	871	6.1	568	763	808
140	20	75.2	886	6.2	586	776	816
147	21	81.4	893	6.25	602	786	824
154	22	87.65	893	6.25	617	795	831
161	23	93.9	893	6.25	631	802	838
168	24	100.15	893	6.25	644	809	843
175	25	106.4	893	6.25	655	815	847

Notice how in this fairly typical example the pigs are slow in getting away after weaning and again how later growing pigs sometimes get 'stuck' at around 16 weeks of age when their potential growth should still continue to escalate by 5% to 6% per week.

Table 3a ACTION LEVEL GROWTH RATES/DAY - IMPERIAL
These can be considered as borderline performance levels

Age		Liveweight (lb)	Pigs growing at (oz/day)	Weight put on per week (lb)	Growth Rate from 21 days (oz/day)	(Out of) Nursery to Slaughter (oz/day)	
Days	Weeks					After 8 weeks in nursery	After 10 weeks in nursery
21	3	12.1	–	–	–	–	–
28	4	14.6	5.5	2.4	5.5	–	–
35	5	17.2	6.0	2.7	5.8	–	–
42	6	20.9	8.6	3.8	6.7	–	–
49	7	25.4	10.1	4.5	7.5	–	–
56	8	31.9	15.1	6.6	9.1	–	–
63	9	39.7	17.6	7.7	10.5	–	–
70	10	48.0	18.9	8.3	11.7	–	–
77	11	56.8	20.1	8.8	12.8	20.1	–
84	12	66.7	22.6	9.9	13.8	21.4	–
91	13	77.2	23.9	10.5	14.8	23.2	23.9
98	14	87.6	26.4	11.6	15.9	24.5	25.1
105	15	99.8	27.6	12.1	16.9	24.8	25.9
112	16	112.5	28.9	12.7	17.7	25.3	27.0
119	17	125.7	30.2	13.2	18.5	25.8	27.7
126	18	138.9	30.5	13.2	19.3	26.4	28.1

Table 3a CONTD - IMPERIAL

Age		Liveweight (lb)	Pigs growing at (oz/day)	Weight put on per week (lb)	Growth Rate from 21 days (oz/day)	(Out of) Nursery to Slaughter (oz/day)	
Days	Weeks					After 8 weeks in nursery	After 10 weeks in nursery
133	19	152.2	30.7	13.5	20.0	26.8	28.5
140	20	165.8	31.2	13.7	20.6	27.3	28.7
147	21	179.5	31.4	13.8	21.2	27.7	29.0
154	22	193.3	31.4	13.8	21.7	27.9	29.3
161	23	207.0	31.4	13.8	22.2	28.2	29.5
168	24	220.8	31.4	13.8	22.7	28.5	29.7
175	25	234.6	31.4	13.8	23.1	28.7	29.8
182	26	248.1	31.4	13.8	23.6	29.1	30.1
189	27	261.2	31.4	13.8	24.2	29.5	31.3
196	28	274.9	31.4	13.8	24.*9	29.8	31.5

Notice how in this fairly typical example the pigs are slow in getting away after weaning and again how later growing pigs sometimes get 'stuck' at around 16 weeks of age when their potential growth should still continue to escalate by 5% to 6% per week.

✓ Check for wrong-mucking / pen-fouling / tailbiting.

✓ Check for unevenness. If distinctly obvious, then pen-splitting rather than re-mixing of individuals is wise.

✓ Straw-based pigs can develop mange at this time which in its early stage can cause low-level but continuous stress and affect food utilisation. Greasy pig can occur at this time but this is more noticeable early on.

✓ Food change? Where food is concerned the specs may be satisfactory but *palatability and texture* may cause reluctance to eat and you should also keep freshness at the back of your mind.

✓ Following on from this, and possibly because the pig's desire to grow is accelerating around this time, *moulds/mycotoxins* could be having a direct or indirect effect (*i.e.* palatability). Producers able to measure feed intake per pen on a daily basis with a C.W.F. pipeline layout have picked this up promptly.

✓ Housing change? Again, direct and indirect effects have been noticed. *Direct* – *i.e.* the new house's environment is not up to standard because the pigs are cold, being placed in *too much spatial volume in relation to their group body weight.* Temporary lids and hovers help here. *Indirect* because the pigs are slow to adjust to new conditions – feed hoppers, dry to wet feed, fewer water points, less dunging area, a change in their socialization pattern. This may take longer-term planning to get better.

35 years ago we did some work which showed *any* change of housing cost about 3 days' growth, but with care this could be halved.

HOW GROWTH RATE AFFECTS MEAT PRODUCED PER TONNE OF FEED (MTF)

Table 4 shows the enormous reduction in output – saleable lean meat in relation to the food fed to obtain it – between the target excellent growth rates in Table 2 and the poor growth shown in Table 3.

Tables 5 and 6 show how cost and income are affected by slower growth over a range of options.

Table 4   REDUCTION IN M.T.F. BETWEEN TOP TARGET AND ACTION LEVEL GROWTH RATES

Top Target (Table 2)	*Action Level (Table 3)*
Days taken 6-100 kg (13-221 lb) 117	Days taken 5.5-100 kg (12-221 lb)157 (+40)

Extra Food required (at 2.25 kg (5 lb)/day in finishing stages)

–	90 kg (199 lb)

Thus expected FCR...	2.4 : 1	3.34 : 1

Assuming a KO% of 73% for both groups of pigs, MTF will be …

323 kg (712 lb) + 92kg (203 lb)	231 kg (509 lb)

And assuming a pig price of 105p/kg deadweight, the equivalent in extra feed costs tonne (PPTE) is

$$92 \times £1.05 = £96.6/tonne$$

Thus, assuming an average feed price for the better performing pigs of £160/tonne, this is equivalent to the poorer performing pigs feed cost rising by 60%!

Of course, these are the extremes, but I still encounter both ends of the spectrum every year, even today.

KEEPING A CHECK ON DAILY GAIN

Keeping an on-going record of liveweight gain in the growing pig is important, yet only perhaps 15 to 20% of producers keep a *constant* finger on this pulse. Most people are content to assess average daily gain (ADG) – also expressed as daily liveweight gain (DLWG) – at the end of each month, or even three months. By then factors which led to any worsening are historical and may be difficult to identify and remedy.

NOT SPOTTING THE CAUSE OF SLOW GROWTH PROMPTLY IS COSTLY

Failure to detect and act on a 10% reduction in daily gain in pigs of 60 kg (132 lb) across a 4 week period results in 15 kg (33 lb) less MTF *i.e.* 15 kg (33 lb) less saleable meat for every tonne fed then and thereafter. This is equivalent to a £15/tonne increase in the cost of grower's food, or about a 10% price rise. Add typical overhead costs to this and the 10% becomes 13% - on some farms 15%.

Table 5   THE COST OF SLOWER GROWTH RATE/PIG
The cost of extra food and overheads from days to slaughter (25-100 kg)

Performance	ADG (g)	Days taken	Extra food needed (kg)	(lb)	Value	Extra overheads incurred	Total per pig	In terms of cost of production of £55/pig
Poor	800	94	39.1	86	£5.46	£2.38	£7.48	+13.6%
Average	850	88	25.3	56	£3.54	£1.54	£5.08	+9.2%
Good	900	83	13.8	31	£1.40	£0.84	£2.24	+4.1%
Very Good	940	80	6.9	15	£0.97	£0.42	£1.39	+2.5%
Exceptional	975	77	–	–	–	–	–	–

Assumptions: Finishing food intake 2.3 kg/day; Finishing feed cost £140/tonne; overheads 14p/pig/day.
Comment: There is no reason why 'Average' producers should not progress to the Very Good category in terms of growth rate. This would, from this performance alone, give them £3.69/pig more margin or a 6.7% improvement in their production costs.
N. American readers should look at the final percentage column as these percentage differences, I find, are fairly similar under their conditions.

Table 6   HOW SLOWER GROWTH RATE AFFECTS INCOME PER TONNE OF FEED FED
The reduction in MTF (25-100 kg liveweight) as a result of poorer growth rate.

Performance	Extra days taken	MTF (kg)	(lb)	Value	Extra overheads incurred/tonne feed	This is equivalent to the producer paying extra per tonne of feed (on a feed cost of £140/t)	
Poor	17	256	565	£281.60	£11.13	£73.83	+53%
Average	11	274	604	£301.14	£7.70	£50.86	+36%
Good	6	291	642	£320.10	£4.46	£28.66	+20%
Very Good	3	301	664	£331.10	£2.40	£15.60	+11%
Exceptional	–	313	690	£334.30	–	–	–

Assumptions: KO% 73% for all pigs. Average pigs ate 200 kg (441 lb) feed. Pig price 110p kg dead. Feed price £140/t (US$196, typical European price).
Comment: Notice the enormous difference in nominal feed cost between, say, the Average producer and the Very Good producer. Moving from the former to the latter in growth rate terms alone, is equivalent to a £35 per tonne (or 25%) reduction in all grower/finisher food and overhead costs between 25 and 100 kg.
N. American readers should regard the final percentage column as fairly typically applicable to them as well

SIMPLE FORMULAE FOR ESTIMATING GROWTH RATE

A variety of formulae exist based on a rolling average inventory usually across three months *i.e.*

Average output weight minus average input weight divided by days in the system for each of the past 3 months ÷ 3.

Table 7 gives an example:–

Table 7   TYPICAL SIMPLE METHOD OF RECORDING DAILY GAIN

A	Jan:	96 kg	– 31 kg	(65,000 g)	÷	77 days	= 844 g/day
	or	212 lb	– 68 lb		÷	77 days	= 1.87 lb/day
B	Feb:	94.5 kg	– 29 kg	(65,500 g)	÷	79 days	= 829 g/day
	or	208 lb	– 64 lb		÷	79 days	= 1.83 lb/day
C	Mar:	92 kg	– 30 kg	(62,000 g)	÷	78 days	= 795 g/day
	or	203 lb	– 66 lb		÷	78 days	= 1.75 lb/day

3 month average (A-C) 823 g/day (1.81 lb/day)

D	Apr:	95 kg	– 30 kg	(65,000 g)	÷	74 days	= 878 g/day
		210 lb	– 66 lb		÷	74 days	= 1.93 lb/day

3 month rolling average (B-D) 834 g/day (1.84 lb/day)

Table 7 immediately reveals the disadvantage of this approach (even if a computer is used).

1.   A month elapses before figures are available.

2.   Casualties and mortality can distort the figures.

3.   As can factors such as underweight marketing.

REPRESENTATIONAL WEIGHING

We need a better system, and while weighing is a highly unpopular task, there is no substitute for representational weighing of pigs. This is because we need to identify as quickly as possible when our test/sample pigs fall below the target daily gain graph.

Because environment can have such an effect on growth rate, it is important to select pigs from a typical piggery.

Ideally, each house containing a different environmental system should have 3 sets of two pens weighed to the schedule below. One pen near the coldest end (or in the tropics the warmest side) and another pen, in cold localities, in the middle of the house.

Suggested representational weighing schedule

First : Feed intake is measured over 12 days in two pens at ***each*** of these weight ranges: 25 to 35 kg; 55 to 65 kg; and 85 to 95 kg, *i.e.* six pens in all. A pen would normally contain a minimum of 10 pigs. In N. America these ranges could be 50-75 lb, 120 to 150 lb and 180 lb to slaughter weight.
Second : Record body weight at the start and finish of each 12 day period.
Third : At the start of each 12 day period, record how old the pigs are in days.
Fourth : Keep an environmental temperature record as a back-up.
You can then plot feed intake/pig in kg/day against observed live bodyweight. From these data you can plot a representational feed intake curve in kg/day (lb/day) and a pig growth curve of bodyweight against days in the piggery.

From this extremely useful farm-specific data the nutritionist can design your diets far more accurately than just supplying formulae off price lists.

He now has at least a guide to ... The genotype you use – the pig's feed intake – their lean growth rate needs – an idea of their current immune status in your circumstances – and what your environmental conditions are from season to season.

Producers should take a decision on how closely to attain this ideal with the workload involved. If this proves difficult right through the growth period, representational weighing is most valuable between 11-13 weeks of age (30-50 kg, 66-110 lb).

I have experience of five producers who do this full programme and their experiences can be summarized as:–

- The extra work involved resulted in an (average) increase of labour load of 12.5% (2 men weighing 30 pigs/week in two houses). On average this extra work increased the cost of producing one finished pig by 0.625% (Range 0.28% to 0.9%).

- While at first it was difficult to convince the staff that this extra chore was worthwhile, within 2 full batches of pig all 5 farms agreed that the information it provided gave a new dimension to their work. However, ***time was put by to do the job*** and it was not forced upon already over-worked stockmen.

Users' comments :

Those farmers who weighed pigs and food representatively on a weekly basis have written to me …

*"How fickle even an experienced eye can be in estimating growth rate. While 'faster' or 'slower' growth is noticeable on a pen basis, we were often wrong in quantifying it. Before measuring it, we **under-**estimated a slow-up in growth by 50% rather too often!"*

"Changes in pen growth on a weekly basis seems largely linked to feed intake" (see below)

*"Plotting weekly growth rates of one pen suggested, from the final graph, that pigs grow in gentle but perceptible waves, often about 14 days in 'wavelength'. If you look **down** a completed growth rate curve from one side and at an angle you often see it."*

"Where we had a slight (not an acute) problem, the drop-off below target was immediately noticeable in the test pens. However, we did pick it up before the weekly weighing in about 50% of the cases, but in the other 50% it made us go back to that pen and look again. This often seemed to occur when the pigs were too warm in summer/too airless on a cold night. Hot pigs can look "puffier" and this fools the eye – but not the weighing machine."

WAS IT WORTH IT ?

The labour cost of weekly representational weighing is equivalent to the cost of food for 1 kg (2.2 lb) of liveweight gain ***in one pig*** – taken as 37.5p (about 50 cents) in this case. Quite modest.

During the two years that the 5 farms have been engaged in the process, the feed cost/kg gain has fallen by an average of 1.8p/kg/2.5 cents (corrected for feed price movements) on 4 of the farms *for each kg gained*. In other words a 90 kg finisher has saved £1.62/pig, while the weighing only cost 37.5p, an REO (payback) of 4.3:1. It is not known how much of this improvement can be put down to alert stockmanship, but these figures from this exercise helps illuminate any cost-benefit position.

DIFFICULT!

I have found it difficult to persuade farmers to test-weigh, especially weekly. In fact the 5 farms concerned represent only a 5% success rate of those approached!

By far the most common objection has been the difficulty in squeezing in this extra job. The next, persuading people that it is worthwhile anyway. If the cost is around 0.6% more on the production cost of a finished pig, and the return is 2.8% more income, it does look to be a good bet.

Speed the day when one pen in the piggery will have a weighing *platform* so that a group of pigs can be driven there and weighed en masse; this is only available for weaners at present.

Calculations have suggested that the device will not save on the 0.625% extra labour cost, as the capital investment will erode much of that, but it will make the concept of representational weighing more practicable.

Alternatively the advent of more wet feeding systems provide evidence that growth rate is tied closely to feed intake. Computerized wet feeding systems can record feed intake very precisely on a daily basis, or even part of a day. If there is a link between growth rate and feed intake then fallaways in daily gain can be picked up within hours, not after one week or more. The basic pen-by-pen system is here with us today in some CWF systems.

THE FUTURE

In future, *each* pig could be electronically tagged and weighed as he stands at the dispenser, so both feed intake and weight gain could be measured and recorded for each pig on a comparative basis. Moreover pen ambient temperature can be measured and controlled by the same device, and temperatures altered as well as nutrient allowance to compensate accordingly.

I have a feeling that by the time this book is published, we will be starting to see this new technology in use on some of the larger pig units and breeding company test farms.

With over 40 direct or indirect causes and effects a growth rate problem is indeed multi-factorial!

We need a more planned approach to a slow growth situation. The following checklist will help

WHAT AFFECTS GROWTH RATE ?
A SUMMARY CHECKLIST

✓	*The pig*	Genetics.
		Age.
		Sex.
		Docility.
✓	*Its food*	Nutritive specifications and balance.
		Raw Material Quality / Availability to the pig.
		Daily intake.
		Palatability.
		Water.
		Growth enhancers.
		Wet/dry & fully wet feeding.
		Adequate trough space.
		Feed texture.
		Access to hoppers troughs.
✓	*Its surroundings*	Temperature over the skin.
		Air speed (draughts).
		Air positioning.
		Humidity.
		Floor surface.
		Bedding.
		Insulation of all surfaces, especially the floor.
		Gases (not necessarily toxic).
		Airborne dust.

✓   *Its companions*   Pen shape.
Position of furniture.
Stocking density.
Group size.
Weight variability.
Docile genes.

✓   *Disease*   Biosecurity.
Diseases present.
The degree of immune protection needed.
Precautionary measures in place.
Curative measures in place.
Veterinary supervision.

✓   *Management*   All-in / All-out (AIAO).
Batch production.
Continuous production.
House changes.
Batching and matching skills.
Weaning weight/size.
Representational weighing/ records.
Computerisation (measuring, monitoring, highlighting action needed).
Monitoring and maintenance of environment equipment.
Weekly (daily?) progress chasing/actioning.

✓   *Stockmanship*   Quality of person.
Time to do the job.
Ongoing education / training / demonstration.
Daily briefing.

PLANNING AND PRIORITISATION – A CONSULTANT'S GUIDELINE

The most effective way to solve a slow-growth problem is, in my experience, carried out like this. While certain sectors below may seem obvious, *it is easy to assume things which may not be true or necessarily evident*. Everything should be checked out so we have a firm base on which the best remedies can be decided.

Knowing how a good adviser wants to work will help both you and him solve the problem.

Approaching the problem

1. Have we a problem?

2. What is the evidence for it?

3. When was the problem noticed, and how did it come to light?

4. Have the assertions you are told about been checked out?

Now we know this . . .

5. How serious is the problem ?

6. What is this costing? (Useful in deciding which action is most cost-effective and which to attend to first).

7. Is there evidence of 'stop-start' growth?

8. So-called compensatory growth?

9. How 'even' do the pigs look?

10. Are any recent veterinary reports/observations available?

    Table 8 provides some evidence of which areas in the preceding checklist are most likely to be involved.

PRIMARY AREAS TO LOOK FOR IN POOR GROWTH RATE

1. Inadequate or unbalanced nutrient intake on a daily basis.

2. Pigs too hot or too cold.

3. Inadequate air movement and poor positioning of air.

4. Disease, stress and likely immune demand.

5. Any changes – of environment, of feed, of comparisons, of management, (possibly) of stockpeople, of outside weather conditions, pen soiling.

6. Unawareness – *i.e.* of weight, temperature, internal climatic fluctuations, feeding times, water supply, and poor batching & matching at weaning or after the nursery stage.

Table 8   AN ANALYSIS OF 137 CASES OF INADEQUATE GROWTH RATE INVESTIGATED OVER SOME 25 YEARS IN 14 TEMPERATE-ZONE OR COLDER COUNTRIES. 90% OF THE CASES WERE FOLLOWED UP WITHIN 6-9 MONTHS (FIGURES IN %)

Area thought to be involved		Resolved or mostly resolved over a period of time	Unsolved
1. Nutrition	a) – dietary imbalance	8	5
	b) – feeding system/ allowance	9	1
	c) – palatability	2	-
2. Temperature		12	4
3. Ventilation		14	8
4. Disease	– pathogenic	4 (*referred to veterinarian*)	5 (*referred to veterinarian*)
	– poor hygiene	6	-
5. Multifactorial, *i.e.* several causes suggested.		7	3
6. Other, *i.e.* outside N°s. 1-4.		8	-
7. Minor or no problem on investigation		2	-
8. Client ignored advice		-	2
		72%	28%

Note: Items 1b and 2; 2 and 3; 3 and 4 were often inter-related

'Other' Includes stocking density, water adequacy, stockmanship errors, too many changes – all thought to be primary factors affecting growth rate on the 8 farms involved.

Comment: Despite being employed as an animal feed firm troubleshooter for 60% of the period, notice that the feed or the way it was fed was only implicated in under 1 in 5 of the slow growth complaints resolved. Notice also the high proportion (28%) of environmental errors responsible. A very different causal list is seen in tropical countries, where genetics/appetite are primary causes.

SOME OBSERVATIONS ON CHECKING UP ON THESE PRIMARY DISTURBANCES TO GROWTH RATE

Nutrition

Compound (ready-made) feeds

Errors in nutritional specifications are much less common than 20 years ago due to computerization in the formulation office and in the feed mill. Mistakes do occur, largely due to errors in assumptions of nutritive value of the raw materials used. However feed manufacturers now analyse raw materials for primary nutrients to a much greater degree than they used to, but energy achievement seems to be a possible weak point, even today.

To a certain extent sub-quality samples have to be used-off in manufacture and it is hoped that this is done ***very*** gradually especially in the case of weather-damaged items or spoilage. If not, a reduction in growth rate is quickly noticed. However short-cuts are (occasionally) taken and farmers should be aware of this possibility. If this is proved – or maybe even suspected, an immediate change of supplier to one more trustworthy is wise.

OVER-USE OF CERTAIN RAW MATERIALS

The writer is old-fashioned enough to believe that a good mix of raw materials to make-up the dietary specifications provides the best growth rate. Conversely however the Americans get excellent results with almost universal corn/soy diets (with a min./vit. supplement) for growing pigs. Such diets are dry and palatable and as a result mycotoxin levels could be lower than in some countries.

Feed compounders do have maximum raw material constraints and purchasers should ask what they are. Over-use of wheat, when price and availability is favourable; wheatfeed, tapioca (manioc), biscuit-meal and rice bran is known.

FARM-SPECIFIC DIETS IMPROVE GROWTH RATE

The feed trade is moving only slowly in the important area of matching nutrient specifications to the immune status of the pigs. All nutritionists are well aware of how this affects performance when it is badly adrift (up to 40% in growth rate, and possibly over 50% in protein deposition), but in the writer's experience some Sales & Finance departments seem to be dragging their heels on cost grounds. Therefore there is a reluctance to change to a radically new way of selling pig food so that immune status can be assessed and the diets supplied to cater for it.

Producers with growing finishing pigs can help themselves by talking directly with their supplier's pig nutritionist, and by providing him with details of the current health and biosecurity status of their grow-out facilities. He can then get specifications closer to being right with a custom-mix. However to utilise the possibilities to the full, such a producer will eventually have to espouse a CWF system (Computerized Wet Feeding) and co-operate with the feed supplier on a "Challenge" or "Test" Feeding programme so that his pigs' lean gain growth curve can be measured and the diet adjusted accordingly. At the time of writing lean growth itself could be a pointer to the immune demand.

PALATABILITY

Sometimes growth rate can be affected by lack of palatability. Most nutritionists have palatability tables and build in the necessary constraints when including likely unpalatable ingredients. However what is less well recognized is the *cumulative* effect of the milder unpalatable ingredients in the diet, especially if other aggravating factors like over-coarse or too-fine grinding, soft pellets, mould residues, added fat levels and some unpalatable chemical additives (*e.g.* nitrofurans), high mineral levels (limestone) are also present. General lack of freshness is also an appetite-reducing factor in small pigs.

In most of the above cases the inclusion of an aromatic taste-enhancer may not work, and most firms selling palatants do not necessarily claim that they do.

Ingredients known to be unpalatable are sorghum (milo), rapeseed, olive pulp/cake, wheatfeed in excess (>40%), any food containing moulds or mould residues (mycotoxins), very hard particles of wheat or maize, most minerals in excess (limestone flour is sometimes over-used), food ingredients which have been oxidised, especially rancid oils/fats, and all ingredients with pungent odours – some confectionery waste can contain such smells. They are not unpleasant to us humans but the pigs definitely object to them.

Suggested maximum inclusion levels are given in Table 9.

If palatability is a suspected cause of slow growth, do not resort simply to flavour enhancers, but immediately change the diet to one known to contain fresh, and a wider variety, of ingredients.

Table 9 SUGGESTED MAXIMUM LEVELS FOR COMMON FEED RAW MATERIALS. ABOVE THESE LEVELS GROWTH COULD BE AFFECTED (% AS FED AND OF GOOD QUALITY)

	Sows	*Weaners*	*Grower*
Ingredient			
Barley	*No limit, but watch dust at high levels*		
Wheat	50	33	40
Maize	40	40	25
Oats	40	10	25
Tapioca	25	10	20
Wheatfeed	35	15	30
Soya	25	30	30
Full fat soya	15	20	20
Meat meal	*(not advised due to risk or fear of BSE in certain countries)*		
Otherwise	7.5	2.5	7.5
Fishmeal	7.5	10	5
Bread waste (63% DM)	40	30	40
Biscuit waste (86% DM)	40	40	30
Cake waste (85% DM)	40	30	40
Confectionery waste (98% DM)	20	7.5	15
Wheat starch syrup (DE12 MJ/kg)	25	15	20
Potato waste (Steamed) (11% DM)	25	15	20
Fodder beet (17.5% DM)	20	5	10
Maize gluten meal (DE 13 MJ/kg)	20	5	10
Rapeseed (DE 12 MJ/kg)	12	5	15
Skim milk (DM 9%, lysine 0.23%)	30	30	30
Whole milk (DM 13%, lysine 0.27%)	60	80	80
Yoghurt waste (DM 14%-20%)	25	15	20
Whey (DM 5.5% DE 0.85 MJ/kg)	25	10	25
Brewers yeast (18% DM)	25	10	10
Extr. Rice Bran (88% DM 16% fibre)	15	5	10
Linseed cake (33% F.P.)	10	2.5	5
Maize Germ Meal (DE 13 MJ/kg)	20	15	10
Distillers Draff (23% DM)	5	0	5 *(after 50 kg)*
Distillers Grains (89% DM)	5	2.5	5
Lupins (88% DM, DE 17 MJ/kg)	10	5	10
Sugar beet pulp (dried) (88% DM)*	10-20*		
Peas	15	10	20
Sorghum	20	0	10

* Care needed due to swelling when moistened.

Some water sources (*e.g.* in central Canada and Mexico) can contain excessive levels of trace elements which may affect appetite.

WHAT TO DO IF YOU SUSPECT THE FOOD

If you suspect your delivered feed is the cause of your slow growth ...

• Immediately take a 2 kg (5 lb) representative sample of the suspect feed.

• Keep at least a 1 kg (2 lb) sample of the food refrigerated, for future use if needed.

• Contact the feed supplier and speak to the company nutritionist. *Use the local representative or sales department staff only as intermediaries*. Generally speaking samples should be taken by them *in a properly representational manner* , half given to you, and a telephoned, then written report by fax or letter sent to you. This should take hours after the arrival of the sample, not days.

• Check that your bulk bins and feed hoppers do not contain obvious mould or mould residues on the 'unpolished' surfaces, *i.e.* how clean they are.

• Assemble and provide proof, if needed, of the batch number of the delivery involved, with all paperwork attached.

• Provide evidence of the performance loss, with dates.

• If requested allow the feed manufacturer full access to your premises.

• Employ an independent pig consultant or pig nutritionist for a second opinion if the feed supplier is dilatory or seems evasive. Allow him also to tour the premises.

Remember : Do not necessarily blame the food. Look again at Table 8. In this survey only one in ten of poor growth problems probably or certainly involved feed quality. This is in contrast to two out of three cases where the food was immediately blamed by the producer for slow growth or poor FCR.

THE ON-FARM MIXER

The three major problems seen with home mixers which affect performance, including growth rate, are:–

1. Failure to provide their nutritionist with details of the expected (or actual) analyses of the home-grown or bought-in raw materials used, especially cereals.

2. Opportunist buying of what appear to be good value raw materials without assurance from the vendor of a nutritive declaration on at least the more important nutrients – "buying blind".

3. Failure to keep the mixing area clean enough, so that the risk of the damaging effects on performance of residual mycotoxins are high.

RAW MATERIAL ANALYSIS

The majority of on-farm mixers use their own grain, or grain from an adjacent farm. Not surprisingly these parcels can vary enormously in nutrient quality and quantity (Table 10).

Table 10 TYPICAL RAW MATERIAL VARIATIONS

	Average Protein	*Range*	
	(%)	*Min (%)*	*Max (%)*
Off-farm			
Barley	11.2	7.8	13.9
Wheat	11.9	8.3	16.4
Bought-in			
Wheatfeed	16.0	13.7	19.7
Soya 44/47	41.1	34.0	47.0
Full Fat Soya	36.1	33.6	45.0
Best fish meal	70.3	64.0	73.5

Source : Dalgety (1994)

While the nutritionist can make fairly close assumptions on bought-in materials, producers should ***always*** provide him with the declared analysis of purchased goods so that he can narrow the variation still further.

Not to do so is negligent, or at best lazy, as the information is often provided free.

THE COST OF GETTING IT WRONG

Many producers think that "taking the rough with the smooth", the quality evens itself out over the lifetime of the growing pig. This is definitely not so. Dealing with feed complaints on protein/ vit./min mixes and where careful analysis revealed the shortfall in formulation, I found the improvement due to subsequent correction was never less than £2/pig ($2.80), and Seemeel in 1993 published a figure of £3 ($4.20).

Even taking the lower figure of £2/pig, at 5 pigs to the tonne of feed, this is a £10/tonne leeway. Assuming the average home-mixer produces 5,000 finishers/year thus consuming 1,000 tonnes feed/year, the shortfall could be £10,000/year or $14,000!

This allows considerable scope for analysing at least the parcels of grain used, if nothing else.

THE COST OF GETTING IT RIGHT

Table 11 suggests current analysis costs . . .

Table 11 CURRENT LAB ANALYSIS COSTS

	£	$
Dry Matter	3	4.20
Protein	6	8.40
Oil	6	8.40
Fibre	7	9.80
Lysine	30-35	42-49
Mycotoxins	25-40	35-56

Several years ago I contacted all the major UK feed compounders and asked them how much they spent per tonne of feed sold on *raw material analyses alone*. The average was £1.27 on feed costing about £145/ tonne or 0.8% (Gadd 1988). In today's prices this would be at least £2/ tonne on a £140/t diet (1.43%).

Figure 1 is interesting as it charts the performance shortfall of grow-out home mixers over grow-out complete feed users across a ten year period in terms of FCR – a similar deficiency occurs in growth rate.

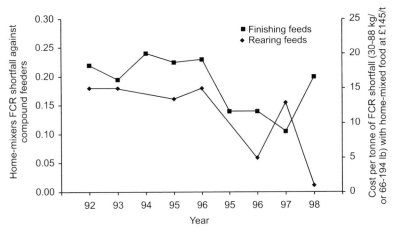

Source: Extrapolated from various recording schemes and compound feed trade data

Comment: While things are getting better, there still seems to be a £7/tonne (6%) difference in poorer performance of home-mixed grow-out feeds to those provided by the feed trade. The fact that the same nutritionists are often involved in the dietary specifications suggests that the home-mixer is adrift - by not knowing the analysis of his on-farm ingredients.

Figure 1 Records show a consistent shortfall in FCR for homemixers v. compounds

How much of this leeway is due to the formulator/nutritionist (the same person in some feed companies) having different conceptions of the raw material analysis he has on hand?

If we assume from Figure 1 that the average shortfall is now £7.50/tonne (at 5 pigs/tonne, £1.50 pig) what can we afford to spend on grain analysis to help the nutritionist to give us a more accurate specification using the bulk raw materials we possess in storage or are buying each month? Producing 5000 finishers/year, even taking this cost at £1/pig this is £5,000, but I find an adequate cost is nearer to only a fifth of that.

Table 12 HOW MUCH TO SPEND ON RAW MATERIAL ANALYSIS?

	12 lysines	at	£38	£420	$588
	15 D.M.s	at	£3	£45	$63
	15 oils	at	£6	£90	$126
	15 fibres	at	£7	£105	$147
Optional …	6 mycotoxins	at	£40	£240	$336
Contingency at nutritionist's request …				£100	$140
				£1000 or 20p/pig	
				($1400 or 34c/pig)	

Spending 20p/pig to recapture what the evidence suggests is a discrepancy of somewhere between £2 and £1.50 pig is an excellent REO – some 7 to 10:1. And if your nutritionist suggests it, you can even afford to pay for more comprehensive raw material analysis!

BUYING BY-PRODUCTS : A CHECKLIST

Many by-products – just because they look cheaper on a feeding cost/day basis (as indeed all are) – are purchased 'blind' or 'on trust'. There can be more variations in critical nutrients than with grain or bought-in straights, especially when the liquid fraction is high, *i.e.* skim and whey. Some by-products are produced to a nutrient specification, which helps greatly in calculating dietary daily intake, but most are not. The following checklist is helpful in getting value for money.

Some 1uestions to ask before buying

A. What is its source (a by-product of what?)

B. If originally designated for the human market why was it rejected? Is there a toxin factor?

C. If past its sell-by date, how stale is it? (Has an anti-oxidant been used?)

D. Is it raw or is it cooked?

E. Is it of a single natural source or is it a mixture? If a mixture, what are its constituents?

F. Is it palatable and digestible?

G. What is its analysis and will it be consistent? Always get a salt level on whey and whey concentrates and a DM declaration on skim and again especially whey. (*See Table 13*).

H. Is there continuity of supply or is it a one-off?

I. Will it be delivered fresh and what is the delivery cost, the cost per tonne of dry matter, the cost per unit of energy and protein?

J. Who else is interested? Who's turned it down?

K. What is it really worth? Do the sums; they matter.

L. **Finally, always consult an independent nutrition specialist.**

VARIATION IN LIQUID BY-PRODUCTS –
HOW IT AFFECTS GROWTH RATE

Both skim and whey are variable products owing to the quantity of water (and in whey's case salt) allowed to enter the collection tanks at the milk and cheese factories.

HOW TO CHECK OUT SKIM MILK

With skim one expects a 9% dry matter content, with a specific gravity of 1.033. Specific gravity (sp.gr.) *is the weight of a substance compared with the weight of an equal amount of some other substance taken as a standard.* For liquids the standard taken is usually water. Thus a sp. gr. for a sample of skim milk of 1.033 means that one kg of skim milk weighs 1.033 times more than water, and can be quickly measured by a hygrometer – a simple and inexpensive instrument.

Table 13 gives a simple on-farm guide to the value of skim milk relative to its sp. gr. Low sp. gr. suggests dilution with washing water at the factory, and the balancer meal/supplement will need nutritive compensation if growth rate is to be maintained, which if these low sp. grs. are consistent, should be drawn to the supplier's attention and appropriate financial compensation claimed for the extra balancer meal required. Talk to your supplier beforehand about these sorts of QC matters. He just wants to dispose of it while you are buying it as a nutrient substitute.

Table 13 RELATIONSHIP OF SPECIFIC GRAVITY, DRY MATTER AND COMPOSITION OF SKIM (AS FED BASIS)

Specific gravity	Measured dry matter %	DE (MJ/kg)	DCP %
1.036	9.5	1.58	3.5
1.033	9.0	1.50	3.3
1.031	8.5	1.42	3.1
1.030	8.0	1.33	2.9
1.028	7.5	1.25	2.8
1.027	7.0	1.17	2.3
1.025	6.5	1.08	2.2

HOW TO CHECK OUT WHEY

Because whey is a high-energy food, and provides a reasonable amount of good quality protein as well, variation in sp.gr. is a very important factor in purchased deliveries. Again a hygrometer is essential to keep abreast of delivery variance and Tables 14a & b show how only small differences in sp.gr. (of only 0.002 sp.gr. or 1% dry matter) per batch can affect nutrient intake, and how a typical whey balancer meal designed by a nutritionist on the ***average*** expectancy of 1.022 sp.gr. (5% DM) needs to be increased or decreased per litre of whey fed to rebalance the nutrient capability of the pig. Tables 14 c&d illustrate this difference on a per pig basis and what failing to do this over quite small movements in sp. gr. from batch to batch costs in lost growth rate and FCR.

Table 14a RELATIONSHIP OF SPECIFIC GRAVITY, DRY MATTER AND COMPOSITION OF WHEY (AS FED BASIS)

Specific gravity	Dry matter of whey	DE MJ/kg	CP g/kg	Lysine g/kg
1.027	6.5	1.07	10.4	0.65
1.025	6.0	0.98	9.6	0.60
1.023	5.5	0.90	8.8	0.55
1.022	5.0	0.82	8.0	0.50
1.021	4.5	0.74	7.2	0.45
1.020	4.0	0.66	6.4	0.40
1.019	3.5	0.57	5.6	0.35
1.018	3.0	0.49	4.8	0.30

Table 14b CHANGE IN BALANCER ALLOWANCE REQUIRED FOR EQUAL NUTRIENT INTAKE WITH VARYING WHEY COMPOSITIONS

Specific gravity	Change in balancer allowance
1.022	No change
1.021	+ 5 kg/1000 litres or 50g/10 litres
1.020	+ 10 kg/1000 litres or 100g/10 litres
1.019	+ 15 kg/1000 litres or 150g/10 litres
1.018	+ 20 kg/1000 litres or 200g/10 litres
1.023	− 5 kg/1000 litres or − 50g/10 litres
1.025	− 10 kg/1000 litres or − 100g/10 litres
1.027	− 15 kg/1000 litres or − 150g/10 litres

Table 14c EXAMPLE (20 PIGS IN A PEN)

| | *S.G. of Whey* | | *1.022* | | *1.018* | |
| | *Whey litres/day* | | *Balancer kg/day* | | *Balancer kg/day* | |
Liveweight kg	*Per pig*	*Per pen*	*Per pig*	*Per pen*	*Per pig*	*Per pen*
25	3.0	60	1.0	20	1.05	21
40	4.5	90	1.4	28	1.50	30
60	7.0	140	1.6	32	1.75	35
80	9.0	180	1.8	36	2.00	40

Table 14d MEASUREMENTS OF GROWTH PERFORMANCE FROM TWO SECTIONS OF ONE FARM WHERE A HYGROMETER WAS OR WAS NOT USED TO READJUST WHEY BALANCER MEAL ACCORDING TO SHORTFALLS IN SPECIFIC GRAVITY READINGS

	ADG 20-88 kg (44-194 lb) (g/day)	*(lb/day)*	*Average variance detected below norm** *DM %*	*Total lysine g/kg*
Hygrometer used and meal allowance increased to Table 14b levels.	759	1.67	1.3%	0.15
Hygrometer not used	721	1.59	–	–

* *On 35 days out of the 90 day period.*

Comment : By adjusting the meal allowance when necessary on a wet-feeding circuit cost 95p/pig more but the quicker growth therefrom saved £1.25/pig food at slaughter. Technically, the producer using the hygrometer, by recording his readings and sending them to the supplier, could claim from the supplier of the whey the 95p/pig spent on extra meal to compensate for the shortfall of 1.3% dry matter below the agreed standard of 5% DM, thus benefitting by the full £1.25/pig. Without a hygrometer no such claim can be substantiated. Always use a hygrometer, and tell your supplier why. At least it may improve the consistency of your deliveries!

NUTRITION AND GROWTH RATE

The correct balance of the primary nutrients in the diet can affect growth rate. Of these lysine:energy ratios are important. While supporting amino acid balance is also critical this has to be left to the nutritionist to get right.

LYSINE: ENERGY RATIOS CHECKLIST

These are often seen as straight ratios averaged out for all three 'sexes' – entires, castrates and gilts. (For example, Table 15).

While Table 15 gives a guide to correct daily nutrient intakes of total lysine and digestible energy for all pigs, further research suggests the true picture is much more complicated. This can be due to:–

✓   Differences between genotypes, often influenced by appetite capability. (Table 16).

✓   Differences within the genotype. (Figure 2).

✓   Difference between the sexes (Table 17).

✓   The effect of immune demand. (see Immunity section).

Table 15 RECOMMENDED OVERALL LYSINE : ENERGY RATIOS AND DAILY INTAKES FOR ALL GROWING/FINISHING PIGS – PAST ADVICE (LATE 1990s).

	Genetically improved stock							
Body weight (kg)	7	12	16	20	40	60	80	100
Body weight (lb)	16	27	35	44	88	132	176	220
Target growth rate/day (g)	250	400	500	650	800	900	950	1000
Target growth rate/day (lb)	0.55	0.88	1.10	1.43	1.76	1.98	2.10	2.20
Lysine needs/day (g)	4.6	7.6	9.6	14.0	20.0	24.4	27.0	28.5
Approx* energy needs/day (MJ DE)				4.5	7.3	9.9	15.5	24.5	30.0	34.0	36.5
Lysine: DE ratio (g/MJ DE)	1.02	1.04	0.96	0.90	0.82	0.81	0.79	0.78

* *Dietary DE can change dependent on appetite, especially of nursery pigs*

Table 16 APPETITE AND LEAN GAIN PER TONNE OF FEED OF 4 MAJOR EUROPEAN BREEDS   (PIGS 55 – 90 kg/ 121-198 lb)

	Appetite			*Lean Produced/Tonne Feed (kg)*		
			Difference to Breed A		*Difference to Breed A*	
	(kg/day)	(lb/day)	%	(kg)	(lb)	%
Breed A	2.84	6.26	–	301	664	–
Breed B	2.78	6.13	- 2.11	268	591	- 1.96
Breed C	2.79	6.15	- 17.76	274	604	-8.97
Breed D	2.51	5.53	- 11.62	253	558	- 15.95

* *Same standard feed fed, but adequate diets fed* ad lib *throughout.*

Econometrics: When the diet of Breed D was adjusted in nutrient density based on a 12% lower appetite potential to that of Breed A, lean per tonne **exceeded** that of Breed A by 2.6%, and while Breed D's diet cost 8% more, the extra yield of lean recouped 80% of this extra dietary cost.

Source: RHM (unpublished)

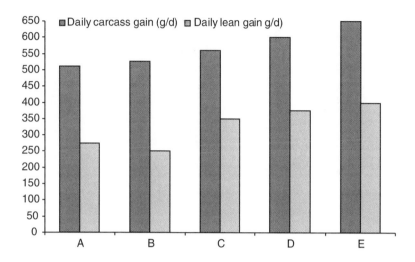

Figure 2 Variation in growth characteristics within a single genotype (Source: Owers, 1994)

And Is There a Tropics Factor ?

I also suggest that certain genotypes ('breeds') may need different lysine:energy ratios when kept under very hot and humid (tropical or near-tropical) conditions, but at present this is surmise on my part but gathered from experience in such conditions. It could be that some high-lean/'low-appetite' genotypes are unsuited to these near-tropical conditions when fed diets formulated to "European" specifications – and this area would repay research. We may even need a 'tropical' genotype of pig with its own dietary specification, and probably do. I come across too many cases of very slow growth in the tropics among genetically-improved, very healthy pigs on clean farms.

Figure 3 suggests that a fourth source of information should be consulted – the seedstock (or semen) supplier. This dialogue is best left to the nutritionist; the producer just needs to check on whether he has done so with the breeding company on the producer's behalf.

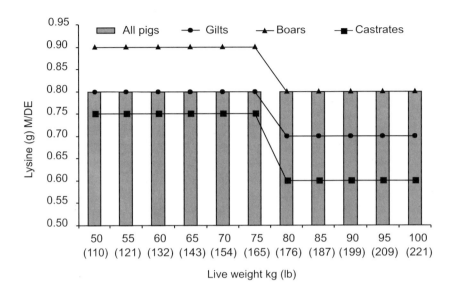

Figure 3   Lysine energy ratios from 50-100 kg (110-221 lb) established by JSR Genetics Research For Their Own Genotype. (Penny 2000)

THE EFFECT OF IMMUNE DEMAND ON GROWTH RATE AND LYSINE NEEDS

Groundbreaking work in America (1993-97) showed us how much of certain nutrients are 'stolen' away from growth rate where the pig is faced with disease challenge and needs to reinforce its immune shield in order to remain healthy (Table 17).

While the Iowa State work in Table 17 could be said to have measured the extremes in the early growth stage, the economics from subsequent trials to slaughter-weight have maybe shown a very serious potential shortfall in income where the growing pig's diet was not altered, particularly in the lysine levels on which supporting amino acids are based, to more closely match the immune demand, or lack of it (Table 18).

Table 17 BY HAVING TO COPE WITH A HIGH DISEASE CHALLENGE, GENETICALLY IMPROVED[†] PIGS 6.3-27.2 kg (14-60 lb) EAT LESS, GROW SLOWER AND HAVE A POORER QUALITY CARCASE

| | *Immune Stimulus Required* | | |
	Low	*High*	*Difference**
VFI (kg/day)	0.97	0.86	12.8% more
ADG (g)	677	477	42% more
FCR	1.44	1.81	25% better
Protein gain (g/day)	105	65[†]	62% more
Fat gain (g/day)	68	63	8% more

Source: Stahly *et al* (1995)

* In favour of low immunity needs

[†] The leaner the genotype the more the protein gain is damaged

Note: Both sets of pigs could be considered "healthy". A high disease challenge is described typically as a 'pig-sick' building and a low disease challenge environment as 'all-in/all-out/multisite' scenario, properly disinfected.

Table 18 THE MONETARY COST OF POORLY MATCHING DIETS TO IMMUNE STATUS (Results averaged from various trials)

		Extra overheads, feed and turnaround costs	
1.	Extra days to 100 kg	18 days more	£ 2.16/pig	($3.02)
2.	[M.T.F. Meat per tonne/food (kg)]	18 kg less	£14.76/tonne PPTE[(2)]]	($20.66)
	Reduced income pig			
3.	From food not utilized		£ 4.64 [(3)]	($6.50)
4.	From needlessly higher-priced food		£ 1.34 [(4)]	($1.88)
	Total reduction in margin/pig (1+3+4)		**£ 8.14**	**($11.40)**

This suggests that just feeding one diet to healthy, high performance gene pigs without taking immune status into account could cost the producer at least £8/ pig ($11)

Notes:
1. At 12p/day overhead-costs/pig.
2. PPTE = Price per Tonne Equivalent. This figure reveals that by getting it wrong the reduced performance is equivalent to a rise in feed costs of £14.76/tonne, or about 12%.
3. At 3.18 pigs/tonne
4. At £4.26/tonne more expensive.

UK Costings at September 1999

Toplis has shows how the genotype's effect on growth rate is magnified when the immune demand is high or low (Table 19).

Table 19 THE EFFECT OF GENOTYPE ON GROWTH IN RESPONSE TO IMMUNE ACTIVATION

	Daily gain, g (lb) to slaughter			
Lean Potential		*Disease Challenge*		
	Low			*High*
Low	680	(1.5)	**599**	**(1.32)**
High	**826**	**(1.82)**	625	(1.38)

Source : Toplis (1999)
British pigs tend to be in the bold sectors, thus we have most to gain.

The very considerable differences in lysine needs throughout the growth period is again illustrated in Figure 4.

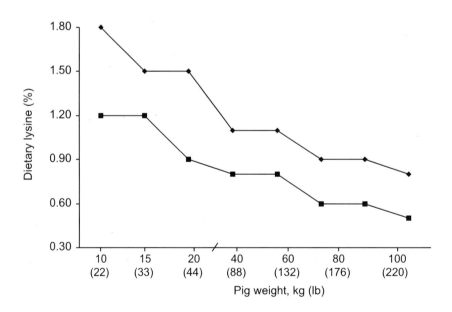

Figure 4 Dietary lysine concentrations to fit immune status. Dietary lysine concentrations to optimise efficiency of feed utilisation in pigs with a low or high level immune system activation. Data derived from castrates with a moderate genetic capacity. (Adapted from Williams *et al.*, 1997)

HOW CAN THE PRODUCER ESTABLISH HIS IMMUNE STATUS ?

Having recognized that tailoring the diet to suit the immune demand of growing pigs is likely to increase profitability substantially, if you can't manage what you can't measure, then how does the farmer measure immune status? Do his pigs need a high protective wall or a low one? If a low one, how much of a better diet do they need? If a high one, how much can lysine be reduced so as not to waste it by using energy to excrete the surplus (de-amination) and so go on to unbalance the correct lysine:energy ratio and lower growth rate still further?

There are three ways of solving the problem. I hasten to add that we are at the start of 'something big' here, and that any suggestions at this early stage are tentative. But the problem is so important to profit that we **must** address it. We need to try the suggestions out and see how they work.

1. **Serology**. This method uses the vet, doing routine blood tests, to determine the level of disease present. Three snags. First, even state-of-the-art serology only covers certain diseases, and so may not include the one doing most of the challenging. Second it is expensive. Third, it could be time-consuming.

    By all means use the vet to disease-profile your herd, especially the breeding herd, to establish correct protocols to prevent disease, but for determining the degree of disease challenge – this is probably not yet an option.

2. **Challenge or test feeding**. Whatever the size of the herd, take 50 pigs and feed them a 'non-dietary-limiting' feed, designed by your nutritionist. Record growth and take periodic lean-scan measurements from the end of the nursery period to slaughter and record carcase measurements after slaughter. This will measure their level of lean accretion and enable the nutritionist to design a diet based on the assumed performance of the whole herd (or section of the farm where the groups are housed) thus automatically taking into account both the disease status as well as the genetic type used.

    If the environment is radically different within the herd, select 50 pigs from each of the difference environments.

    The snag lies in the expensive deep-scanning equipment needed to measure lean deposition, so it is really the province of a feed manufacturer who sells his food on a herd-specific basis – a marketing strategy for the future, though a few firms are already trying it in different ways.

3. **Measure growth rate**. Recording daily gain accurately, while in theory less accurate than Challenge Feeding, seems from initial reports to be worth trying as there may be a simple correlation (linkage) between immune status and growth rate. Of course, the snag here is that other things besides immune demand (environment, stress, appetite, wet-feeding, the design of the diet itself, *etc.*) can alter speed of growth, but this possibility is so simple and farmer-friendly that is must be explored.

ACHIEVING A LOW IMMUNE STATUS A CHECKLIST

Action	*Further information*
✓ **Minimize Buying in of Stock.**	*Weaner gilts*
✓ Keep visitors away.	
✓ **Adopt AIAO and batch rearing.**	*See relevant section*
✓ Establish strict reliable vehicle sanitation and driver discipline.	*Biosecurity section*
✓ Establish farm delivery and collection areas well away from pigs.	
✓ *Never* allow a knacker/casualty disposal firm to cross the farm boundary.	
✓ Incinerate all casualties.	
✓ **Put in place a complete biosecurity system.**	*Biosecurity section*
✓ Appoint vermin control/bird control stockpeople with responsibility for these areas.	
✓ Realise the value of routine 'fogging' to keep respiratory disease as low as possible.	*Biosecurity section*
✓ **Induce an adequately-long and well-planned induction protocol for any new breeding stock.**	*Empty Day section*

Action	*Further information*

✓ Avoid a 'needle-happy' vet.

✓ Use a veterinarian experienced
in the techniques of naturally-
acquired immune protection.

✓ Have isolation facilities, preferably   *Choosing a gilt*
off-farm, for all new stock.   *section*

✓ **Do frequent stress audits.** Stress   *Stress section*
lowers resistance

✓ Realise how important it is to wash   *Pen Soiling section*
away looseness and scour immediately
it is noticed.

✓ Check all diets are adequately
provided with zinc, especially organic
zinc.

✓ Get your ventilation checked over and   *Ventilation Section*
then establish a monitoring procedure
to ensure fans, etc. are kept up to
scratch.

✓ Does your slurry drainage flow ***out***
of the unit or through the unit?

✓ Farrowing house and nursery slurry
pits also need sanitation when AIAO
is practised.

✓ Site dung heaps off-perimeter and
cover them.

✓ Have a clean and tidy rest room.

✓ Never allow farm staff to consume
pork/meat products on-farm.

✓ Replenish foot dips frequently and   *Biosecurity Section*
use the correct disinfectant.

✓ Have a laundry system for clothes
and washing kit.

✓ On outdoor units, the match is a vital disinfectant.

✓ On outdoor units, beware overuse
of land, *e.g.* duplicating hut runs
for weaners.

	Action	*Further information*
✓	If you have more than one farm unit, do not exchange tools/vehicles etc., and change *your* clothes when visiting.	
✓	Monitor space allowances frequently to avoid over-crowding creeping up on you.	*Stocking density*
✓	Keep domestic and farm animals out especially sheep and chickens.	
✓	Cats and dogs should be kept inside the unit and not encouraged to roam adjacent fields.	

THE POST WEANING CHECK TO GROWTH

If we 'abrupt wean' between 16 days and 28 days all the weaners will slow up to a certain extent in growth rate. Post-weaning nutrition is primarily involved. Disease, poor management and incorrect housing are also concerned but the **primary cause is nutritional**.

This is covered in the section on 'Problems at Weaning – Nutrition' where several checklists will be found.

COMPENSATORY GROWTH

One eminent pig researcher has stated …

"There is no such thing as compensatory growth. It is the last refuge of producers who have not managed pigs properly. Experiments have shown that there is no regaining of lost time and no improvement in growth efficiency in piglets whose growth has been hampered."

My field experience leads me to suggest modifying these forthright opinions! I don't disagree with the 'last refuge' statement but I have found on several farms that if the postweaning check is not too severe, say around 5 days of measurable growth reduction compared to other weaners, these pigs can finish within a day of the less affected pigs at slaughter weight. I know because I weighed them all on the same day! This can only mean that in this (and other cases reported to me by clients) the pigs *did* seem to catch up in liveweight terms. Certainly it does not seem to happen often. How often I know not.

However, when the producer and I examined the killing-out percentages of the respective 'checked' and 'non-checked' pigs at weaning, the processor's returns showed, without fail, a reduction in the KO% of the checked pigs compared to the unchecked (or less-checked) pigs. The KO% was down between 0.2% and 1.06% suggesting that in **liveweight terms** the (moderately) checked pigs may indeed catch up, but in deadweight terms (which is the measurement which matters in profit terms) they do not do so? **Lean gain** doesn't seem to catch up or compensate - which is the result which mattered to us, and of course, matters just as much to you as a producer.

Whether there was any difference in liveweight or deadweight feed efficiency I don't know, as it was not possible to measure it under farm conditions, but we were meticulous over our weighing in the week after weaning and at slaughter, and the elapsed days were identical between checked and non-checked pigs.

The fact that the checked pigs had less dressed-out meat suggests to me that their FCR would be worse, as food into lean in weight terms is a good bargain, meat containing as it does so much water.

All this may be semantics as the end result in profit or income terms is the same, but the subject needs airing, I guess. Surely the correct statement is that "*compensatory **lean gain** is a myth*" not "*compensatory growth is a myth*".

FOOD CONVERSION 18

TERMS

Food Conversion Ratio (FCR) Food Conversion Efficiency (FCE)

Liveweight Food Conversion (LFC) *is the number of kilograms of food required to produce one kg of liveweight, e.g. kg feed per kg liveweight gain (or lb feed per lb liveweight gain, USA).*

Deadweight Food Conversion (DFC) *is the number of kilograms of food required to produce one kg of dressed carcase weight i.e. whole carcase less head, selected internal organs and digesta, and in the case of entires, the sexual organs.*

Lean Tissue Food Conversion (LTFC) *applies the same principle to* ***lean tissue*** *formation as distinct from deadweight, and is used principally in research.*

SUGGESTED TARGETS FOR GROWING/FINISHING (TO 100kg)

Note: Live FCR.
Targets for ad lib fed genetically-improved pigs kept within their LCT/ECT temperature zones

Age (weeks)	Target FCR in that week	Overall, from 3 week weaning	Action level
10 (30 kg) (66 lb)	1.90	1.4	In the present
11	2.00	1.5	economic climate a
12	2.10	1.6	10% worsening of
13	2.15	1.7	the overall FCR
14*	2.20	1.80	(30-100 kg) (66-220 lb)
15*	2.30	1.90	figure opposite could
16	2.40	2.00	be considered an
17	2.45	2.10	action level *i.e.*
18	2.50	2.20	more than 2.5:1
19	2.55	2.25	(6-100 kg) (13-220 lb)
20	2.60	2.30	or over 2.2:1 (6-65 kg)
21 (100 kg) (220 lb)	2.65	2.35	(13-143 lb)

Overall FCR 2.35:1 (30-100kg) (66-220 lb)
* Assumes a change of house or pen-size in these weeks

SOME THOUGHTS ABOUT
FOOD CONVERSION EFFICIENCY

- Food conversion worsens as the pig ages – gets heavier – due to older pigs needing more food for body maintenance (Table 1).

Table 1 CHANGE IN FEED CONVERSION WITH LIVEWEIGHT

	Feed consumed/day (kg) (lb)	Used for maintenance (kg) (lb)	%
60 kg (132 lb) pig	1.75 (3.85)	0.7 (1.54)	40%
90 kg (198 lb) pig	2.50 (5.50)	1.10 (2.42)	44%
120 kg (264 lb) pig	2.9 (6.38)	1.35 (2.97)	47%

- The effect of higher lean tissue growth deposition due to generations of genetic improvement is changing the traditional profile of the FCR graph from more of a 'V' to more of an 'L' shape. This means that the producer has more scope for *ad-lib* feeding to higher liveweights before grading suffers.

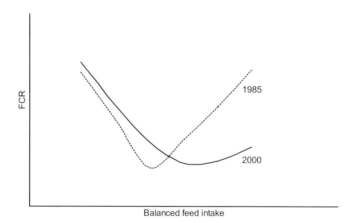

Figure 1 How FCR is changing on-farm

- In Figure 2 the longer, flatter shape to the FCR line also means a wider range of feed given is possible (A) over which little change in FCR is apparent compared to less improved strains (B).

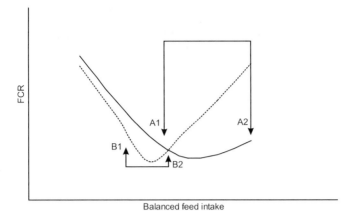

Figure 2 Perceptible changes in FCR allied to feed intake. Note that under modern genetic conditions, the difference in FCR variation is over a much wider spread of balanced feed intake (A1-A2) than over the same FCR variation in the past (B1-B2)

- While FCR is a very important measurement, it must be taken into account along with daily gain and carcase grading – all three affect profitability.

  Nutritionists now have predictive computer models, which *if given accurate information* can put these three important factors into econometric perspective.

  This means that a worsening FCR must always be referred to a competent pig nutritionist.

- One problem with FCR is that it is difficult to measure accurately enough on a busy farm (Table 2). While this table is over 25 years old now, recent analysis of FCR complaints concerning feed suggest that any error is only a little less, if at all. 3 recent investigations (1999-2000) revealed 0.2 LFCR differences above or below what was claimed. This is equivalent to a 16% rise or fall in the current price of feed.

- If FCR is poorly estimated under real conditions, the MTF measure (Saleable Meat Produced Per Tonne of Food Fed) is a better yardstick – it is easier to measure and thus likely to be more accurate.

  This is fully described in the Chapter on New Terminology. Use both measurements if you wish, but you will soon find yourself preferring MTF if you do, leaving FCR to the research worker who can measure it accurately.

Table 2   FARMERS CALCULATED AND ACTUAL FCR'S TAKEN FROM 5 COMPLAINTS OVER POOR PERFORMANCE (1975 - 1977)

Farm	Weight range (kg)	(lb)	Farmer's estimated FCR	Actual FCR based on careful measurements on-farm	Likely reason for error
1	6-28	13-62	2.9	2.71	Input food not weighed
2	20-91	44-200	3.2	2.86	Poor recording
3	30-90	66-198	2.9	2.81	Mistake over input batches
4	25-86	55-190	2.6	2.92	Guesswork (!)
5	30-64	66-141	2.6	2.45	Poor recording.

Average error of 0.22 FCR over 48 kg (106 lb) LWG, an 8% error

Source : RHM Agriculture (Unpublished) 1977

FACTORS WHICH AFFECT FOOD CONVERSION RATIO (FCR)

IMPROVE IT

- Younger pigs
- Genetically superior strains especially terminal boars /AI
- Nutrient-dense food
- Correct amino-acid/energy balance
- Adequate clean water
- Pigs within their LCT/UCT comfort zone
- Appetite – as food intake rises in fast-growing pigs less food (in % terms) is used for maintenance
- Freedom from stress of many forms
- Sex – entires have better FCR
- Wet feeding in correct environmental conditions
- Well made pellets over meal
- Low immune demand
- Segregation and batch rearing
- More available trace elements in feed
- Good biosecurity protocol
- Growth enhancers

WORSEN IT

- Age
- As fat deposition increases
- Overfeeding energy
- Overfeeding protein
- Too few or unbalanced amino-acids
- Castration
- Overstocking & aggression
- *Ad lib* feeding too long in later growth period when appetite is high
- Pigs of low genetic merit (lean tissue deposition potential poor)
- Cold (seasonal changes a clue). Also cold draughts
- Too dry or too humid air
- Ill health
- High immune demand
- Feed wastage
- Too dilute wet feed/by-product feeds

Further factors which can worsen FCR

- Particle size
- Incorrect feed
- Floor feeding
- Dusty feed
- High gas levels
- Continuous occupation of pens
- Poor cleaning & disinfection
- Poor control of self-feeder and wet/dry feeder settings
- Mycotoxins

DEALING WITH AN APPARENT FCR PROBLEM

A PRELIMINARY CHECKLIST

To save the embarrassment of the adviser finding your figures/assumptions are incorrect, double-check the FCR situation. Common errors discovered on-farm have been...

✓ Input food dockets not matching with output pig numbers or weight.

✓ Bulk bin residues not taken into account.

✓ Surges or dips in pig-flow not taken account of in monthly output averages.

✓ Double-counting of pigs and food deliveries.

✓ Inaccurate recording by staff – cross-check yourself just to be sure. This includes weights of pigs and foods.

✓ If feeding by volume *e.g.* when pipeline-feeding, failing to take account of feed bulk density – see the relevant paragraphs in this chapter.

✓ Computer inputting errors.

✓ Failure to record mortality removals from the tally.

ARE YOU FEEDING WHAT YOU THINK YOU ARE ?

Investigations into complaints of poor FCR have revealed that one in five producers are simply using the wrong – or a less beneficial – diet!

At the time of writing most producers are feeding on the '3-step' principle – Post Weaner (or Starter) / Grower / Finisher – which in future will eventually change to a 5 Phase or even Multiphase system under tighter control *e.g.* automatic dry feeding or fully wet feeding, with pigs raised in batches *i.e.* in one particular environment.

Until that desirable situation occurs it is vital to check that the correct diet is chosen for the class of stock kept, and their age or weight at transfer from one environment to another.

In an ideal world there are four advisors who should be consulted, two of whom should have on-the-spot knowledge of the farm.

The Geneticist : Your breeding stock supplier should be able to give you target nutrient requirements for his genetic lines. This must be used as a basis for dietary design and daily intake (appetite is a critical factor these days when *ad lib* feeding).

But this basic data needs further qualification from:–

The Nutritionist : Pig producers do not consult the person designing their feeds anything like frequently enough. Twice a year is minimal. Only the nutritionist knows the analysis of the raw materials he adds to his feeds and these change. If on-farm mixing is practised he needs to know, from you, what the critical analyses of the on-farm ingredients are, so at least 2% of feed costs must be invested by the farmer in getting the less-established ingredients (especially cereals) analysed.
The nutritionist doesn't need to visit the farm but needs far more information than he is currently being given. I estimate that the average nutritionist, because he is working on assumptions – and this includes overages to protect performance – costs his clients 0.15 on their FCR (6-100 kg) in so doing. This is not his fault – *it happens because he hasn't sufficient dietary analysis information to practise the precision nutrition of which he is capable*.

The Veterinarian : The effect of disease challenge on FCR is underestimated. In my experience it can be as much as 20-30% and this is confirmed by workers like Stahly (1997) (25%). The concept of the drain on nutrients to build a protective immune wall in healthy but challanged pigs is discussed in the checklist section on Immunity.

So the presence of an experienced pig veterinarian reporting back to the feed design team at least twice a year on the likely immune status of the herd is vital, as it rises and falls across time. Where this has been done effectively gross margin has risen by £8/pig for an extra veterinary cost of £1.30/pig, including the tests needed – an REO of 6:1.

The Environmentalist : He is the second expert who should visit any pig farm at least twice a year.

First : To analyse any environmental deficiencies, tell you what they are costing you and to suggest cost-effective improvements so you can prioritize your capital expenditure needs.

Second : To revisit periodically and review progress, suggesting short term adjustments if needs be.

The effect of substandard environment on the FCR of the growing/finishing pig is considerable, and like the problem of matching nutrient intake to immune challenge, is generally underestimated. This is covered in the Growth Rate section of this book, but Tables 3 and 4 give two examples of how poor environment affects FCR.

Table 3 THE COST OF INADEQUATE TEMPERATURE

TOO COLD : Pigs kept 1°C (2°F) below LCT (100 sows' progeny)

	Extra feed/day		Extra feed/year		Extra days to slaughter
Weaners 6 to 20 kg	8 g	0.28 oz	680 kg	1500 lb	2
Growers 20 to 100 kg	25 g	0.88 oz	6000 kg	13228 lb	3.5
Total penalty (if not increased)	Food conversion worsens by 0.03:1		6.68 tonnes (14727 lb) for the herd		5.5 days overheads for each pig

Many pigs are kept 3 to 4°C (5.5-7.2°F) below LCT, especially at night, costing at least 0.1 higher FCR   (6-100 kg) (13-220 lb)
Poor control of temperature costs typically 8 kg of lean meat per tonne of feed (8 kg (17.6 lb) MTF)   (6-100 kg) (13-220 lb)

Table 4 THE COST OF POOR VENTILATION ON PIGS FROM 6-100 kg (13-220 lb)

Depression re:	Not achieving UCT or LCT	Appetite	Stress*	Health	Total
Food conversion	0.1	0.1	0.05?	0.20	0.4 worse
Daily gain, g/day (oz)	30 (1.06)	20 (0.71)	10? (0.35)	50 (1.76)	110 g/day (3.88) slower

*   *Difficult to measure, minimal estimate only. Poor ventilation alone can cost 30 kg (66 lb) MTF (6-100 kg, 13-220 lb) i.e. lean meat per tonne feed.*

REGULAR VISITS ESSENTIAL

Like a veterinarian who can improve MTF in the growout pigs by around 10 kg (22 lb) by minimizing the drag on FCR through better health and advice on more closely matching dietary design to immune needs, the environmentalist can set in train improvements which can provide *another* 15 kg (33 lb) MTF.

Thus at current feed and pig prices, an extra 25 kg (55 lb) of MTF is equivalent to a reduction in all nursery and grower-finisher feed price of 20%. Most of these savings are in lower FCRs.

While a regular supervisory veterinary visit is requested by at least 25% of typical producers, the figure for an environmentalist (to monitor thermodynamics and air movement) world-wide is less than 1%. In my view a serious oversight.

A PLAN OF ACTION

After double-checking the degree of the problem and the facts from which it is calculated, ***contact the qualified nutritionist who designed your feed***. Beware of itinerant feed salespeople and 'specialists'; use them solely as communication links. If you have FCR problems, you need specific scientific advice based on fact and calculations, not 'opinion' or 'sales experience'.

A FOOD CONVERSION CHECKLIST

The 10 Essentials : Provide the nutritionist with the following :

✓   ***Type of diet fed plus expected analyses.***
      He may – or should – ask for check-analyses.

✓   ***Type and method of feeding system used.***

✓   ***Proof of amount fed.*** The nutritionist works on expected daily nutrient intake details, so the bulk density of your feed could be critical, especially if wet feeding. See the section on *bulk density* in this chapter.

✓   *Weight, numbers and ages of the pigs fed* and if segregated by sex.

✓   *Your current daily dietary changeover points* and whether associated with other changes, like housing.

✓   *The genotype(s) used.* This information must come from your breeder/supplier and/or AI source. Only 20% contact their breeder when an FCR problem occurs.

✓   *The health status of the pigs.* This information must come from your veterinarian.

✓   *The environmental details.* An environmentalist would supply these data, but in his absence, the nutritionist needs to have an idea from you of:–

- Floor type; bedded, concrete, slats, etc.
- Wetness of floor or bedding.
- Stocking density allied to weight/age.
- Roof and wall insulation.
- Ventilation adequacy *i.e.* air changes/hr, air speed over the pig's back at the hottest/coldest part of the 24 hrs, air flow directional plan (cross section), typical temperature range. You cannot see air movement. Use smoke generators or tubes to do this.
- In summer, cooling facilities available.
- In winter, likelihood of draughts, especially nocturnal. (Be honest about this; draughts affect overall FCR considerably).
- Likelihood of dust.

✓   *The degree of biosecurity.* This means the method of cleaning, disinfection, mycotoxin and vermin control. Again, assessment can be made from expert biosecurity firms like Antec International and a veterinarian is a help here too.

✓   *Type of feed receptacles, positioning and use.*

✓   *Estimate of wasted food.*

THE FEED'S TO BLAME !

90% of FCR problems will be either wholly or partly in these 'non-feed' areas. The other 10%? Regrettably it does have to be poor quality feed ingredients or spoilation in manufacture. Many farmers, when complaining about poor performance, suspect it to be the other way round. It is not; the 90:10 ratio is about right, in my experience.

Some other factors which can impinge on FCR are:

Deamination

Deamination occurs when the pig has eaten an excess of amino-acids in relation to energy intake and so cannot use all of them. They therefore have to be reprocessed by the pig to be excreted as nitrogen and this takes up food energy. The resultant alteration to the amino acid:energy ratio worsens food conversion efficiency to a surprising extent in pigs which otherwise look healthy enough.

Thus it is essential to get the proportion of essential amino acids to digestible energy level correct according to the pig's age and weight, otherwise FCR suffers.

This is a nutritionist's job. Your job is to ensure you *are* feeding what he thinks you are!

Bulk density of feed

As this can vary substantially in both dry and especially wet feeding, FCR can be affected.

Table 5   DOES FEED VARY IN BULK DENSITY ?

Product	Mean density (g per l)	Range between deliveries
Pig breeding and finishing pellets	646	616 to 700
Pig breeding and finishing meals	479	456 to 497
Cattle nuts	611	532 to 675
Broiler pellets	645	560 to 696
Layer's meals	561	481 to 728

The overriding influences in the case of pressings are the grinding and pelleting conditions: in the case of meals, the choice of raw materials and grinding conditions. (Figures supplied by one feed compounder)

More producers are adopting fully wet-feeding and fail to allow for changes in bulk density. Two vital checks are needed on an ongoing basis.

CHECK THE DRY MATTER CONTENT OF THE LIQUID DILUENT (IF NOT PLAIN WATER)

Both skim milk, but especially whey, can vary considerably from delivery to delivery, with other factory carbohydrate by-products (such as yoghurt) being less of a problem. It is essential to do a simple hygrometer test on each delivery after obtaining from the supplier an expected range of dry matter density, thus securing an average. If this varies by more than 10% batch-on-batch, then errors in predicted or target FCR are likely. In any case, if 10% below expected average, an equivalent price adjustment can be requested for that delivery.

Specific gravity is the weight of a substance compared with the weight of an equal amount of another substance – in this case water. If the specific gravity of water is 1, and if a sample of whey shows an SG of 1.022 this means that whey is 1.022 times heavier than water, the extra weight in the case of whey is primarily the nutrients it contains.

Table 6 shows typical specific gravity differences in whey deliveries across a year and around a SG 1.022 norm.

Table 6 RELATIONSHIP OF SPECIFIC GRAVITY DRY MATTER AND COMPOSITION OF LIQUID WHEY (AS FED BASIS)

Specific Gravity	Dry Matter of Whey (%)	DE (MJ/kg)	Lysine (g/kg)
1.027	6.5	1.07	0.65
1.022	5.0	0.82	0.50
1.018	3.0	0.49	0.30

Liquid whey needs a special balancer solid diet. If the SG of whey drops from 1.022 to 1.018 and a 60 kg pig is fed 7 litres of whey/day, then a typical balancer meal allowance must be increased from 1.6 to 1.75 kg/day otherwise the overall FCR will increase by 0.05.

BULK DENSITY OF DRY FEED, ESPECIALLY PELLETS

The bulk density of feed ingredients varies considerably – we have often seen this when an empty 15 tonne bulk bin barely holds a 12-tonne bulk delivery!

Table 5 shows by how much individual deliveries of bulk meal (wet feeding) and pellets can vary. 5% variation is common and if the food is fed volumetrically either 5% nutrients are wasted, thus affecting F.C.R. by 5%, or 5% are under-supplied, leading to a 5% reduction in daily gain and perhaps a 4% reduction in FCR.

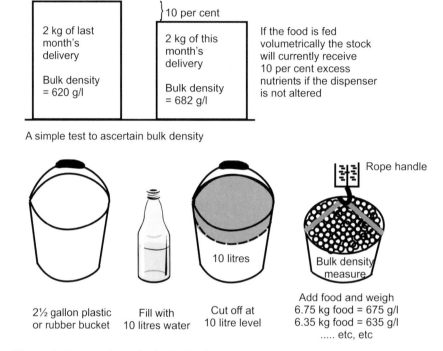

A simple test to ascertain bulk density

Figure 3 How to check for bulk density of dry feed

SINGLE SPACE FEEDER OPERATION

Evidence continues to accumulate on the advantages of correctly-operated wet/dry or single-space feeders over free-access dry feeders. From 18 trials surveyed in the late 90's the benefits lay between a 0.15 and 0.28 FCR improvement, mostly due, it is thought, to less waste.

However, when the single-space feeder is not frequently adjusted to meet individual pens of pigs' feeding style, or differences in feed texture, then the drawbacks can be considerable.

Research done in 1994 (Table 7) suggests that the difference between 'adequate' and 'too little' flow-plate adjustment can be as much as 18 kg MTF (30-100 kg, 66-220 lb), equivalent to a 14% rise in feed price at the time. The difference in FCR was 0.23, or 6.6% worse, showing that careless or lazy attention to single space feeder settings can completely negate the economic advantage of the wet/dry feeder over the simpler and cheaper dry feeder.

The correct setting depends on the design, so consult the manufacturer for guidance. In general each pen of pigs should be made to 'work' for food, and this extra action does not seem to carry a performance penalty at all.

Table 7   THE EFFECT OF FEED SETTINGS ON PIG PERFORMANCE

	Low	*Medium*	*High*
Food intake, kg/day (lb/day)	1.97 (4.33)	2.14 (4.71)	2.21 (4.86)
Liveweight gain, g/day (lb /day)	727 (1.60)	797 (1.75)	845 (1.86)
Food conversion efficiency			
of carcass gain	3.70	3.58	3.47
Backfat thickness at P_2, mm (in)	10.6 (0.42)	11.1 (0.44)	12.1 (0.48)

From Walker & Morrow (1994)

The effect of poor feeder setting on pig behaviour and pig performance is given in Table 8 also from Walker and Morrow's work and Table 9, extrapolated from the same research.

Table 8   THE EFFECT OF FEEDER SETTINGS ON PIG BEHAVIOR

	Low	*Medium*	*High*
N°. of feeder entries/pigs/24 hr	51.5	45.6	42.2
Feeding time/pig/24 hr (minutes)	110	78	87
Queuing incidents/pig/24 hr	70	45	26

Table 9   THE PENALTY OF INCORRECT FEEDER ADJUSTMENT (PIGS 30 – 100 kg) (66-220 lb)

Feeder gap	*Too restricted*	*Ideal*	*Too Generous*
Days to 100 kg (220lb)	95.3	87.88	81.3
Food eaten/pig (kg) (lb)	190 (4.18)	188 (414)	207 (455)
MTF (kg) (lb)	270 (594)	279 (614)	248 (546)
Av. backfat at P_2 (mm) (in)	10.4 (0.41)	11.0 (0.43)	13.2 (0.52)

Figure 4, from video-recorded data, shows where pigs tend to rest or congregate in a pen with a single space feeder set laterally. Compare this activity pattern with the diagrams in the stocking density section.

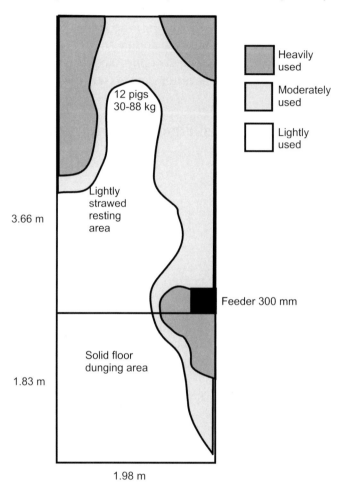

Video'd activity as:

	Mean %	Range %
Asleep/dozing	4.95	20-80
Feeding/drinking	9.9	4 - 18.4
Trying to feed	2.6	0 - 15
Social activity	30.2	6 - 48.5
Fighting/playing	7.8	0 - 25

Figure 4 Good single space feeder pen

FURTHER READING

For information on other factors which affect food conversion, check out the sections on:–

- Stocking density.
- Temperature, air movement and draughts (Growth Rate and Ventilation sections).
- Dust and gases.
- Biosecurity (cleaning, disinfection and mycotoxins).
- Stress and stressors.
- The post-weaning check.

REFERENCES

Walker, N. and Morrow, A. (1994) Some observations on single space hopper feeders for finishing pigs. Hillsborough Research Station Report. Quoted in *NAC Pig Unit Review*.

STOCKING DENSITY 19

The minimum amount of space required by pigs so as not to restrain performance or aggravate aggression, and maintain welfare.

A COMPLEX SUBJECT

The space requirements of pigs can be divided into various groups :

Body-occupation space	-	a pig lying down takes more space than one standing up, a supine pig takes up more space than one in the semi-sternum position.
Body-activity space	-	the space required by body posture changes, like getting up or down, lying supine, turning round/grooming itself.
Social space	-	the space required for socialisation with other pigs or access by stockpersons. An important part of this – often undervalued – is 'fleeing space' to mitigate aggression.
System space	-	the space required by different management systems *e.g.* straw v. slats, gestation stalls, groups in yards, wet feeding v. dry hoppers, etc.
'Dead' space	-	the space required for partitions, passages, corners and pen furniture.

Thus the ideal commercially viable space requirement of any pig is complex and variable. Most of the categories above are two-dimensional, but three-dimensional space – volumetric space – should be borne in mind as low ceilings often predispose towards respiratory disease and may merit fewer pigs housed per square metre (or the use of improved air flow) to combat the build-up of infection.

TARGETS

SPACE REQUIREMENTS FOR GROWING FINISHING PIGS

(for creep-housed piglets see p330; for breeding stock see pages 300 and 301, at the end of this section.)

One has to start somewhere, and Table 1 is based on advice from several countries.

Table 1 RECOMMENDED SLATTED (LEFT) AND STRAW-BASED (RIGHT) SPACE ALLOWANCES FOR NURSERY-TO-FINISH PIGS. COMPILED FROM VARIOUS SOURCES WORLD-WIDE. PER PIG HOUSED

| Slatted & partially slatted/dung scraped | | | | Straw-based* | |
kg	lb	m_2	ft_2	m_2	ft_2
5	11	0.1	1.1	0.25	2.7
10	22	0.15	1.6	0.4	4.3
15	33	0.175	1.9	0.45	4.8
20	44	0.2	2.2	0.5	5.4
25	55	0.25	2.7	0.75	8.1
30	66	0.3	3.3	0.8	8.6
35	77	0.325	3.5	0.95	10.2
40	88	0.35	3.8	1.0	10.8
45	99	0.375	4.0	1.1	11.8
50	110	0.4	4.3	1.2	12.9
55	121	0.425	4.6	1.25	13.5
60	132	0.45	4.8	1.3	14.0
65	143	0.475	5.1	1.4	15.1
70	154	0.5	5.4	1.5	16.1
75	165	0.525	5.7	1.55	16.7
80	176	0.55	5.9	1.6	17.2
85	187	0.575	6.2	1.65	17.8
90	198	0.6	6.5	1.7	18.3
95	209	0.65	7.0	1.85	20.0
100	220	0.7	7.5	2.0	21.5
105	232	0.72	7.75	2.1	22.6
110**	243	0.74	8.0	2.2	23.7
115	254	0.75	8.1	2.25	24.2
120	265	0.76	8.2	2.3	24.8

Notes

- *Assumes pens are no longer than 2½ x 2, trough space not less than 100 mm (4 in)/pig and not more than 20 pigs per single-space feeder.*
- *Many producers on solid floors with no or minimal bedding would allow 10% more space to accommodate ease of dung removal (except for Straw Flow designs).*
- *Some research suggests no improvement in performance was obtained from the same space allowances on straw and on slats. However, in the author's experience this is not tenable in many or even most farm circumstances, and deep-bedded straw-based pigs in groups should have at least twice to up to 3 times the advised spatial levels for those on slatted floors.*

* Minimum depth 10 cm (4") ** UK Welfare Regulations require 1 m^2 (10.8 ft^2) per pig over 110 kg.

WELFARE MINIMUM STANDARDS

The guidelines in *Table 1* can be considered just that – guidelines. They match or exceed the Minimum Welfare Standards which are already in place in several countries like the U.K., Sweden, Australia, Denmark and Canada. For example the U.K. Welfare of Livestock Regulations (1994) implement the E.C. Directive 91/630 which lays down minimum standards for the protection of pigs are as follows:–

Annex Schedule 3 of the U.K. Welfare & Livestock Regulations (1994)

This states ...

The unobstructed floor area available to each weaner or rearing pig reared in a group must be at least:

- 0.15 square metres for each pig where the average weight of the pigs in the group is 10 kg or less.

- 0.20 square metres for each pig where the average weight of the pigs in the group is more than 10 kg but less than or equal to 20 kg.

- 0.30 square metres for each pig where the average weight of the pigs in the group is more than 20 kg but less than or equal to 30 kg.

- 0.40 square metres for each pig where the average weight of the pigs in the group is more than 30 kg but less than or equal to 50 kg.

- 0.55 square metres for each pig where the average weight of the pigs in the group is more than 50 kg but less than or equal to 85 kg.

- 0.65 square metres for each pig where the average weight of the pigs in the group is more than 85 kg but less than or equal to 110 kg.

- 1.00 square metres for each pig where the average weight of the pigs in the group is more than 110 kg.

This method of expression in bands or 'jumps' has encountered criticism in being too rigid and "Is at odds with the fact that pigs grow continuously and not in steps. It is illogical that a pig weighing 20 kg needs 0.2 m²/pig while a pig weighing 21 kg suddenly needs 0.3 m²." (Morgan, 1997)

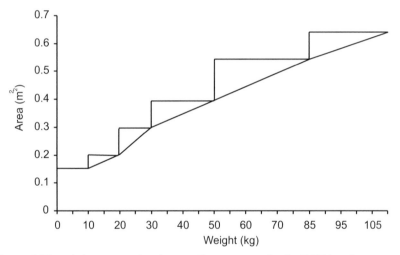

Figure 1 The minimum permitted space allowances under the EC Directive compared to the average expected weight of the pigs in a group. This is depicted as a rising line.

There are also economic drawbacks to this method – a stepped system means that in order to comply with the Regulations, producers may have to move pigs regardless of whether their routines and pen sizes allow them to do so.

For example : Under the Welfare Regulations, if pigs stay in a pen to any weight over 50 kg they must then be given the same space as 85 kg pigs. If pigs are normally taken out of a pen at 55 kg (a little over the 50 kg mandate of 0.4 m²/pig *i.e.*, say 0.42 m²/pig) then this 0.42 m²/pig allowance is adequate. But the Directive says each such pig must be given 0.55 m²/pig – a difference of 0.13 m²/pig. This would be an extra capital cost of £26/pig ($42) if new pens are needed to satisfy this destocking level below the needs from a graphical curve rather than a stepped system. On some units operating an all-in/all-out policy the management system could be disrupted and group sizes altered. In addition, such unnecessary destocking could increase the need for heat input in cold weather or at night which, if not done, could then compromise the pig's welfare or health!

HOW MUCH DOES OVERSTOCKING COST?

In the 1980's I recorded the pen measurements and stocking densities of all the grower/finisher houses I entered. In general some 38% were overstocked by 15% or more, which is putting 14 pigs into a pen designed to hold only 12 – obviously quite an easy thing to do.

We then carried out a carefully measured trial on three farms where we deliberately destocked half the pens on each farm to the correct densities as listed in Table 1. One farm had spare accommodation to take the surplus, and it was summer in the case of the other two, so these surplus pigs could go into yards or outside kennels.

Table 2 shows a typical result.

Table 2   LIKELY COSTS INCURRED BY OVERSTOCKING A NURSERY AND A FINISHING HOUSE BY 15 PER CENT

| | Pigs 6-35 kg | | Pigs 36-100 kg | |
	Correct density	*+15 per cent*	*Correct density*	*+15 per cent*
Daily gain, g (lb)	518 (1.14)	480 (1.05)	844 (1.85)	848 (1.86)
Days in pen	56	60	77	77
Overhead costs @ 16p/day (£)	8.96	9.60	12.32	12.32*
FCR	2.02	2.12	2.42	2.63
Total food eaten in period (kg)	58.6	61.5	157.3	171.0
Total food cost (24p and 17.2p/kg) (£)	14.06	14.76	27.06	29.41
Extra costs/pig (£)	*1.34 plus*	*2.35*	*Total 3.69*	

Savings in 15% less housing cost per pig (at £4.26/pig)**Savings 64p ($1.03)**
Costs £3.69 ($5.94)

The average REO for deliberately destocking to guideline levels on all three farms was a fraction under 6:1.

Factors affecting stocking density in growing pigs

Check the following …

Pen shape : Long and narrow versus square. A ratio of 2 to 1½ is satisfactory, and 3:1 is not good if the shorter length comprises the feeding or dunging space. This is often seen in wet-fed pens where tailbiting can arise.

Temperature : Allow for 15% more space in hot weather (24°C or 75°F) dependent on airflow and cooling devices. An example is that Boon (NIAE) showed that given the option, the lying space occupied by growing pigs increases by some 15% as air temperatures rises by 6°C from their LCT.

*Draughts :*A pen with adequate space allowances but which causes pigs to huddle to keep warm can cause the same drag on performance as overcrowding.

Feeder space : Penny (JSR Genetics, 2000) reports … "Offering pigs additional feed area can override the negative effects of reducing floor space. Looking at the performance of 396 grower pigs from 20-40 kg (44-88 lb) over a 28 day trial period, reducing pen space from 0.4 m^2 per pig to 0.3m^2 (4.3 to 3.2 ft^2), both with 50 mm (2 in) per pig feeding space allowance, can affect performance." . . . "A decrease in floor area results in poorer liveweight gains and lower feed intake, but this can be overcome by offering more feeding opportunities. Raising feed areas to 100 mm (4 in) per pig can override many negative responses from increasing stocking density."

Solid floor dunging : sleeping space ratio

If the pig's sleeping area leaves room only for the dunging passage space/slatted area, the pigs are technically overstocked. In the author's opinion the ratio of sleeping, plus socializing/feeding to dunging space should not be less than 3:1, or 25% more than the resting area alone. This is borne out by Edwards (1987) who writes:– "An experiment (was effected) where the size of pen was changed each week in relation to the weight of the pigs. At the lowest space allowance the pigs were given only as much space as was necessary for them all to lie down. This was compared with three progressively more generous space allowances, giving 12, 25 or 42% additional area". The results are shown in Table 3.

Table 3 EFFECT OF DIFFERENT SPACE ALLOWANCES ON PIG PERFORMANCES

| | Lying space only | Other space allowance | | |
		+ 12%	+25%	+42%
Daily gain, g (lb)	844 (1.85)	862 (1.89)	882 (1.94)	897 (1.97)
Feed conversion	2.70	2.56	2.60	2.59

(from Edwards, Armsby & Spechter, 1987)

Even with constant group size, good temperature control, *ad libitum* feeding with plentiful hopper space and ready access to drinkers, individual pig performance improved as space allowance increased.

When the economics were examined, it was found that the higher feed cost per kg of carcase produced more than offset any saving in housing cost at the lowest space allowance. Since this trial was carried out in such good conditions it is likely that even greater effects would be seen in day-to-day practice on commercial farms."

Edwards recommends +25% as an advised level.

Pen shape

Many dung passages are not broad or deep enough as this saves construction and cleaning costs. This is especially true of some wet feed pen designs where pens are much too long and narrow. The 5 to 7:1 length/breadth ratios allow a large number of pigs to feed at one time but pens remain wet from misplaced voidings and wet-feed spillage, so the pigs lie uncomfortably/lose more body heat to the floor.
This is why wet-fed pens to the Suffolk design (trough to a broad front) tend to give better results than the traditional long/narrow design with the wet feed trough to one or both sides. However Suffolk pens tend to be 18-24% more expensive in capital and running costs, so such long-term housing costs money.

Uneven pigs. Differences in bodyweight of more than 3 to 5% per group may need destocking. However pigs are very adaptable; but providing greater spatial allowance during such disparities will help minimize check to growth. In nurseries the author finds a 15% easement beneficial. In older pigs the provision of screener boards may help but only makes matters worse in overstocked pens where fleeing space/ hiding ability is constrained as a result.

Pen furniture

In pen groups which follow the minimum stocking densities outlined in Table 1, careful positioning of pen furniture will help reduce aggression and improve performance. Figure 2 shows two pen layouts using a single feeding point.

A
Incorrect (restless) pen design?

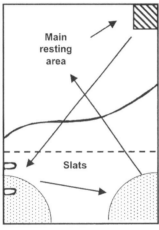

B
Pen design causing less aggression

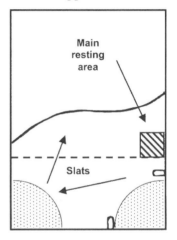

1. Feeder causes disturbance inside resting area.
2. Drinkers on wrong side of pen / too close together / too close to corner which is elimination area.
3. Pigs cross routes – more aggression

Rotary pattern established causing less disturbance, less aggression.

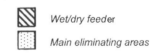

Wet/dry feeder
Main eliminating areas

The arrows refer to pig movement sequences

Figure 2 Even if stocking density and pen shape are correct - the positioning of pen furniture can calm or inflame agression.

Table 4 illustrates what happened when the remainder of a new finishing house was built to the left hand in place of the preferred right hand design (to make hand-filling the feeders easier).

Table 4   PERFORMANCE COMPARISON OF PENS A AND B (12 PER SIDE)

(*All-in/all-out*)	(A)	(B)
Daily Gain, g (35-90 kg, 77-198 lb)	567 (1.24 lb)	608 (1.33 lb)
LFCR (35-90 kg)	3.0	2.86
MTF (kg)	394 (869 lb)	416* (917 lb)
Stockman's estimate of pig's time spent resting during daylight hours	*c.*60%-70%	*c.*80%-85%

*The extra 22 kg (48 lb) meat sold/tonne feed was equivalent to a 14% reduction in feed cost at the time.

Misconceptions about Stocking Density

"Pigs aggress more when overstocked."

Not necessarily. Pigs can adapt to overstocking and recover performance levels given time.

Research suggests that up to a certain degree of crowding all measures move in the expected direction and then a reverse trend occurs. The animals seem to adapt to over-crowding. Even so, there is always performance loss (or worse, *i.e.* tail-biting) until they do.

"Giving pigs more room improves performance."

This depends on how overstocked the pigs were initially. Several trials have shown that while LFCR and ADG improved between decreased stocking densities of, say, 0.8m²/pig and 0.5m²/pig (8.6 and 5.4ft²), ***the meat sold per house was greater at the higher-stocked level***. Always measure physical performance on an MTF basis as a counter-check to physical performance, as it is profit that counts (Table 5).

Table 5   EFFECT OF DIFFERENT AVERAGE PEN SPACE PER PIG PER HOUSE

Mean space allowance/pig		*Daily gain* (*g*)	(*lb*)	*Feed intake* (*kg*)	(*lb*)	*F.C.R.*	*Weight of meat sold per house* (*Tonnes*)
0.49m²	5.27ft²	640	1.4	2.23	4.9	3.49	125
0.66m²	7.10ft²	681	1.5	2.31	5.1	3.38	98
0.81m²	8.71ft²	731	1.6	2.39	5.3	3.31	82

Source : Powell & Brown (1996)

"Stress, as measured by plasma cortisol, increases with overcrowding."

It often doesn't. Again, the pig is very adaptable. Stress is very difficult to measure scientifically.

"Antibiotic (in fact all) growth promoters work better when pigs are overcrowded."

Research suggests the effects are similar.

"Overstocking is a behavioural problem."

Sure, stress must play its part. Much more likely is that overcrowded pigs eat less, are indeed more stressed which may or may not show behaviourally and the transmission of disease is increased, possibly by a reduction in immune status. The combination of less feed intake with a greater and altered demand for nutrients due to the need to raise the immune defences in overstocked pigs combine to give the performance reduction. Also, once the immune system readjusts/adapts, the performance may recover even though the stocking density remains high.

Correct stocking density and understocking.

Some people suggest the deliberate understocking (within reason), while increasing the cost per m², improves performance to such an extent that income is actually increased. In my experience this happens only rarely.

Look again at the Canadian work in Table 5 above. The extra liveweight produced on the 'accepted' space allowance of 0.49m²/pig (5.3ft²) which in fact still gives a 22.5% overage from the 0.4m²/pig (4.3ft²) minimum welfare standard on an average herd weight of 50 kg (5-100 kg) – is dramatic. Compared to 0.81m²/pig (8.7ft²) but using the following reasonable assumptions and basing the performance comparisons on an econometric basis, Table 6, this time expressed in costs per m², reveals a much less dramatic picture, even though the conclusion is the same – it does not pay, in general, to understock. In fact any more than it does to overstock (*see* Table 2).

Table 6 shows how much it pays, under good management to adhere to the correct target stocking density minimum quoted in Table 1.

Table 6   THE ECONOMICS OF UNDERSTOCKING

Using the following assumptions based on current U.K. economics, the performance results from Powell & Brunn's work reveals the following ...

<u>Assumptions</u> : *1000 pig spaces, 2.31 batches year (5-100 kg; 11-221 lb). 4 days for batch cleandown and shipping lag. 72% dressed carcase weight. Av. food cost 15p/kg. Av. overhead costs 16p/day. Pig price (deadweight dcw) 110p/ kg. Housing cost £23/m². Maximum space available for production 490 m².*

	Savings/m² from improved performance			Costs of lower throughput/m²		
	From food	*From overheads*	*Total/m²*	*Sales/m²*	*Difference*	*Nett shortfall/m²*
0.49m² (5.27ft²)	–	–	–	£358	–	–
0.66m² (7.10ft²)	£ 9.13	£ 2.24	£11.37	£233	£125	- £113.63 (-32%)
0.81m² (8.71ft²)	£14.00	£ 8.21	£22.21	£127	£231	- £208.79 (-58%)

Conclusion Despite improved performance, radical destocking to 35%-65% below the recommended levels given in Table 1 could prove extremely expensive in lower throughput from the more generous space allowances quoted.

Variable geometry – the crusher board

The crusher board is cheap, flexible, is simple to use, easy to clean and the need for adjustment makes the stockperson look at the comfort and condition of their pigs more frequently.

It is used as a simple divider board in a passageway creep area (to discourage the newborn piglets from wrong-mucking in a heated creep) right up to pigs as heavy as 65 kg/143 lb (to provide a rising scale of cost-effective area per square metre of space available).

Moreover, it cuts down the workload; pens don't get fouled so frequently. In technical terms:

• By maintaining the correct social space, it keeps young pigs warmer (many pigs are still kept below their LCT in winter or at night).

• So they grow faster.

- It keeps pens cleaner (as young pigs with too much room may dung at the back and too-hot pigs will dung/urinate around themselves to provide an indoor wallow).

- So they are less stressed, less prone to aggressive behaviour and thus convert better.

Some suggestions:

The most effective housing in which to use the crusher board idea is the kennel or bungalow, especially if the design is long and narrow (like the monopitches) or wide and shallow (like the conventional Suffolk design – not the 'zig-zag' version – and especially the wet-feeding variety with a long outside trough). Conventionally one stocks these simple pens at 30 pigs and thins down to 20's or 15/16's. Even so, for 40 per cent of the time the pigs are either understocked or overstocked to a greater or lesser degree – hard on the pigs at some times, or on the pocket at others.

- Even so, crusher boards do not allow you to put more pigs into what space you have available but they help you manage the temperature better and more evenly. Thus for reasonably low-lidded pens a laterally-moved crusher board should reach up to the lid itself so as to cut down on the air volume. This allows for a reduction in the air volume to be circulated as well as reducing the floor space to be occupied by the young pigs. This assumes correct air movement within the air space provided.

- If the roof line is high (and for young pigs it shouldn't be, allowing for ease of inspection) then a rick-sheet batten-frame slotted in to the upper part of the crusher board cuts down on weight and cost markedly – as long as it is out of the pig's reach, of course.

- Always locate the board to the pen division, but use hooks or pegs only on the *board*, and round eyes or recessed lugs on the *pen wall*, in order to avoid damage to the pigs' skin as the board moves back up the pen leaving the fixtures exposed.

- Do not use the board to such an extent that late-to-bed pigs are forced to lie right in a kennel doorway and thus are exposed to a night-time draught. Move it one notch backwards to accommodate them and/or use flexiflaps in the pophole itself.

- Try not to angle the board diagonally across the pen – pigs may dung in any acute-angle deadspace you create. A right angle allied to a correctly measured space allowance is best.

Crusher boards can be templated to fixed equipment, like troughs and hopper lengths, ***provided the inside of the feeder or trough is also shuttered off.*** This is not onerous or expensive; for example a wet feed trough can have a moulded moveable concrete plug (with handle) placed at the junction with the crusher board. Stale feed negates the whole exercise, so seal off the unused portion of any feed receptacle. Finally, a plea to the housing manufacturers. Could you design, as an optional-extra, a crusher board device into each pen/kennel division? Only a very few in the world seem to do so. And also give more thought to variable-geometry pen divisions? We haven't explored this cost-saving idea nearly enough. With intensive housing costs now rising past 15% of the cost of producing a finished pig, we must attempt to make better use of spatial investment.

Exercise : 25 x 15 kg (33 lb) second stage weaners to be thinned down at 25 kg (55 lb) to 12's, then grown on to 70 kg (154 lb) if required. Pens are 3m x 2m (say 10ft x 6.6ft).

Using Table 1, 25 x 15 kg growers need 25 x 0.175 = 4.375m² (say 47ft²) on entry and will occupy approx 75% of the space available, so place the crusher board at 1.5m (say 5ft) from the side wall (Figure 3) and reduce progressively until all the 6m² (65ft²) space is occupied, in this case at 25 kg (55 lb).

Then (Figure 4) thin down to 12's or 15's, the former allowing occupation until 70 kg (154 lb), the latter to 50 kg (110 lb).

In the case of 12's the crusher board, after cleaning both it and the pen thoroughly, is positioned at 3 metres (say 10ft) from the side wall, and in the case of pens of 15, at 2 m (or 6½ft) (as approximately only 2/3rds of the initial space is taken up) and again moved back progressively until all the 6m² (65ft²) space is taken up at 50 kg (110 lb)liveweight – when the group is rehoused in the grow-out facility.

Value of crusher boards

Table 7 gives an indication of the value of a simple crusher board used in nursery kennels.

How to use crusher boards

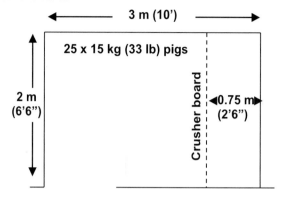

Figure 3 Position of crusher board on entry of weaner pigs.

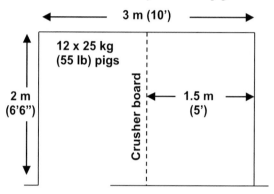

Figure 4 Crusher boards re-introduced on thinning down. In squarer shaped pens crusher boards can run either way - up or down/side-to-side.
Conversion: 1 metre = 3.281ft

Table 7 PERFORMANCE AND ECONOMETRICS TO SLAUGHTER FROM USING CRUSHER BOARDS (12-88 kg/26-194 lb)

	Before		*After*	
LFCR to slaughter	2.94		2.91	
ADG to slaughter, g (lb)	613	(1.35)	631	(1.38)
MTF, kg (lb)	291	(642)	293	(646)
Value of extra meat and saved overheads (2 days) (per pig)	–		£0.89	($1.43)

Pigs put through crusher board regime/year 820
Extra value from use of crusher boards 820 x £0.89 = £730 ($1175)
Cost of 10 crusher boards (home-made) £400 ($644)
Payback therefore 0.54 years – under 7 months.

STOCKING DENSITY FOR BREEDING STOCK - GILTS

Most gilts are selected or bought-in from 85 to 100 kg (187-220 lb) and placed into a wide variety of housing varying from kennelled yards on concrete to bedded straw in groups.

For examples of pen design (plans) see Brent, "*Housing the Pig*" (1986), a textbook which has stood the test of time and experience excellently. Numbers vary, usually from 4 in a pen to 20 in a yard; I prefer a maximum of 6 housed together.

There are a number of basic precepts to follow so as to allow the animals to settle-in rapidly, avoid injuring each other and ensure good signs of oestrus.

CHECKLIST ON GILT ACCOMMODATION SPACE REQUIREMENTS

✓ No gilt pen should have a side less than 2m long (6.6ft).

✓ Lying area alone for new entrants at 85-100 kg (187-220 lb) should be not less 0.6m²/gilts (6.5ft²).

✓ If kept in groups until first farrowing allow 1m²/gilt (10.8ft²) lying area.

✓ Many tractor-scraped solid dunging areas allow a 1:1 ratio with the resting area: this itself is generous but very satisfactory.

✓ If the pen includes a slatted dunging area, this must be 25% of the lying area given above, and this slatted area have not less than a 1m side (3ft 3 in). However, from experience, the author finds a minimal total area of 2.8m²/gilt (as for sows) to be preferable (30ft²).

✓ Providing individual feeder spaces is desirable. If so their width will depend on the body shape, especially the shoulder width, of the genotype chosen, but generally 450 mm to 540 mm (17.7 to 21.3 inches) is adequate. Gilts will get larger in future, as longer induction times catch on.

✓   Gilts may have to be moved to the boar. If so, doorways must be a minimum of 900 mm (3ft) and have rubber or plastic corner-protectors if made from brick or block walls. Passageways should be 1.2m (4ft).

✓   Wherever possible do not overcrowd gilts – remember the 2.8m²/animal advised (30ft²). Allow adequate 'fleeing-space', this helps reduce aggression and makes first service easier and more effective.

SOWS

As with gilts the range of satisfactory accommodation for sows is enormous – and this includes gestation stalls and the mistakenly maligned farrowing crate!

Table 8 may help breeders who require some minimal standards where considering building alterations and design. Those marked * are taken from the *UK Pig Animal Welfare Group* publications listed below.

TARGETS - SOWS

Table 8 SOME BASIC TOTAL SPACE ALLOWANCES – SOWS

		m²/animal	*Approx ft²/animal*
	Sows on slats, loose-housed	2.8	30
	Stalled sow	1.5	16
PAWG N°.4	*Cubicle & free-access stalls	2.3 to 2.9	25-31
PAWG N°.5	Yards and Individual Feeders	3.26 to 3.73	35-40
	Farrowing crate & pen	4.6 to 4.8	50-52
PAWG N°.6	*Trickle fed or wet fed, yards or kennels	2.3 to 2.79	25-31
PAWG N°.7	*Floor fed yards or kennels	2.33 to 3.73	25-40
PAWG N°.9	Single Yard Electronic Sow Feeders	2.66 to 3.18	29-34
PAWG N°.9	Twin Yard Electronic Sow Feeders	2.7	29
	Group mixing pen	3.5	38
	Outdoor Sows	15 to 20 sows per hectare	37-50 per hectare

For further information see Mixing Pigs section.

BOARS & SERVICE ACCOMMODATION SPACE REQUIREMENTS

Many boars are badly housed in pens which are too small, badly positioned, too cold and uncomfortable.

Table 9 SPACE ALLOWANCE TARGETS - BOARS

- ✓ Minimal area per boar 7.5m² (81ft²).
- ✓ Preferred area per boar 9.0m² (97ft²).
- ✓ Sides should preferably be 3m long; with a minimal height of 1.5m (5ft).
- ✓ Service pen (and boar accommodation if combined with a service pen, not advisable due to slipperiness) to allow adequate movement and minimize abrasion – 10.56m², i.e. 3.25m x 3.25m (114ft²; i.e. 10.7 x 10.7ft).
- ✓ Allow for 350mm (14") gussetted corners, *i.e.* blanked-off corners, not a 90° right angle.
- ✓ Boar pens should preferably be fitted with a stockperson's escape pole, which duals as a boar rubbing post. This should be close enough to the corner not to allow the boar to get jammed in.

BABY PIGLETS

For forward creep areas and their design, see the *Wrong Mucking Section*.

REFERENCES

Brent, G. (1986) *Housing the Pig*. Farming Press (UK).

Edwards, S., Armsby, A. and Spechter, M. (1988) Effects of floor area on performance of growing pigs kept on fully-slatted floors. *Animal Production*. **46**, 553-459.

Morgan, D. (1997) Beware, the Devil is in the Detail... **Pig Farming**, 16-17.

Penny, P. (1999) JSR Genetics Technical Conference. Press Reports.

PROBLEMS WITH HOT WEATHER 20

Heat Stress: The body temperature of the pig must remain between certain limits to maximise production, safeguard welfare and resist disease.

When the pig's temperature rises beyond the upper limit it starts to pant. This is known as the Evaporative Critical Temperature (ECT). Panting increases the evaporative heat loss from the lungs as the pig starts to try to control its body temperature. Normally the pig breathes at 20 to 30 breaths per minute. ECT is estimated at around 50 to 60 breaths/minute, but this can rise to 200 breaths/minute as it crosses towards UCT – see below. The start of panting is a good indicator of heat stress and is a definite and urgent action level.

Upper Critical Temperature (UCT) is generally taken as being approximately 3 to 5°C above the warning signs of ECT.

Crossing beyond the UCT threshold immediately affects the pig's metabolism, seriously reduces appetite, lowers production and can jeopardise its defences against disease; it also compromises it's welfare.

As with LCT (Lower Critical Temperature) both ECT and UCT can vary across a 5 to 6°C ambient range dependent on, for example, the pig's age and weight, fat-cover, food and type of energy intake, protection from solar radiation, floor type, airspeed and skin wetness.

Table 1 gives an approximation of ECT and UCT for the different conditions shown.

Advice: I advise the adoption of ECT as a warning threshold and UCT as crossing a definite danger boundary.

Table 1   THE RELATIONSHIP BETWEEN ECT AND UCT UNDER DIFFERING CONDITIONS

Stock	Age (wks)	Wt (kg)	(lb)	Floor type	Air speed (m/sec)	Energy intake (MJ/d)	Skin* wetness (%)	Ambient temperature (°C) ECT	UCT	(°F) ECT	UCT
Piglets	1	2	4.4	Mesh	Still	ad lib	15	35	41	95	106
	4	5	11					33	39	91	102
Weaners	5	7	15	Mesh	0.1	ad lib	15	35	41	95	106
	6	10	22					33	39	91	102
	8	16	35					30	37	86	99
	5	7	15	Concrete	0.1	ad lib	15	36	42	97	108
	6	10	22					34	40	93	104
	8	16	35					31	38	88	100
Growers	9	20	44	Concrete	0.1	ad lib	15	30	38	86	100
	15	50	110					28	36	82	97
	21	90	199					27	36	81	97
	9	20	44	Concrete	0.5	ad lib	15	32	39	90	102
	15	50	110					30	37	86	99
	21	90	199					29	36	84	97
	9	20	44	Concrete	0.5	ad lib	60	34	42	93	108
	15	50	110	& spray				32	40	98	104
	21	90	199					31	39	88	102
Dry Sows		150	331	Concrete	Still	27	15	27	36	81	97
				(one sow)	0.3	27	15	29	38	84	100
					+ spray	27	60	33	40	91	104
				Concrete	Still	27	15	26	35	79	95
				(group of	0.3	27	15	28	38	82	100
				5 sows)	+ spray	27	60	32	40	90	104
Lac. sows		150	331	Mesh	Still	ad lib	15	22	32	72	90
					Still+drip		30	26	33	79	91
				Concrete	Still	ad lib	15	23	33	73	91
					Still+drip		30	25	34	77	93

Note : •   This table is not a set of recommendations. Use the values above as guidelines only and carefully observe pig behaviour.

•   UCT data are estimates only. Beware of air temperatures approaching these UCTs. They signal danger; deaths could occur at temperatures at or above them.

*Skin Wetness   15% is normal wetness due to drinkers.

30% is typical wetness of pigs during drip cooling.

60% is an average wetness of pigs during spray cooling.

Based on advice from the Australian Pig Research & Development Cooperation, (who are arguably the world's leading experts on hot weather pig production) and from other sources.

INITIAL CHECKLIST
HOT WEATHER STRESS

✓ ECT (panting) is a warning sign of the commencement of profit loss.

✓ Check the ambient temperature over the pig with a clean thermometer.

✓ Consider the list of recommended actions show in the following tables.

Remember

If the temperature climbs 2 to 3°C above that measured when you notice panting, acute performance loss is likely. At UCT your pigs may enter the death zone. See Table 1. UCT is made much of by ag. environmentalists, however, *I consider ECT to be a more useful threshold than UCT* for two reasons:–

✓ ECT can often be identified by panting and respiration rate encouraging you (in fact instructing you) *to take remedial action at once*. Many times I have passed panting pigs at 60-80 breaths/min and the stockman did nothing. On instructions to wet the animal(s) or redirect airflow *now*, the stockman said "Do you mean *now*?" The answer is "Certainly. UCT is approaching or has been crossed and the danger point has arrived. Do not delay one minute."

✓ If the pigs have reached or exceeded UCT it could already be too late. Their metabolism is compromised and may have been damaged, taking some while to repair. Loss of productivity has been suffered, which in certain places in the breeding cycle for both boars and gilts/sows can be severe, *i.e.* litter size and farrowing rate in particular.

HEAT LOSS

The pig can to a certain extend defend itself against the effects of heat stress by the following:–

Radiation (radiant energy): **The emission of energy waves from a source**, *e.g.* heat from a roof heated by the sun then striking the pig's body surface. Conversely the pig's body surface can radiate energy to the air around it which eventually warms the surfaces (roof, walls) around it, thus losing body heat.

Radiation typically accounts for 20% of the pig's heat loss in hot-weather – but if the building surface temperatures are above that of the animal, there will be a net heat gain.

Convection : **The rise and removal of warm air from around a hot surface**. This is helped by moving the air over the pig's skin surface – as long as the air moves fast enough to dislodge the layer of still air which lies close the skin. This is an effective means of cooling; but without wetting only if the air velocity is at least 1m/sec and the temperature of the air 3°C below the pig's normal body temperature of 38.9°C (102°F). Thus if the ambient temperature is between 26°C to 33°C the pig can dissipate about 30% of its body heat to vigorous dry air movement of this standard.

Evaporation : **The conversion of liquid into vapour**. As it is the fastest moving molecules of heated liquid (in this case water) which escape from the pig's hot skin surface, the stored (kinetic) energy of the remaining water molecules is reduced, and therefore evaporation causes cooling. Air movement increases evaporative loss, so it is the combination **of air speed and wetting** which contributes most to skin cooling not just air movement alone, or wetting alone.

Pigs can create their own evaporative heat loss by panting, wallowing and pen soiling, which are all early indicators of ECT.

For every 100 cc of water evaporated, 220 Btu of heat is required. This means that in an ambient temperature of 27°C (81°F), panting can in theory account for about 40% of a growing pig's heat loss. However considerable and prolonged rapid respiration is needed to evaporate 100 cc of lung moisture, which is why skin wetting is preferable.

Conduction : **Heat transmission from places of higher to places of lower temperature**. The pig can alter its posture to provide greater contact with a cooler or wetter floor surface. But don't overrate it; conduction usually only accounts for 5% to 10% of the heat loss in hot weather as only 20% of the animal's skin can be in contact with a cooler floor surface.

WAYS OF KEEPING PIGS COOLER

Reducing radiant heat

Producers in hot countries, as well as those of us with more temperate climates now beginning to experience the effects of global warming, do not generally appreciate the value of insulation to anything like the extent of those who have to deal with cold weather.

Piggeries in the tropics often have little or no insulation, relying on air movement, solid floors and wetting to cool the pigs. An adequate insulation layer together with a white-painted outer surface to the roof alone can reduce the still air temperature inside by as much as 3°C. *Just painting the outside surfaces white can reduce internal solar radiation by 30%.*

Any benefits from other cooling methods follow on from this head start. A formula I brought back from Australia which is cheaper than white paint is a mixture of 4.5 litres (one Imp. gallon) of PVA emulsion, 22 litres of water (about 4.9 Imp. gallons), 20 kg (44 lb) hydrated lime and one handful of cement. Important: add the PVA to the water first and the lime afterwards. This solution needs to be kept stirred.

Even in temperate zones it is surprising how much solar heat can penetrate a roof if the insulation is deficient. Recent measurements have shown solar gains of up to 30 Watts per square metre (W/m^2) coming through old roofs, and as much as 85 W/m^2 through single skin roofs. It is important to check the standard of your insulation and to bring it up to modern recommended standard U value [*] of 0.4 $W/m^2/°C$.

Shade cooling

Shades shield out the sun's rays (solar radiation) and provide a cool ground surface for the pigs to lie on. Shades can cut out the radiant heat from the sun by as much as 40%. Various commercial materials are available which can be set up in spring and taken down in autumn, but they need to be well-anchored and, as we can see in the section on Seasonal Infertility, *need to be erected soon enough* to catch the clear bright days of 10 hours sunlight in early Spring – in this case not so much to produce shade cover but to distil the effects of bright light on reproduction and reduce sunburn.

[*] U value is the heat in Watts which passes through each square metre for every 1°C difference between inside and outside temperatures.

Shades should if possible be sited on high ground to catch the breeze. On lower ground, locating them 50 metres (yards) downwind from a wood or lush vegetation helps cool what breeze there is. If shades are set high and sloped, eg from 2m rising to 2.5m (6½ to 8¼ ft) with the higher end facing *away* from the sun's travel, this will help maximise radiant heat loss from the animals by exposure to the 'cooler' northern (or southern) sky, according to what hemisphere they are in.

Always check the sun's angle when planning roof overhangs towards the sunward side of any tropical building where the side walls/blinds are removable to assist airflow (Figure 1). As well as protection from sunburn, the radiant heat emission from a concrete walkway fully exposed to the sun is considerable and either a permanent overhang or an extendable sunshade must be provided. Never use blinds to provide shade (except in a crisis sunburn situation) as this will disturb cross-flow movement of air, and risk considerable elevations of ambient temperature despite the sun's exclusion.

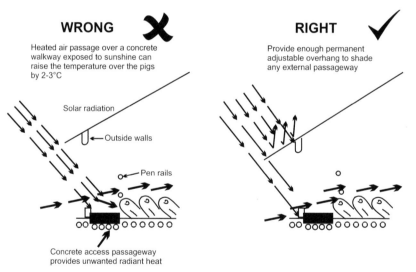

WRONG ✗   **RIGHT** ✓

Heated air passage over a concrete walkway exposed to sunshine can raise the temperature over the pigs by 2-3°C

Provide enough permanent adjustable overhang to shade any external passageway

Solar radiation

Outside walls

Pen rails

Concrete access passageway provides unwanted radiant heat

Figure 1 Shading to prevent sunburn and lower house temperature

Ventilation in hot weather

Rapid air movement over the animal helps both convective and evaporative heat loss. Airspeed below 0.1m/sec is considered to be 'still air', and, as we have seen, an airspeed about 10 times more (1m/sec) and higher provides a cooling benefit for an unwetted skin surface. Thus *ventilation rate* and *air positioning* are the two main components of cooling by air movement.

Correct ventilation rate for hot conditions which is preferably measured in cubic metres moved per hour (m³/hr) is relatively easy to calculate (Table 2) and is often adequately incorporated into a hot weather ventilation system involving propeller fans. The converse is true where natural ventilation or Automatically Controlled Natural Ventilation (ACNV) is involved; these can rarely cope with outside temperatures over 28°C (82°F), where the heat produced by the pigs can raise the **Temperature Lift** – *the difference in temperature between the inside and outside (climatic) temperature* – within the building by 4 to 5°C, which often approaches or exceeds ECT and even UCT in some circumstances.

In practice a temperature lift of 3 to 4°C is aimed for in hot weather conditions. However, the smaller the rise in lift the greater is the airflow needed to achieve it; in a given situation a 3°C lift requires about a 32% increase in airflow than for 4°C.

Table 2 HOW TO CALCULATE MAXIMUM VENTILATION RATE

Maximum Ventilation Needed	=	Heat output from the pigs in Watts
		Difference in temperature target between inside and outside x 0.35

Reference data : Heat production in watts

Weight (kg)		(lb)	At ECT (watts)
1		2.2	4
6	preweaning	13	24
6	postweaning	13	13
10		22	35
20		44	51
40		88	94
60		132	121
80		176	144
100		220	163
Dry sow 170	kg	375	142
Farrowed sow 170	kg	375	272

Worked example: First stage grower to contain 500 growing pigs at maximum weight of 60kg. Ventilation to be designed to keep temperature lift down to 3°C.

Heat output of 500 x 60 kg (132 lb) pigs   is 500 x 121 watts = 60,500 W.

Max Ventilation Rate needed is:– $\dfrac{65,500}{3 \times 0.35} = 62,381$ m³/hr

(1040 m³/min, or 36,327 ft³/min)

This sounds a lot, but remember 500 x 60kg (132 lb) pigs are involved!

We now know how much air must be put into the building to maintain temperature lift no higher than the pig's ECT, assuming the pig's skin is not deliberately wetted.

AIR POSITIONING

But what if the outside air is hot? There is then a limit to the degree of cooling which can be achieved by ventilating with outside air. The air must now be directed over the pigs themselves, so *air positioning criteria* are needed.

The principle is to have the airspeed over the pig's back at between 0.75 and 1m/sec dependent on the current ECT of the pigs – generally the higher level is advised but beyond 0.75m/sec, a lower airspeed with skin-wetting is more effective and uses less power.

In contrast to air positioning in cold conditions where we require a long air travel, visiting the passageway and dunging area first, good air mingling, then passing across the resting pig's back last (at 0.15 to 0.2 m/sec) – in warm weather we need to direct the air directly over the resting pigs first and exiting over the dunging area (or in some cases the roof vent) last. Figure 2 in the Ventilation Section shows the two contrasting airflow patterns.

HOW TO USE EVAPORATIVE COOLING

There are two types of evaporative cooling for pigs.

Spray or mist cooling for growers over 30 kg (66 lb), finishers and dry sows

and

Drip cooling for farrowing sows, and also nursery pigs, say to 30 kg (66 lb).

Spray or mist cooling is used for whole house cooling of pigs if the temperature is above 26°C (79°F) for one hour, where the risk of chilling baby pigs and weaners is not present. Drip cooling, being more particulate, generally cuts in at 22°C (72°F) for lactating sows and 30°C (86°F) for weaners.

Spray cooling, with water droplets up to 20 times bigger than mist droplets, need not necessarily use more water than misting (a common sales-point of misting over spraying) if it is controlled properly as described in Table 3 and when so, I find it much more effective. However many piggeries where spray cooling is used are badly operated. Here is a check-list on what to avoid.

Table 3   SPRAY COOLING GROWERS, FINISHERS, DRY SOWS AND BOARS

Application Rate	330 ml per hour per pig
Cycle time	5 min on, 45 min off
Nozzle Flow Rate	* 3 litres/hr x number of pigs/nozzle
Switch on temperature	26-28°C (Lower than this, increase ventilation rate)

* *Examples*	1 spray nozzle in pen of 10 growers requires a nozzle flow rate of 30 l/hour while 2 spray nozzles in a pen of 10 growers requires a nozzle flow rate of 15 litres/hour

Source : Kruger, Taylor and Crosling (1992)

CHECKLIST ON EVAPORATIVE COOLING ERRORS

✓   Inadequate or over-use (waste) of water.

✓   Wetting and thus chilling (especially at night) sucklers and many newly-weaned pigs.

✓   **Spray**-wetting lactating sows. These should be **drip**-cooled.

✓   Not applying forced air movement (under control) over wetted pigs in the hottest conditions.

✓   Not maintaining spray or drip cooling equipment.

✓   Not adjusting wetting times and evaporative intervals between to suit climatic changes.

✓   Using spray cooling under 23° to 25°C for growers (73-77°F). In this temperature band increase the ventilation rate or raise/direct the airspeed more closely over the pigs.

WHAT SPRAY COOLING DOES

Dependent on relative humidity and the amount of air movement over the body, a wetted pig takes up to an hour to dry. If its skin temperature is lowered by a short burst of spray of around 5 minutes followed by a break of 45 minutes to enable evaporation to take place during this time, even though the pig is living in a temperature of 25°C - 27°C (77-81°F), it *feels* as if it is only at 20°C (68°F) and its metabolism responds accordingly.

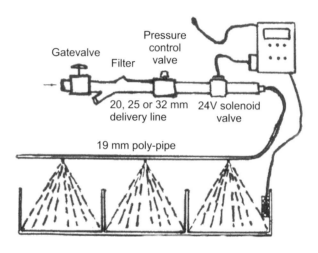

Figure 2 Spray cooling installation (Source: PRDC, see references)

ADJUSTMENT MAY BE NEEDED

Individual producers may need to adjust the frequency and length of time the spray nozzles operate. Adjustments will be affected by several factors: type of building, ventilation, stocking density, and local climatic conditions. In areas of high humidity, water will take longer to evaporate from the pig's skin, therefore the interval between spraying may need to be increased slightly.

AN INSTALLATION CHECKLIST
(SPRAY COOLING ONLY)

✓ Make sure the water flow can be turned on or off.

✓ Have one day's reserve water supply in header tanks.

✓ Keep water source as cool as possible, as spraying with **conduction-heated** water stresses pigs.

✓ Water pressure is normally 140 kPa (20 *psi*)

✓ Use a 200 micron water filter capable of a flow rate of up to 1.4 litres/sec.

✓ Houses of up to 40 metres will need 20 mm diameter delivery lines.

> 40 – 60m will need 25 mm delivery lines.
>
> 60 – 100m will need 32 mm delivery lines.

✓ For lateral lines use 19mm diameter tubes.

✓ Spray nozzles should deliver a uniform distribution of **large** droplets directed straight down.

✓ It is better not to use **mist** or **fog** nozzles as these increase humidity, have much less effect on shed temperature and are distorted by air movement.

✓ Do not use manually-controlled systems. Use solenoid valve automatic controllers.

✓ Use a control system (*e.g.* Farmex) which can integrate with curtain or shutter openings.

It is important to stress that water cooling will not be effective unless there is some air movement over the pig. Water cooling must be done in conjunction with adequate ventilation. *A minimum air speed over the pig's back of 0.2 metres per second (5 seconds to cross one metre) is necessary.*

Checkpoint

A spray cooling system working properly will result in pigs reverting to 20 to 30 breaths/minute within 25 to 30 minutes after the 'off' period. If respiration rate is climbing to twice as much at this time, shorten spray off-time and check airspeed/air positioning over that batch of pigs.

DRIP COOLING

Drip cooling in farrowing crates where the sow can be wetted, leaving the sucklers largely untouched, is a cost-effective and efficient way to cool lactating sows in hot weather. Table 4.

Table 4   THE BENEFITS OF DRIP COOLING ON PERFORMANCE

Sows 27 to 34°C (81-93°F)	*Drip*		*No Drip*	
Breaths per minute	28.5		63.6	
Weaned per litter, kg/lb	56.21	(124)	50.91	(112)
Sow weight loss in lactation, kg/lb	3.79	(8.4)	38.53	(85)
Daily feed intake, kg/ lb	5.74	(12.7)	4.79	(11.0)
Return to estrus (days)	5.0		5.0	

Drip at 4 litres per hour per sow

Source: Murphy *et al* (1988)

Table 4 is typical of the benefits in a hot country – piglets are weaned at heavier weight and sows maintain body condition better than non-cooled sows.

In more temperate climates where farrowing house summer temperatures cause the sow to be at ECT or near UCT levels for more than 12 hours/day the benefits have been 0.5 more piglets born alive, 300g (0.66 lb) heavier weaning weights and 1 kg/day (2.2 lb) more sow lactation food eaten.

Table 5   EFFECT OF DRIP COOLING ON SOWS INCORPORATING STILLBORNS AND MORTALITY

Still borns reduced 0.1 / litter	$p < 0.05$
24 hour deaths reduced by 0.13 per litter	$p < 0.01$
Weaning weight increased by 400g / piglet (0.88 lb)	$p < 0.001$
ADG increased by 14g/day to weaning (approx ½oz)	$p < 0.001$

Payback on equipment cost from these figures 1.69 years.
Year = two hottest months of summer *i.e.* payback 3.38 months use

Source : Cutler (1989)

Even allowing for the need for drip cooling in the four hottest summer months, the payback over cost from 290 kg (640 lb) more liveweight sold per sow/year at weaning and 2 days quicker growth to slaughter was … "At current prices this paid back the capital and installation costs on a 100 sow unit in one year" (Maxwell 1989).

Table 6 details the design requirements for drip cooling sows and weaners and Figure 3 gives installation guidelines.

Table 6 DESIGN REQUIREMENTS FOR DRIP COOLING

	Lactating Sows	*Weaners (only in extreme temperatures)*
Application rate	330 ml per hour per pig	65 ml per hour per weaner (5 weaners/dripper)
Cycle time	1 min on, 10 min off	1 min on, 10 min off
Dripper flow rate	3 to 3.5 litres/hr	3 to 3.5 litres/hr
Switch on temperature	22-24°C (72-75°F)	32°C (large weaners only) (90°F) 35°C (small weaners) (95°F) Avoid use in high air speeds

Source: Kruger, Taylor & Crosling (1990)

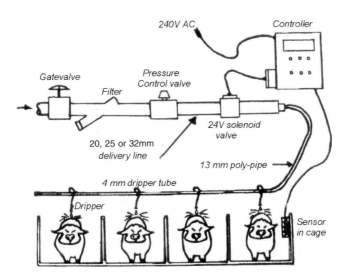

Figure 3 Drip cooling system (Source: PRDC, see references)

A DRIP-COOLING CHECKLIST

General

✓ Provide ventilation. 0.2 m/sec seems adequate. (5 seconds to cross one yard).

✓ A filter is wise, capable of dealing with a flow rate of 0.8 l/sec at a 200 microns mesh.

✓ A pressure reducer is advisable to maintain even flow.

Lactating sows

✓ Location is very important to minimise wetting the litter and the sow's feed.

✓ Fix the 4 mm dripper tube to the *top* of the 13 mm delivery pipe so that the 4 mm tube rises up and bends down. This ensures that when the water is turned off the dripper line empties quickly and there is no drip 'run-on'.

✓ Locate the drip over the shoulder within the circle dimensions described below – this assumes a perforated/slatted bed. If the bed is solid, to minimise the chance of mastitis, less-good but advisable is to locate the drip over the rump where the rear mesh / slats are.

✓ Do not instal drippers where the water can flow into the creep area.

✓ Water must be able to be increased or decreased and turned off when the pen is empty or on cool nights.

Studies on how often the sow lies out flat in the farrowing crate show that she occupies this position for 94% of the time. In a normal 575 to 600 mm-wide (23-24 inch) farrowing crate with equal space outside the side rails, sows seem to have a preference and will normally lie on one side for 70% of the time and about 30% on the other. The head is displaced laterally by some 180 mm (7 in)between these two positions – which is not much. Thus the target area for wetting is a circle approximately 300 mm (12 in) radius, with the drip set at least 500 mm (20 in) back from the edge of the feed trough.

Weaners

✓ Use one dripper for every 5 weaners, dripping over the slats.

✓   Avoid use in high air speeds, this will chill them – a maximum of 0.2 m/sec even in high temperatures.

Producers may care to use 'piggy showers' rather than drips, where a special shower area is available in each pen. This, while excellent, tends to be expensive, with paybacks calculated from installations of nearer 3 years.

EVAPORATIVE PAD COOLERS

These are growing in popularity in hot countries especially where the climate is *hot, dry and dusty* in summer. The system works less well or hardly at all in areas of high humidity. The benefits in dry climates can be remarkable (+ 38 kg MTF or 56 lb/ton, 12-105 kg or 27-232 lb) but *installation and design are critical*. Air is drawn through wet fibrous pads. Water is supplied to the pads by a water sump and overhead distribution system. Rigid plastic pipes or open rain gutters with evenly spaced holes allow water to drip uniformly over the pads. Pipe size, hole size, and spacing depend on water flow-rate and are sized for each system. For best evaporative efficiency, more water is supplied than is evaporated. To conserve unevaporated water, provide a return to the sump. A sloped gutter below the pads collects and conveys unevaporated water back to the sump. Control the make-up water line with a shut-off float valve.

Protect the water distribution system from insects and debris. Filter recirculated water before it returns to the sump. Also, install a filter between the pump and distribution pipe or gutter. Control the system with a thermostat set to begin wetting pads at the desired cut-in temperature *i.e.* ECT. To reduce algae growth on the pad, stop the pump several minutes before the fans to dry pads after each use.
Figure 4 illustrates one such design incorporating a cooling chamber.

It is vital to have the installation designed by a ventilation engineer. There are formulae for water requirements, pad size and power loading needed, as well as the need to control the internal ventilation carefully between operant and non-operant times. There are home-made systems, often incorporating a cork pad taking up most of the end wall, with the house's ventilation system drawing air through. These often cause problems and soon get clogged up which lowers efficiency, raises power costs and under-ventilates the far pens in the room.

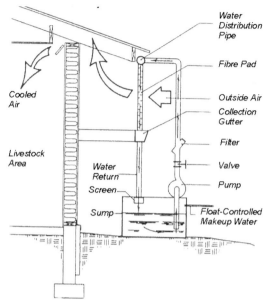

Ventilating exhaust fans pull hot air through wet fibre pads.
Air heat evaporates the water and lowers air temperatures.

Figure 4 Evaporative pad cooling system, based on PRDC advice

SNOUT COOLING
(Sometimes also called Zone Cooling)

This is an excellent low cost system to improve appetite in hot weather
and is almost always confined to lactating sows who can suffer very
badly from lack of appetite in hot conditions, especially hot humid
conditions.

There seem to be two principles involved.

1.   Freshening the air over the feed trough

Here air from the outside – which may well be at 30°C (86°F) or more
– is positioned accurately and deposited to fall gently over the feed
trough at about 0.25 to 0.3 m/sec. The idea is not to cool the sow at all
but to keep the space near the sow's head, especially over the feed
trough, free of gases. A 15-20 cm (6-8 inch) plastic pipe conveys outside
air, pushed in by a lightweight propeller fan, to 8 cm (3 inch) downpipes
fixed to the front end of the farrowing crate. The downpipe contains an
adjustable damper to alter exit air speed and stops over the feed trough
just out of the sow's reach.

EVAPORATIVE COOLING
A BASIC CHECKLIST

Evaporative cooling systems require regular maintenance for proper operation. Develop a maintenance schedule from the following guidelines:

✓ Replace woven fibres annually, so a swing-away wire mesh front is advisable.

✓ If the pad settles, add more pad material so air does not short-circuit.

✓ Hose pads off at least one every two months to wash away dust and sediment.

✓ Control algae build-up with a copper sulphate solution in the wash water. Light-tight enclosures around pads and water sump also help control algae.

✓ As water evaporates, salts and other impurities build-up. Bleed off 5%-10% of the water continuously to remove salts or flush the entire system every month. *Caution*, bleed-off water can be toxic – dispose of it properly.

✓ Remember, power costs rise by almost 25% during operation due to the pad's resistance to the drawn-in air. The engineer may advise extra fans to ensure adequate air is drawn in *and circulated*.

✓ For a 50 sow farrowing house or 200 pig nursery about 2 litres/minute flow rate (about ½ gallon/minute) will be needed around the system, which will have at least 5 m^2 of pad area to be wetted, *ie* about 2.25 x 2.25 metres in size (7.5 x 7.5 ft).

Figures from farms who adopted this form of air freshening show, without question, lactation feed intake improves by 1 kg/day (2.2. lb) in temperate zone summers and up to 2 kg/day (4.4 lb) in the tropics.

Remember, *snout air freshening does not cool the sow*, it encourages her to eat more and so improves body condition at weaning.

2.   Zone cooling

In hot weather, most animals lose 60%-70% of their heat through evaporation from the respiratory tract. Zone cooling the area around the head helps improve cooling. Zone cooling is generally used for stall- or crate-restrained animals and occasionally for animals in individual pens such as a boar pen. In farrowing buildings, zone cooling makes the sow more comfortable while allowing higher temperatures in the pig creep.

Zone cooling however does not satisfy all of the hot weather ventilation needs, and a conventional system sized to provide adequate ventilation rate is still needed.

In theory zone cooling will not work so well if the outside air is very humid. This is because the high moisture content of the air hinders effective dissipation of the respired moisture from the lungs. However, the air freshening aspect is still there, and the higher air velocity to the head plays its part. Because of this higher velocity (Table 7) great care must be taken not to cause a draught in any side creep area, as the air can 'skid' around solid farrowing pen divisions and chill the piglets.

Table 7   RECOMMENDED AIR FLOW RATES FOR ZONE COOLING OF BREEDING PIGS, m³/MINUTE (cu ft/min)

Type of animal	*Ventilation rate* m^3/min (ft^3/min)	
Farrowed sow, 200 kg (441 lb)	2	(71)
Dry sow, 150 kg (331 lb)	1	(35)
Boar, 250 kg (551 lb)	1.6	(57)

*Zone cooling systems should still be ventilated at the hot weather ventilation rates given in Table 1

Source: Jones *et al* (1990)

A ventilation engineer will have internal distribution duct and downpipe duct dimensions to accommodate the air velocity recommendations in the above tables, as well as the fan power loadings needed.

THE USE OF CURTAINS (BLINDS) IN HOT WEATHER

Vertically rising reinforced canvas curtains or blinds are a common method of ventilating hot weather buildings. Their design and operation leave much to be desired, however.

Curtains on 90% of the farms with daytime temperatures usually between 26°C and 34°C (79-93°F) are designed for such a daytime temperature band, where a reasonable flow of air is required across the pigs, along with spray cooling.

However, thought is rarely given to late evening and night-time temperatures which can often fall sufficiently to cause chilling. This is especially dangerous with young pigs in the 20-40 kg band (44-88 lb). While there is a problem with food conversion –pigs growing between 20-40 kg will eat 18g food more/day (0.28 lb/week) for each 1°C below their LCT to maintain growth. So if appetite is limited, growth is 12g/day slower, which results in 2 days longer to slaughter.

Even so the real danger from slightly low temperatures is chilling. Chilling causes stress and stress lowers the animal's immune defences to many diseases, especially respiratory ones. Chilling is an additional effect to the surrounding air temperature, as the floors are often wet from the day's spraying to keep the pigs cool. Wet floors increase thermal conductivity so the pig loses additional heat, not only to the cooler air over it, but also to the cold surface under it. Some critical organs (liver and kidneys) are in direct contact with this too-cold surface.

Manual curtain alteration is never frequent enough

As the warm evening turns to a cooler, chilly or even cold night, the curtain is the main device to retain the correct inside air temperature – just above the pigs' LCT. These, being manually operated, are usually lowered too late; as a result the pigs are chilled and stressed as the outside temperature falls.

I have measured such falls in the tropics between 7 pm and midnight. In each case I asked the owner what he thought the range of fall over this period was, the reply being "4°-5°C", yet the instruments showed it was between 8° and 12°C. Only one curtain adjustment meant that all pigs were well outside the correct range across 4 hours, and from midnight onwards were frequently below LCT by 2°C (older pigs) and as much as 5°C below LCT for weaners.

If you get more virus and respiratory disease than you would like, then crude and inefficient curtain operation is probably a major reason.

Solving the problem

1.   Automate the curtain

No stockman, however diligent, can keep pace with the diurnal temperature changes twice a day (morning and evening), and maybe further changes during the day or night if the weather changes. An automatic controller makes over 30 adjustments in a normal day, and sometimes more.

A series of temperature sensors (normally 3 for a building 10 metres wide x 30 to 40 metres long) linked to a controller which instructs a curtain motor, (or motors, see below) will keep the temperature bands close to the pigs LCT (this can be preset into the controller according to the age of the pig, type of floor surface, degree of roof insulation and feed density etc) so that the pig is not chilled even if the outside temperature falls either quickly or slowly. (Figure 5).

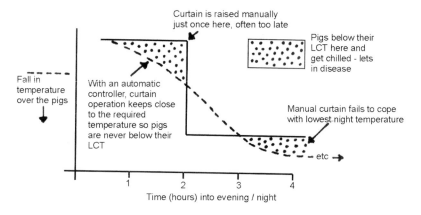

Figure 5 The disadvantage of manually operated curtains

2.   Make the curtains fit snugly

The top should be rigid and fit into a slot. The sides should 'fold' into a box, or if on rollers should rise and fall via a narrow slot. Both are to reduce draughts, and this is especially important for weaners to about 20 kg (44 lb).

In non-tropical areas solid louvres are better than curtains, but are some three times more expensive to construct.

3.   Quartile the building

'Quartiling' is to divide the building into four squares or broad rectangles served by *two* curtains down the side of the building, not just one.

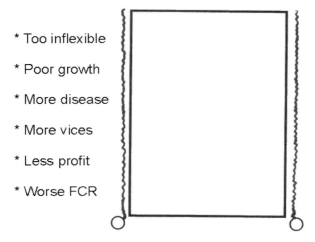

* Too inflexible

* Poor growth

* More disease

* More vices

* Less profit

* Worse FCR

Figure 6 Wrong: Two long blinds running all down one side of the building

Just having one blind down one side is too inflexible as it:

• Cannot cope with the sun's daytime movement from horizon to horizon.

• Cannot sufficiently cope with changes in wind direction.

• Curtains are often too long, so the upper gap tends, especially after time, to be unequal down its length.

This is not important during the day, but can be critical at dusk and during the night. Maximum operational length at the time of writing is about 30-35 metres. If the building is longer, use several blinds (Figure 7).

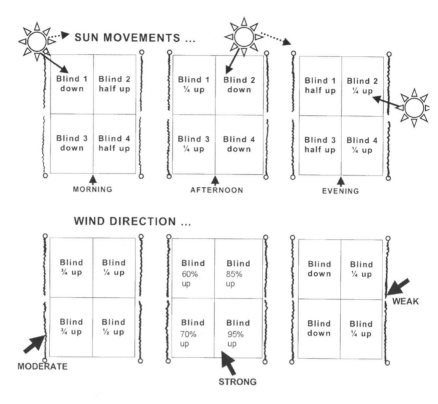

Figure 7 Correct: 4 shorter blinds quartiling building

THE BOTTOM LINE

Table 8 shows that in growing pigs, due to faster growth and better food conversion, the food saved was 26.4 kg/pig (58.2 lb) mainly due to faster growth. This translates into a massive 162 tons of feed/year for this 300-sow unit.

Table 8 EFFECT OF QUARTILING CURTAINS

	Before	*After quartiling*
FCR 20-90 kg (44-199 lb)	3.2	3.0
Days to 90 kg (199 lb)	161	153
% Dirtying Pens	36% (at 29.3°C/85°F)	12% (at 29.2°C)
Food saved on 300-sow farm/year (t)	–	162
After deduction for one-off cost of auto installation (t equiv)		130
PAYBACK in months	10 months	

Source: Clients records

The only difference in the above table was that two buildings (nursery and finishing) were converted from manually operated curtains, to quartiled automatically-controlled curtains. The cost of conversion was 80% of the cost of the food saved, thus the payback was in 10 months. Not surprisingly the farm is now fully converted to automatic curtaining. Further exercises have shown the payback to be 1.2 to 2.1 years on food savings alone, not counting the cost of reduced disease. One farm reported a marked improvement in the mortality and morbidity level of nursery pigs after 1½ years. This alone paid for the alterations in 6½ months.

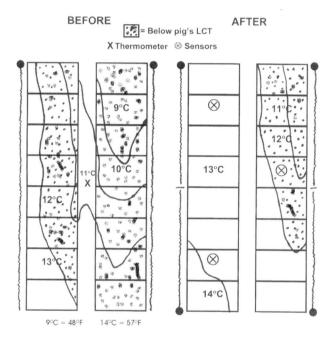

Figure 8 Temperature bands before and after quartiling (nightime) (9°C = 48°F; 14°C = 57°F)

THE BOAR IN HOT WEATHER

The testis is usually 2.5°C cooler than normal body temperature, and this temperature must be maintained for optimum boar fertility. The most significant effect of "overheating" is a reduction in the numbers of motile sperm ejaculated at mating. Motility enables sperm to reach and fertilise eggs after mating, with 95% of ejaculated sperm normally motile. This percentage declines at air temperatures above 30°C (86°F), to the extent that on one trial, motility of sperm collected from boars heated to 40°C (104°F) fell to *below5%*.

So, at air temperatures above 33°C (91°F), in addition to depressed motility, other influences on fertility such as total sperm numbers, percentage of live sperm and percentage of abnormal sperm ejaculated can be expected to deteriorate; ***this effect appears between 2-5 weeks after heat exposure***, with the shorter time period likely to apply to those boars mating more frequently.

Boar libido

Australian research has demonstrated boar sex drive doesn't appear to become depressed until air temperatures approach 38°C (100°F), with no residual effect on libido following the period of heat stress.

Boar libido plays an important role in enabling boars to actively seek out females in-oestrus, and display an adequate level of stimulatory courtship behaviour prior to completing an active, quality mating. So hot weather may reduce preliminary courtship.

The activity of "flank nosing" by boars within the first 30 seconds of mounting females appears to be an important behavioural trait to ensure good fertility. Boars displaying high levels of nosing activity during courting generally have higher conception rates than those boars exhibiting lower nosing activity (Table 9). It is thought that increased nosing activity by boars may stimulate the release of oxytocin in the female, which in turn may lead to increased chances of fertilisation occurring through increases in uterine contractions and sperm transport along the oviduct.

Table 9 SERVING – LENGTH OF ACT AND CONTENTMENT

Score	Conception rate %	'Courting' (mins)	Mounting attempts	Intromission (mins)
1. Poor 2. Moderate	80.5 (75-86%)	1.8	1.9	2.1
3. Good 4. Excellent	83.4 (75-91.8%)	0.42	1.1	3.1*

* Litter size was increased by 0.48 piglets for each minute's increase in duration of intromission

Correct receptivity : Correct surroundings : Quiet unhurried handling.

Source: Rikard-Bell (1994)

WORKLOADS

The effects of summer heat on boars may be lowered by reducing the frequency of boars mating during periods of hot weather. The effects of heat on sperm production and the determination of the optimum number of sperm for fertilisation to occur show that following short-term heat exposure, boars can maintain fertility provided they are used less frequently.

The recommendations below were drawn from Australian work and are:-

Table 10 MATING FREQUENCY GUIDELINES IN HOT WEATHER

Temperature to which boars are exposed for a single 12 hour period		*Maximum mating frequency 1-6 weeks following exposure*
Below 30°C	(86°F)	Twice/day
30°	(86°F)	Once/day
33°	(91°F)	Four times/week
36°	(97°F)	Twice/week
40°	(104°F)	Infertile for most of the period

It is possible that some boars will be capable of maintaining their fertility at higher mating frequencies than those recommended. All other things being equal, I suspect these boars are likely to be those with larger testicles, but this needs scientific confirmation.

Regardless of workload, it is recommended whenever possible that matings be conducted early in the day (when temperatures are likely to be lower), and ***prior to feeding***, in order to maintain boar libido. (Table 11)

Table 11 COMPARISON OF EARLY MORNING *v* ALL DAY SERVICE ROUTINE IN HOT, HUMID WEATHER

	Sows served between 5 am – 7 am (Temperature 24°-26°C; 75-79°F)	*Sows served all day (Temperature 27°-34°C; 81-93°F)*
Farrowing rate (%)	88	72
% repeats	13	23
Born alive/litter	9.87	8.91

REFERENCES & FURTHER READING

Assistance with this section is gratefully acknowledged from the Pig Research & Development Corporation, Australia, whom I consider the world's leading experts on hot weather pig production. Their publication in the Australian Pig Housing Series "Summer Cooling" (1992) ISBN 07305 9892 6, editors, Kruger, Taylor & Crosling, is a major source of valuable practical information (*cf* Figure 2 and 3) and I am grateful to the PRDA for permission to quote from selected advice therein. I am also obliged to a variety of commercial equipment firms who have sent diagrams and advised on technical specifications.

Jones, D.D., *et al* (eds) 1990; 'Heating Cooling and Tempering Air'. Monograph published by Iowa State Univ, USA (Mid West Plan Service) pps 12-15 and 24-28.

Murphy, Nichols & Robbins (Kansas State University) 1989; 'Drip Cooling of Lactating Sows', Pigs (Elsevier Publications) May 1989 pps 8-9.

WRONG MUCKING / PEN FOULING 21

Pigs soiling their resting/sleeping area with urine and/or faeces.

TARGET

A dirty pen floor is one of the most annoying and time-consuming problems facing pig producers. I see bad cases on one in twenty of all farm visits – with a third of them left unattended when a simple solution is all that is needed to rectify matters. The target must be zero.

Pen fouling can occur at any time – from sucklers voiding in their 'creep' areas where they are supposed to lie warm and dry in between meal-times, to finishing pigs making a quagmire out of their sleeping and socialising areas. Dry sows in stalls can mess up their underbelly aera too.

Strange as it may seem, I have always enjoyed being asked to deal with a dirty pen floor problem as it is relatively easy to cure. If you are observant, the reasons are always there to be seen.

THE BABY PIGLET

Creep-fouling

Nearly always due to giving the newborn pigs too much space in the shut-off creep area. Faced with a lying space which is too generous, they naturally treat one end of it as a lavatory.

331

Sure cure: allow no more than 1½ times the supine[1] body area of the total number of piglets in the litter; or 2 times the area once a creep hopper is placed in the creep. Vary the area by putting in a wooden dividing board, a plastic can or wooden box to shut off or take up the unwanted space until the pigs get bigger (Figure 1).

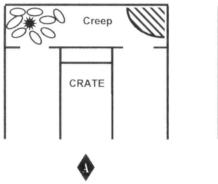
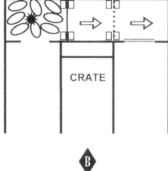

Giving sucklers too much room in a forward creep encourages them to use the excess space as a lavatory.

Remedy: Close off one pophole until the end of the suckling period. Fix vertical battens at 1/3 and 2/3 of the creep box, and place a crusher-board at either location as the piglets grow.

By training them to accept that the only area to rest in is that allowed for sleeping supine, they will not foul the box – unless the space under the lamp or over the heated pad is uncomfortably hot.

Figure 1 Dealing with a fouled forward creep box in a farrowing pen

Fouling near to the sow's head

With a solid floor this area – on either side of the sow's head – can be filled in with a concrete rounded 'hill', rather like a woman's breast, which will deter elimination. Alternatively put a grille or drainage area in these 'dead' places in the farrowing pen. If the sow spills a lot of water in this position, consider how this can be lessened – provide a good deep 4-litre (one gallon or more) drinking trough, for example. Another solution – but at the farrowing house design stage – is to eliminate this 'dead' space by allowing the creep area in the adjacent farrowing pen to occupy it. Remember, all pigs prefer to eliminate in corners, so don't have obvious corners at the sow's head – only down at the other wet end of the pen.

[1] Supine = The space occupied by the head, body and extended legs of the piglet when lying on its side as distinct from the sternum position where the pig lies full-length on its chest.

Adequate temperature gradient

Even a newborn pig can be instinctively drawn to eliminate in a colder area. Trouble is it needs an adequate temperature gradient between the resting/ exercising area and the elimination areas to encourage it to do so – my own experience suggests a drop of over 3°C, at least. This is often not provided in the open-plan farrowing house, as insufficient use is made of flaps or deflectors to 'knock down' some of the cooler incoming air towards the sow's rear end. Not a draught, but just a gentle downward drift of cooler incoming air. This is easily achieved with flaps set at the correct angle up on the ceiling, together with the use of simple Draeger smoke test tubes to see where the incoming air is going and how fast it falls. But hardly anyone uses this essential aid!

OLDER PIGS

Anthropomorphism (the attribution of human characteristics to non-human objects) is common among pig producers, probably because the pig is, in many ways, rather similar to us. *We* regard wrong-mucking as a 'bad habit', but the pig probably considers it to be a natural reaction to alleviate an uncomfortable situation it finds itself in, often through no fault of its own. Or it could be just pure laziness! Let me explain.

There seems to me to be five main reasons why a pig eliminates in the wrong place.

1. **Because it is too hot, and wishes to start to cool itself by lying in urine and faeces to conduct away more heat from its body.**

    Examples:

a) The house and ventilation design which cannot get rid of an excessive temperature rise towards and beyond the pig's ECT (Evaporative Critical Temperature).

b) Overstocking, which causes a localised temperature rise.

c) Poor ventilation, especially in centre pens, pens exposed to mid-day sunshine, and troughs or solid partitions which deflect cooling airflow.

2. **Because the pig has too much space in which to void.** We may provide a suitable elimination area, but if the free space between its

preferred (dry) resting area and where we wish it to void is too generous, it may decide to relocate its voiding area closer to 'home'. Anthropomorphically we call it 'laziness', to the pig it is merely common-sense!

3.   **Because it is difficult to get to the voiding area.**   The above problem is made much worse by making the access to the dunging area difficult (narrow pass-ways, excessive change of floor height (steps), slippery surfaces, insufficient voiding area as the pig gets longer with age). Providing unpleasant conditions where *we* want the pig to dung can persuade the pig to disagree with us. For example poor slat dimensions and cracked or pitted solid floors, darkness in the slatted area and excessive temperature differences between resting and voiding areas.

4.   **Overstocking :**   Overcrowded pigs can also be forced to eliminate wrongly because they are prevented from establishing their usual social pattern of behaviour. What they see is no clear voiding area, or other pigs in the way preventing the lower orders in the group from getting to it. Usually overcrowding results in over-hot or airless, gas-filled conditions and the pig's natural answer is to wet (cool) itself which is a purely defensive reaction, and urine is the most convenient source of liquid available to it.

5.   **Habit :**   The classic case is where a young pig has been kept in a hot nursery with a totally wire or slatted floor and is then moved to a ***solid or part-solid*** pen.   Not surprisingly it has been encouraged to void where it likes in the all-wire pen, especially if there is poor air placement. It may continue to do so in the new quarters. This can be overcome by ensuring that :–

a)   There is a distinct air pattern provided in the all-wire room (often a flat-deck) where cooler air falls on to the preferred dunging area (as far away from the resting/feeding areas as convenient and with drinkers sited within this cooler zone).   The temperature difference between lying and voiding area should be at least 3°C dependent on the LCT (Lower Critical Temperature) of the pigs.

b)   Training nursery pigs on an all-wire floor not to void on a solid area when they are eventually moved by putting a solid comfort-board in front of the feeder in the all-wire pen so that they get used to solidity being a 'clean' surface.

c)   The space allowed in the part-solid follow-on pens should not be too generous.   It often is, as small pigs are put in pens big enough for them at finish weight. The answer is to use crusher boards or variable-geometry pen fronts to close the space at first so as to allow the pigs to grow *with* the space available rather than grow *into* it. Housing designers – please think about this!

d) The slatted/void to solid/resting ratio should be adequate, at 1 to 3 or 1 to 4 maximum. Again, it often isn't, especially in the long narrow pens seen often with pipeline wet-feed units or in broad finishing piggeries where too narrow wet/dry feeder pens make best economic use of total area.

CHECKLIST
THINGS TO LOOK FOR WHEN NOTICING DIRTY PENS

✓ *Where* are they in relation to end/outside walls, centre of the building, ventilation inlets/ extractor fans. Where are the *dry* ones? What's the difference – this will help you decide what the cause is.

✓ Is there anything making the pigs in dirty pens too hot? Or too gaseous ? (Use a gas detector phial.)

✓ Have you lost control of air movement – summer/winter or day/night ? In other words, can you measure it? (Use a smoke 'candle'.)

✓ Check stocking density. If you've got dirty pens up to maximum stocking density, try destocking by 15% (1 in 7).

✓ Have you got the tools to assist you? A proper clean calibration-checked thermometer set just above the pigs back? Smoke tubes? Draeger ammonia phials and suction bulb (10-50 ppm NH_3 range is adequate using ammonia as a marker gas). Deflector boards hung on wire or string? An airbag? And lastly – elbow grease plus a pail of water and yard brush!

CHECKLIST WHEN WRONG-MUCKING OCCURS IN PART-SLATTED GROWER/ FINISHER PENS

Pigs prefer to eliminate in corners – we need to tell them which corners!

✓ Check stocking density for summer/winter conditions, sometimes day/night conditions. (See stocking density section).

✓ Dunging area 3°C colder than lying area? In the tropics a difference of 2°C is usually adequate.

✓ Pens should not be longer than x2 the width, with 1½ times better. Use a crusher board to adjust balance, *i.e.* shape.

✓ Feeders in among, or beyond the lying area are problems – check that they are quite close to drinkers in dunging areas as long as there is at least 1.65m (almost 5½ ft) gap to the opposite division across the pen – otherwise move the feeders to the front wall/rails.

✓ Always site pen furniture so that the pigs follow this primary natural pattern – awakening, feeding, drinking, eliminating, social congress, resting – in a clockwise or anti-clockwise direction. Too many crossed paths may result in vices, aggravated by high stocking density and inadequate ventilation control.

✓ The dunging area should be along the short side of the pen, of sufficient depth to accommodate a finished pig's length (1 metre or 3' 3"). Do not close the dunging:sleeping ratio to less than 1:3.

✓ High temperatures - Is it hot or airless at the following temperatures ?

Up to 23°C* (73°F)

Encourage better ventilation – provide natural cross-flow ventilation *over* kennels, for example, to remove the airlessness inside. Put an offset hip to any kennel roof to encourage rotary airflow inside.

23°C to 26°C* (73-79°F)

To put in air-bag ventilation (so as to position greater air movement/turbulence over the pig's resting area). This can be shut off at night if required (see Ventilation Section on how to do this).

26°C+* (79°F)

To merit spraying if the circumstances allow.

*   *Dependent on the E.C.T. of the pigs in question. You should consult a ventilationist as it can vary by as much as about 4°C from these approximate problem thresholds.*

DEEP STRAWED PENS

Groups of pigs have to eliminate somewhere in a yard, so provide a concrete scrape-through area. This is commonly seen, but what is almost never done is *to provide downflow ventilation over it as well*. Generally a simple airbag hung on wire loops with a single row of holes on the underside is adequate. Growing pigs especially will often also eliminate in the strawed corners, and some sows will too. The airbag provides that 3°-5°C vital difference, even in summer, when in the hottest weather the bag can be replaced with one possessing a row of dual outlets (called sipes – see Ventilation section) set at 45 degrees on either side of the bag to push air across both the scrape-through passage *and* strawed resting yards. In this way the animals are less stressed and are encouraged to dung/urinate in the customary areas.

Clients have often been impressed by this simple, low-cost precaution, which in cold weather can be shut off at night. Many pig producers have ignored the airbag concept – a serious error. I give details on how to make your own in the Ventilation section.

PIG MIXING 22

Pigs may have to be mixed so as to facilitate pig flow, reduce stocking density or make maximum economic use of pig space.

TARGETS - GROWING PIGS

In growing/finishing pigs it is far preferable never to have to mix separate pens together, in view of the antagonism this causes. Howard Hill, one of the world's greatest pig managers, once famously said:

"Don't mix those damned pigs!"

MIXING GROWING PIGS

However this has to be done when batching and matching weaners into or out of the nursery or when a production bulge necessitates some thinning down, and also when laggards need penning together so as to free-up pens in the final few days before the slower growers become heavy enough to be shipped.

The older the pigs, the more disparate their weight, the greater the difference in the new environment, the more fighting is likely to occur, especially when there is insufficient movement space and where feeding and watering is radically different. Good stockmanship and forward planning minimises the need for mixing, which can hold up growth surprisingly. (Tables 1 and 2)

Table 1   ENFORCED MIXING OF PIGS FROM 10 DAYS BEFORE SHIPPING CAN SLOW GROWTH RATE CONSIDERABLY

Pigs varying in weight by 10.8 kg or 24 lb (av. wt. 82.1 kg or 181 lb) growing at an average of 760 g/day (1.67 lb) were mixed from 4 pens into one pen of 15 pigs until average shipping weight of 92.2 kg (203 lb) (contract 90 kg), and were compared to pigs of similar weight which remained in pen groups until shipping.

	Mixed pen		Unmixed pens		
Av. daily gain, g (lb)	696	(1.53)	805	(1.77)	(+13.5 %)
Av. feed intake/day, kg (lb)	2.05	(4.52)	2.21	(4.87)	(+ 7.2 %)
Av. FCR (25.1 to 92.1 kg)	2.94:1		2.73:1	(− 7.7 %)	
(55 to 203 lb)					

Source : Clients' records

Comment: The marked reduction in growth rate of the mixed pigs seems due to the lower-end pigs in the new peck-order eating less, thus pulling the pen average growth rate down badly, probably combined with increased anxiety stress reducing the FCR of the food they did eat.

Alternative options are to leave the laggards in the 4 pens (ensuring they were warm enough). At today's capital costs amortized over 10 years, locking up the growing space thereby was a roughly similar cost at 0.9p/kg deadweight compared to the extra costs of food and overheads from mixing the 4 groups – at 0.8p/kg in this trial. On the other hand, selling the laggards along with the others at 82 kg (181 lb) instead of 92 kg (203 lb) reduced income in this trial by as much as 9p/kg dead (2.8c/lb) and would not be an option, as every producer knows.

Table 2   MIXING WEANERS IMMEDIATELY PENALISES GROWTH ADG 20 DAYS POSTWEANING, g (lb)

Litters kept together, not mixed with other litters	350	(0.77)
Litters mixed	240	(0.53)

Varley(2001)

WHAT DOES MIXING COST?

Pope (1996) reports, compared to unmixed pigs (controls), average daily gain over 2 weeks was 19% less when the pigs were mixed at 84 kg just before sale. Mixed pigs spent 17% less time eating than the controls on the first day and 24% less during the second.

19% less daily gain from 84 kg (185 lb) to sale weight at 96 kg (212 lb) reduces the saleable lean produced/tonne feed (MTF) by 5.45 kg (12 lb), equivalent to a rise in feed cost over that final period of 4%.

Mixing young pigs – as distinct from the postweaning check to growth where the gut surface is damaged – in the author's experience, does not usually penalise days to slaughter very much, and sometimes not at all, as the mixed pens if remaining healthy often display compensatory *liveweight* growth by slaughter despite what the experts say. However, careful measurements of saleable lean from the processors' returns I have seen where this had occurred (to the producer's relief) suggest, however, that the compensatory growth after the **mixing** check to growth suffers in respect of **lean dressing percent** ranging from 0.5% to 0.72%, average 0.61%. At 5 pigs/tonne feed and a typical 74% dressing percent this is a reduction in MTF of 3.05 kg (6¾ lb), equivalent to a 2.2% price rise in all grower finisher food fed from nursery to slaughter.

As Howard Hill says, 'Try not to mix pigs' so . . .

* **Minimise the need to mix pigs in your pig flow plan.** So can you put them elsewhere – in temporary but satisfactory accommodation? Can you secure a satisfactory market for your lighter pigs? This need not then cause so much of an income loss as expected. I have records of several case-histories where the loss of income from selling lighter pigs averaged only 4p/kg deadweight, but mixing them with pens of heavier pigs and selling them at the normal sale weight cost 5p/kg. This was due to a check to growth in both the newcomers and the indigenous pigs, plus aggression damage (scarring and tailbiting) causing condemnations which accounted for a substantial part of the reduced income.

CHECKLIST
IF YOU ARE FORCED TO MIX PIGS

✓ The big error made by most stockpeople is to **fail to allow sufficient space** for submissives to avoid dominants, and for dominants in each group to get the challenges over with minimal damage and stress to either party.

✓   The answer to this is **not** to assume that if the to-be mixed groups have existed satisfactorily at an acceptable and approved stocking density, that they will be able to so on being mixed. You must allow them more space, and at least +20% is suggested. This is why smaller groups settle down better than larger groups in relatively constricted pens, such as in a typical flat-decked nursery and why much larger groups can often be mixed with no discernible trouble in straw-based yards where there is much more getaway space, of if needs be 'sleepaway' space for a night or two.

✓   'Sufficient space' certainly involves voiding/exercise and resting areas **but also feeding space** and is certainly one reason why two drinkers (or more) in a pen is always preferable to just one. The majority of aggressive incidents occur at or near feeders and this is why an extra feeder or temporary trough/hopper seems to help.

✓   If practicable, reduce the feed allowance slightly for the to-be-mixed and recipient pigs, say by one-third, from the morning of mixing day. Do not overdo this as it will increase aggression at mixing. Just get all the pigs slightly hungry, but not overly so.

✓   If withholding some food prior to mixing, extra trough space is vital as two-thirds of aggressive acts are directed at pigs trying to approach the feeder. Feeder seeking is so important that subordinate pigs can attack dominants. Put in a temporary feeder/trough so the chased-away pigs can gain access to it.

✓   Introduce the pigs just before dusk, certainly by late-afternoon.

✓   Move pigs quickly but **gently**. Stressed pigs aggress more especially in new surroundings.

✓   Allow ample food once the pigs are moved.

✓   Immediately spray all the pigs with lavatory freshener aerosol or detergent. Sump oil is much too messy.

✓   If using bedding, re-bed amply just before mixing.

✓   Place some of the to-be-mixed pigs faeces in the voiding area.

✓   Check drinker flow and adequacy. Submissives will be last to eat – they tend to tank-up on water and this itself may induce tailbiting.

✓   Do not batch **too** evenly, especially young pigs out of the nursery.

A 4 kg (9 lb) range is better than identical weights (Lean, 1985) for smaller pigs and maybe up to a 6 kg (13 lb) difference for older pigs. This allows dominance to get established more quickly and the pen settled down sooner. At 21-28 days, a 1 kg (2¼ lb) range may be good within each batch of, say 20.

✓ Check you aren't exceeding stocking density – a common error. In fact, try to put in say, one less pig per pen if practicable, *i.e.* *destock* by 15%. Remember mixed pigs need fleeing space.

✓ In typical intensive nurseries small groups (<20) mix better than large ones, and 12 is better still.

✓ In larger groups, *i.e.* 'big pens', now gaining favour, once group size reaches 50-100 animals the drop off in weight gains after mixing begins to disappear.

✓ Recheck ventilation. Getting the two groups to 'rub together' in the first night is important, so an ample warm and dry sleeping area helps this natural socialization. Submissives tend to 'sleep out', get cold and stressed and may tailbite next day. A temporary cover or lid across the pen has been known to help, but don't place it flat – raise one end by a few centimetres to circulate the air over the sleeping pigs.

✓ Observe, observe, observe. Mixing can awaken a latent persistent tailbiter – often, I find, a small female. Watch for this pig, which must be removed and penned with others displaying a similar trend where, in large units, a sedative can be given. Such pigs rarely convert well, and whether to keep such individuals is debatable.

MIXING TARGETS - SOWS

Sooner or later, even on the best-run batch-served breeding unit, sows have to be regrouped. No producer needs to be told how much damage such strong and heavy animals can do to each other and every attempt must be made to minimise the need to mix groups together or introduce individuals into established groups. The writer has evidence that on some farms new to grouping skills when the gestation stalls had to be replaced with yarded systems, extra culling rates of from 5% to 11% were directly attributable to fighting and/or vulva biting.

MIXING SOWS

Because Swedish and British breeders were in the van of moving from individual stalls to sow groups in yards under mandatory welfare legislation imposed in their countries, much valuable information on grouping sows has been accomplished by research farms in both nations, in particular Cambac, MLC (Stotfold), MAFF, ADAS (Terrington) and others in the UK, and the Swedish Pig Research Centre (Svalov). Recently the Danes and the French are contributing to our findings.

My own experience, like many breeders, has been largely by trial-and-error. My check-list below doubtless will be improved upon as the years progress and we all share our experiences, but it will serve to illustrate the current advice available at the time this section was written.

MIXING SOWS – A GENERAL CHECKLIST

✓ Cater for the least dominant sow in the group. If you get the conditions right for her, not only will she benefit materially but the whole group will settle down more quickly. "Farm for the most timid sow" is as sound advice as Peter English's "Farm for the smallest pig in the litter."

✓ Allow a minimum of 3.5m² (37½ ft²) per sow (MAFF 1993).

✓ Adequate fleeing space seems vital. MAFF reports that 75% of aggressive sows will not chase another sow beyond 2.5 metres (a little over 8ft) (range 0-20 metres). While the jury is still out on ideal pen shape, it could be that a larger, narrow pen could allow more fleeing space than the same stocking density in a square or near-square shape. While the latter is generally favoured, if you give them enough space I guess it doesn't matter.

✓ A specially designed mixing pen (Figure 1) may be a useful idea, especially where a gilt has to be introduced into an established group. The pen should be straw-bedded and floors provide adequate grip *i.e.* not a soft, spongy under-surface.

✓ The mixing pen can be used for about 24 hours. Aggression is worst in the first 4 hours from introduction, so periodic suspension is wise during this time, say once an hour. (Extrapolated from Kay, 1999.)

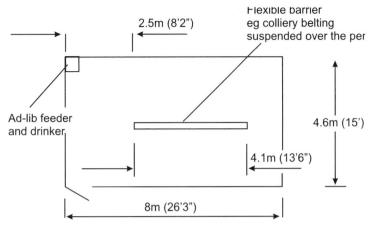

Note: All measurements internal

Figure 1 A specially designed mixing pen

✓ Pen size should be matched to age and weight of the animals housed.

✓ A mixing pen adds 6% to the breeding units capital cost; about 0.5% to total production cost/ year (500 sows).

✓ Ad-lib feeding reduces aggression. (Peet, 1993 and ADAS/SAC n.d.)

✓ Competition for food is a significant trigger. If you can, provide sugar beet pulp from single space feeders. (Figure 2).

✓ Both wet-feeding and to a lesser extent, trickle feeding (Hunter, 1998) minimise aggression (Table 3).

✓ Despite careful measurement of subsequent performance in one trial (Burfoot & Kay n.d.) when sows were mixed from 1 week to 6 weeks after service (to cover the implantation period in depth) there were no significant differences in total born, born-alives or total litter weight born.

✓ *However*, it is still current advice ***not*** to mix sows during the implantation period, and the writer can quote several cases where returns to service and poor litter size responded quickly when it was suggested that it was a good idea to move batch introduction of new gilts into the herd's farrowing profile, rather than their introduction into what is called dynamic breeding groups (*see below*). This could be valuable in outdoor breeding paddocks even though, paradoxically, there is much more room for escape.

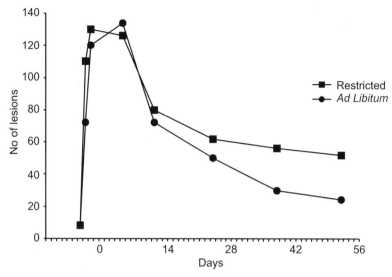

Figure 2 Number of skin lesions on gilts fed *ad libitum* or fed restricted from an ESF feeder

Table 3 PERCENTAGE SOWS SETTLED ONE HOUR POST FEED-DROP

Study number			Percentage settled (one hour hour post feed-drop)
1	Wet fed (conversion)	10	56%
2	Wet fed (purpose-built)	10	75%
3	Trickle (converted)	6	56%
4	Trickle (purpose-built)	5	52%
5	Dump fed (converted)	55	22%
6	Dump fed (purpose-built)	20	21%
7	Dump fed (service house)	12	52%
8	Drop (converted)	6	15%
9	Floor fed	72	15%

Source: Hunter (Cambac Group) (1998)

✓   Mixing in the evening may be beneficial.

✓   Adding fresh straw at mixing tends to delay the settling of peck-order as the more dominant sow's attention is distracted for a while. *However* adding fresh straw at evening or late-afternoon may be beneficial in aiding socialization – it depends in my experience on whether the sow genotype is basically placid, the space available and the frequent presence of a familiar stockperson. If some or all are borderline, fresh straw sure helps!

✓ Avoid re-mixing sows whenever possible, if so, a mixing pen as described is very worthwhile.

IF YOU HAVE TO FARM IN DYNAMIC GROUPS . . .

Dynamic groups are large groups of 20 to 30 sows upwards, where sows are added and removed on a regular basis, usually weekly, on the demands of farrowing, weaning or service.

A DYNAMIC SOW GROUP CHECKLIST

✓ Introduce *more than* 3 animals in the sub-group, whatever is the size of the main group.

✓ It is vital to have adequate feeding, sleeping and watering space for all the sows. Providing extra *ad lib* feeders is now considered essential. Remember, "farm for the most timid sow/gilt."

✓ Introduce after implantation is complete, technically 28 to 35 days after service, but after 28 days in my experience seems satisfactory.

✓ Take time to introduce gilts to a dynamic group, and preferably grow them towards 130 kg (287 lb) before this is attempted. Ensure the gilts are not bullied off their feed by partially bucket feeding them and standing by for a few minutes; this helps in flushing anyway.

✓ Individuals in the gilt sub-group should have been introduced to each other in electronic feeder training pens, a specially-designed mixing pen, or a holding pen near the main sow group. Gilts are excitable pre-oestrus and need familiarization with each other so as to minimise stress rather than excitement.

✓   The use of sedatives (*e.g.* amperozide) seems to me to merely delay aggression, though this maybe needs further exploration.

✓   The presence of a boar in a dynamic group does not seem to reduce inter-female aggression.

✓   Sows may be able to remember each other's place in the social hierarchy for 4 weeks, so within this time-scale "individuals may be returned to groups of sows with which they have previously been housed without any in-fighting" (Arey, 1998).

✓   If you have trouble in introduction, try penning off an area and keeping the new entrants there for a few days. Pen that area off away from access by main group sows for a couple of days before the sub-group is introduced. (This is a very useful idea.)

✓   Breaking up the lying area with divisions is sometimes advised. I recommend dividers are essential with electronic sow feeders on hard standings as seen in the continent of Europe, but my experience suggests that this is unwise in deep straw-bedded yards (see Figure 3), as it tends to disturb the natural territorial occupancy of the chosen resting area by the various sub-groups.

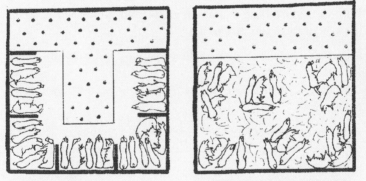

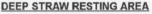

HARD FLOORS

Divisions are essential. Sows will lie facing the feeding area many in the 'Dog-Crouch', semi-sternum position

DEEP STRAW RESTING AREA

Sows lie in family groups, the most dominant group(s) nearest to the feed area. Divisions will tend to hinder this. Notice many more sows are in the fully recumbant position.

Feeding and socialising areas

Figure 3 Two yards ... but two totally different layouts

✓   Be careful not to have protruding fixtures (feeders, drinkers) or sharp corners, especially on mixing pens where there is much sudden movement.

The author recognises that much about sow group behaviour is far from being fully understood and that the advice given in these pages may be modified as experience accumulates.

SOME TIPS FROM A LIFETIME'S EXPERIENCE

I first kept sows in yards in 1949. We found that . . .

- A docile breed was a godsend! Saddlebacks and Berkshires, then Accredicross (Seghers) were easy to manage. (They still are!)

- We soon discovered the benefit of ample space.

- And of the straw yard.

- But both were expensive, so we used old converted (but warm) buildings and I suggest using moveable partitions to alter divisions. Docile sows and good stockmanship need nothing fancy, so save your money!

- There was always the timid sow or gilt. We had a couple of feeding stalls for these, in which they could be shut in daily. It took more time and trouble, but was always worth it – up to 4 more pigs per year for these ladies.

- After a while we used extra *ad lib* feeders as routine, as is recommended now some 50 years later! Things go full circle, don't they!

- Sows in yards need much better stockmanship than sows in stalls – and more time.

- With electronic feeders I much prefer the double-yard system as once a day you get all your sows up and moving past you. (Figure 4) Before they settle down again you have time to examine things like udders and vulvas and mark any which need an eye kept on them subsequently. Also those which are slow to rise can be noticed and catered for.

- Enough time, enough space and careful planning are the three routes to trouble-free grouping of sows, whichever breed you choose or in whatever climate and housing system you raise them.

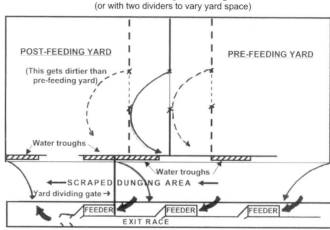

Daily feeding cycle starts at noon or 5 pm when all sows are allowed through from post-feeding yard to pre-feeding yard

Figure 4 Sketch of rotary station feeding layout for 120-150 sows

REFERENCES

Arey, D. and Turner, S. (1998) *Housing pigs in large groups*. SAC Report.

Burfoot, A. and Kay. R. (1995) *Agression between sows mixed in small stable groups*. ADAS Report.

Hill, M. (1997) Allen D Leman Swine Conference. University of Minnesota.

Hunter, L. (1998) Behaviour and welfare of sows in group housing. CAMBAC/JMA Research Report.

Kay, R. (1999) Sow agression under spotlight. *Farmers Weekly*, 8th October.

Lean, I. (1995) Matching for size increases fighting. *Pig Farming* (April). pp 68.

MAFF (1993) *Pig welfare advisory group booklet series*. UK Ministry of Agriculture (now DEFRA). London.

Peet, B. (1993) Sow pen design agression and feed. *Farm Building Progress*, October, pp 3.

Pope, G. (1996) The cost of mixing pigs. *Pig Industry News* (Australia).

Varley, M. (2001) More space boosts piglet performance. *Farmers Weekly*, 22nd June.

Whittaker, X. and Spoolder, M. (1990) *The effect of ad-libitum feeding on a high fibre (sow) diet*. ADAS Research Report.

TAILBITING 23

(Flank and bar gnawing, PINT, ear chewing, vulva biting, urine drinking)

Chewing of the tail end which, if unchecked, can lead to severe erosion of the whole tail, sepsis, stress and even death.

TARGETS

Incidence should be nil. It has been suggested that 14% of all condemnations in Europe are due to tailbiting, so the problem is a major one. I see evidence of tailbiting (past or present) on 1 in 5 of the farms I visit.

CHECKLIST

The checklist gives an indication of what are the most common causes. (This is just one man's experience of witnessing or being asked to help alleviate several hundred cases)

Over 38 years I have seen a trend away from nutritional causes to those of environmental effects and possibly that of genetics.

We are feeding pigs better now, but maybe overcrowding them too much. Our pigs are growing very fast, and could be less docile than they were decades ago.

These are the main areas you should examine when tailbiting occurs.

Analyses of 236 outbreaks 1961 – 1999*

Areas of Attention			Importance Rating
✔	Overstocking		60%
✔	Ventilation	- inadequate	50%
		- wrongly positioned	50%
		- gases (CO_2, NH_3)	15%
		- low speed cold draughts at night	20%
✔	Badly placed pen furniture causing aggression		10%[1]
✔	Uneven mixing		18%[1]
✔	Poor trough design		15%[1]
✔	Sick pigs not removed promptly		60%
✔	Genetics (tendency to lose docility?)		15%[1]
✔	Nutritional	- salt	20%[2]
		- diet which pigs dislike	15%[3]
✔	Water inadequacy (all variants of)		20%[1]

At least 10 nutritional factors are said to be involved (like low protein, etc.) but I classify them still as largely unproven/undemonstrable. We feed pigs well these days – look elsewhere!

[1]   These cases/incidences I suspect are more important than the rankings indicate, at least since 1985.

[2]   Salt is strange. While the salt levels in tail-bitten pigs' diets are found to be adequate (0.4%), raising the level to 0.8% (Muirhead, 1989) *with plenty of water* does help cure outbreaks. But why? Could it be the presence of other unpalatable ingredients masked by an increase in salt?

[3]   Includes sudden change, and/or mycotoxin presence.

*   These are subjective findings : what one man's experience has found to have stopped the outbreaks when they were attended to, in order of approximate ranking when followed up later. However, how many of these cases would have cleared up anyway?

PRIMARY CAUSE – SOMETHING TO DO

Pigs have a natural rooting instinct. They are inquisitive, curious and aware animals, so when not eating, drinking and sleeping are looking for 'something to do'.

Yet we keep them in 'overcrowded' and 'unnatural' conditions. Modern production means we have to do this – compared to the wild, anyway. We have to ensure that we do not overstep the *overcrowding* or *unnaturalness* boundaries.

Both cause restlessness or low-level stress (anxiety). Happy pigs rest for up to 82% of their time, dozing away the hours, so 18% of their time is spent up-and-about being inquisitive and looking for something to do. And if they can't find something to do, then they'll look for trouble!

So we must give any confined grower something to do – and if rooting comes naturally, something to root – a ball, old 6" diameter ball-cocks from lavatory cisterns, an old tyre, a heavy log, some greens, even a sod of earth (very short-lived!) Not chains, these swing about, slap other pigs in the face and tend to raise the restlessness level! In the wild, pigs do not graze branches like cattle – they *root*. So give them 'rootability'. A very good diversion is a 2 metre (2 yard) long piece of toughened alkathene tubing, as pigs like to chew across an object as well as pushing it.

Next, keep the pigs resting – comfort and a feeling of well-being helps enormously. *Contented pigs do not tailbite!* I've seen it time and time again. If your pigs aren't dozing away 20 hours out of the 24 you should hold a restlessness audit! As for yourself, what you are doing, by mistake, to raise the restlessness level?

HOW TO APPROACH A TAILBITING OUTBREAK

While the effect of tailbiting is restlessness and stress, the causes could be many and interrelated. Because of this combination of factors it is difficult for the stockperson to pinpoint the one reason which has pushed the pigs over the brink.

Most producers are haphazard in their approach. The problem is all too visible and needs urgent attention, so a wide variety of remedies are applied at once. Instead (1) Remove all bitten pigs immediately. (2) Try to prioritise the *most likely* causes from the advice given in the following pages and attend to them first. (3) Only try one or two solutions at a time, and allow time – about 2 days, for any effects to show. (4) Record dates, pen numbers, weather changes, numbers of pigs affected and any noticeable changes in behaviour.

The purpose of this planned approach is to locate the cause and solve the problem permanently. It works!

THE FOLLOWING CAUSES HAVE ALL BEEN IMPLICATED IN TAILBITING

Overstocking

Correct stocking density for pigs (this is covered in detail in the Stocking Density section) is unfortunately not as simple as the recommended published allowances suggest. This is because the pig uses some parts of the pen as a living area and other parts which he will tend to avoid, such as wet or mucky floors, cold, draughty spots and areas where he feels constricted and may find it difficult to move away from if attacked. Long narrow pens are an example – fewer pigs (it is suggested –15%) must be housed in pens which are more than 2½ times as long as they are broad. A square or a 1½:1 rectangle is preferable. In any case, do not exceed the rule of thumb of 115 kg liveweight per m^2 (approx. 25lb/ft^2); this is particularly important when penning entires together, which can be more prone to tailbiting when overcrowded.

Table 1 indicates the effect of overstocking.

Pen furniture

It has been suggested that pigs, on awakening or from rest, tend to eat first, then drink, then urinate/ defecate, then 'socialise', then return to rest. It is perhaps best to locate these various areas so that the pigs 'rotate' around the pen to left or right rather than cross across each other and so invite antagonism (Fig. 1). How important this is in reducing tailbiting is not known, and it is probably less important than ensuring the correct stocking density level.

Even if stocking density and pen shape are correct – the positioning of pen furniture can calm or inflame aggression

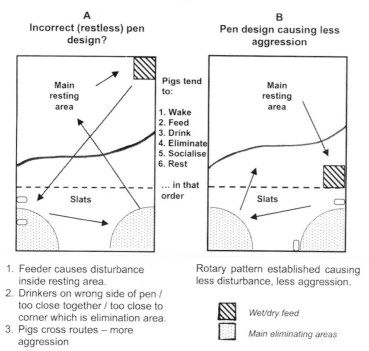

A
Incorrect (restless) pen design?

B
Pen design causing less aggression

Pigs tend to:

1. Wake
2. Feed
3. Drink
4. Eliminate
5. Socialise
6. Rest

... in that order

Main resting area

Slats

1. Feeder causes disturbance inside resting area.
2. Drinkers on wrong side of pen / too close together / too close to corner which is elimination area.
3. Pigs cross routes – more aggression

Rotary pattern established causing less disturbance, less aggression.

⬛ Wet/dry feed

▫ Main eliminating areas

Figure 1 Arrows indicate progression within the pen

Table 1   LIKELY COSTS INCURRED BY OVERSTOCKING A NURSERY AND A FINISHING HOUSE BY 15%

| | Pigs 6-35 kg (13-77 lb) | | Pigs 36-100 kg (79-221 lb) | |
	Correct density	+ 15%	Correct density	+15%
Daily gain, g (lb)	518 (1.14)	480 (1.06)	844 (1.86)	848 (1.87)
Days in pen	56	60	77	77
Overhead costs @ 16p/day (£)	8.96	9.60	12.32	12.32
FCR	2.02	2.12	2.42	2.63
Total food eaten in period, kg (lb)	58.6 (129)	61.5 (136)	157.3 (347)	171.0 (377)
Total food cost	14.06	14.76	27.06	29.41
Extra costs/pig (£)	**1.34**	**Plus**	**2.35**	**Total 3.69**

Savings in 15% less housing cost per pig (at £4.26 pig)**Savings 64pCosts £3.69**

Source: Client's records (trial)

Temperature and environmental factors

In my experience tailbiting does not necessarily follow from the pigs being obviously too hot or obviously too cold. Much more likely is the overall number of other stressors associated with too hot/too cold conditions. Again a stress audit is advisable – stocking density, temperature fluctuations, sufficient air changes, presence of new companions, dietary change, unpalatable/stale food, presence of a cold draught (especially at night) can themselves precipitate tailbiting when if absent, sheer temperature errors alone may not.

Check temperature variation across 24 hours

One of the commonest causes is diurnal (day to night) temperature *fluctuation*. The pig has a 'comfort zone' inside which there is little thermal stress if any. The comfort zone is the temperature above the LCT (Lower Critical Temperature) and below the ECT (Evaporative Critical Temperature). This band varies in width from 5C° in the weaner to about 11C° in the finisher so is quite narrow, and I have found regular excursions outside the relevant band for the age/weight of the pigs housed over a 24 hours period often triggers tailbiting. This may be the reason why runt, thin or 'nervy' weaners/early growers often instigate the process, as they are just not warm enough.

Always check the pigs are within their comfort zone *day and night* so as to avoid too much temperature fluctuation.

Draughts

Robertson (1999) recommends that to contain tailbiting an air speed of 0.15 to 0.3 m/sec is needed in buildings with temperatures below 20°C (68°F), and at above 28°C (82°F) the velocity should be higher, from between 0.74 to 1.3 m/sec.

In winter and especially at night, walls tend to be cold, so cold, heavy air will flow downwards on to a resting pig lying close up against them. In volume of (cold) air and in downward air speed these are not great, but over a period of time are sufficient to cause sufficient discomfort (stress) to precipitate tailbiting in the affected pigs which become restless. Nailing a 3 cm (1½ inch) wooden batten, triangular in cross-section, at 1 to 1.5 m

(3 to 5 ft) internally on such (outside) walls deflects the cold air onto the rising stream of warm air over the resting group and the draught is dissipated. Several puzzling cases of tailbiting have been cured once this simple and cheap device was installed.

Gases

High ammonia (NH_3) and carbon dioxide (CO_2) levels cause irritable behaviour, including tailbiting. This is seen in the cold winter months when ventilation adequacy is reduced in order to keep the pigs warmer/lower power bills. Pigs can be affected at NH_3 levels of only 10 ppm which is very easy to exceed. Tests have shown that winter month levels of 25 to 30 ppm NH_3 are quite commonplace in some grower houses/nurseries and a reduction to 12 ppm has stopped tailbiting.

So apart from attention to ventilation adequacy, keeping pens cleaner and drier (especially bedding), ensuring slurry channels are less than two-thirds full, *i.e.* 15 cm (6") from the slat under-surface, having slurry pit baffle boards to prevent cold up-draughts and the inclusion of a yucca ammonia-inhibitor like DeOdorase (Alltech) in the feed or in the slurry itself will all help. Dust is also a vehicle for ammonia particles (see Ventilation Section).

Tailbiting in kennels

Most kennels provide a dry and warm sleeping area yet are often the source of tailbiting. Why? Most are designed for cheapness and so have flat roofs, which contain a small ventilator and/or can be raised at the front. (Figure 2) In cold weather however kennel lids are shut down and only the ventilator used. This is insufficient as the air does not circulate, but hangs around at lid-height accumulating gases. The air depends mainly on the pigs' movement to agitate it, which at night can be insufficient or even non-existent over the sleeping animals.

Far better is to spend (about 18%) more on a staggered hipped roof. (Figure 3) This gets the air rotating naturally and with the ventilation open or part-open, the gases eventually disperse. Table 2 demonstrates the value over the years of such a design *à propos* tailbiting. The same sort of figures are probably equally true of streptococcal meningitis, another problem brought out by hot, airless conditions.

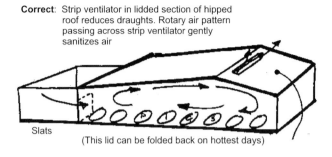

Correct: Strip ventilator in lidded section of hipped
roof reduces draughts. Rotary air pattern
passing across strip ventilator gently
sanitizes air

Slats

(This lid can be folded back on hottest days)

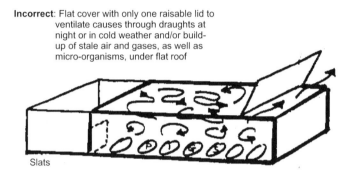

Incorrect: Flat cover with only one raisable lid to
ventilate causes through draughts at
night or in cold weather and/or build-
up of stale air and gases, as well as
micro-organisms, under flat roof

Slats

Figure 2 Natural ventilation of kennels.

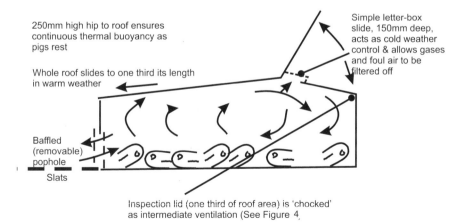

250mm high hip to roof ensures
continuous thermal buoyancy as
pigs rest

Whole roof slides to one third its length
in warm weather

Simple letter-box
slide, 150mm deep,
acts as cold weather
control & allows gases
and foul air to be
filtered off

Baffled
(removable)
pophole

Slats

Inspection lid (one third of roof area) is 'chocked'
as intermediate ventilation (See Figure 4)

Figure 3 Avoiding stale air in kennels.

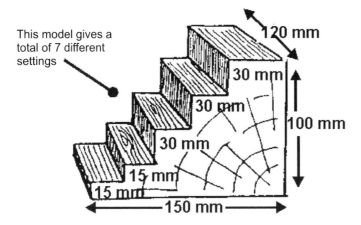

This model gives a total of 7 different settings

120 mm

30 mm

30 mm

30 mm

100 mm

15 mm

15 mm

150 mm

Figure 4 Example of a 'chock'.

Table 2 INCIDENCE OF TAILBITING IN FLAT AND HIPPED ROOFED KENNELS

	Flat roof used	*Hipped roof used*
Number of pens affected	27	5
Where roof line was changed to hipped	15	2

Clients' records (1971-1995)

Conclusions

(1) Tailbiting from whatever cause could be up to 5 times more likely in flat-roofed kennels.
(2) Where the roofline was changed to hipped, tailbiting was reduced sevenfold.

Lighting

Does bright light or very low light encourage tailbiting? Reduced light intensity has helped to reduce vices in poultry, and the same may be true of pigs (not proven). Pigs do seem to need about a 4 to 6 hour dark period every 24 hours (<15 lux), while continuous light over 60 lux may cause irritability? A largely unexplored area. EU welfare regulations will stipulate 40 lux for 8 hours/day in the near future.

Feeding

Nutritional errors have long been thought to be a precursor of tailbiting but this has probably been overrated. Some 40 years ago diets were less well balanced than they are now and manufacture is more consistent.
The following nutritional areas may be implicated:–

Salt

Check that the level is at least 0.5% (0.2% sodium).
If tailbiting is present, raising the salt level to 0.75% or even 0.8% is permissible for a time (5-10 days) ***providing adequate water is available***.

Other minerals

·   Total phosphorus should be greater than 0.5%.

·   Ca:P ratio should be less than 1.25:1

·   Magnesium could be increased to 0.08.% (American data). They use magnesium oxide at 1 to 2 kg/t, about 2 to 4 lb/ton, I'm told.

·   A proportion of the trace element provision should be of bioplex origin (mineral proteinates).

·   Some people use salt and special anti-tailbiting blocks (Frank Wright Ltd.)

Other findings are

·   Low soya bean meal (<5%) may aggravate it.

·   Spent mushroom compost (surface casein layer removed) cured outbreaks at 0.5 kg pig/day (about 1 lb) placed behind a wire mesh grid (Hillsborough, 2000). Some people have used fibrous peat.

·   Energy deficiency/imbalance could be involved in young fast-growing pigs; provide extra energy (Kyriazakis, 1996).

·   Increase tryptophan level (Aherne 1997).

Single space feeders

Queues at the SSFs are a common contributory cause, stimulating frustration and anxiety. Robertson (1999) suggests up to 70% of feeding visits can be cut short by forced withdrawal from the feeder. The writer has observed such action carefully and noted mild tail-chewing within 3 minutes among those pigs which were chased away.

Fitting dividers or 'stalls' to feeding positions "causes a significant decrease in this effect and had a subsequent effect on reducing the incidence of tailbiting," says Robertson.

In the case of the SSF, providing an extra feed station is the only alternative, apart from destocking below the accepted levels, which is often impractical, and can be costly.

Tail-docking

In Europe approximately 80% of intensively-housed pigs are tail-docked, where 75% or more of the tail is removed soon after birth, up to about 3 days old, but not in piglets more than six hours old so as to allow the piglet to have at least 4 good, uninhibited, colostral suckles. Alternatively, but less-favoured, in older pigs approximately 1-2 cms of the tail-tip only is removed. Cambac, a UK pig research consortium, after surveying tail-docking on over 40,000 pigs in a postal survey, concluded that from this evidence that tailbiting is three times more likely where tails have not been docked – 9.4% of undocked pigs suffered incidents of tail-biting while 3.3% of docked pigs were affected. We do not know how future EU welfare regulations will view tail-docking.

Veterinarians believe pain following tail-docking of very young pigs is transitory and research evidence (Noonan et al, 1994) suggests that done correctly it is no more distressing than simple handling.

Adverse effects may result from poor technique however, therefore stockpersons performing this task must be properly trained, preferably by a veterinarian.

TAIL-DOCKING CHECKLIST

✔   Two people are better than one.

✔   6 to 16 hours old to 3 days. 12-48 hours is best in the author's experience.

✔   Get trained in the technique and materials to use.

✔   Leave approximately 16 mm of tail length (Muirhead, 1997), about 2/3rds of an inch.

✔   The instrument must be sharp. Alternatively use a gas cauterizer.

Do *not* use a burdizzo ring as for lambs' tails.

✔   Preferably dip the wound in iodine cow teat dip antiseptic. In less developed countries a mixture of chemist-grade French Chalk and sulphamezathine powder 50/50 can be used, but do not let the powder mix get congealed, so renew frequently.

✔   Bleeding, if any, should stop in 30 seconds.

✔   Check for bleeding after 5 minutes. If it continues, apply a strong tourniquet – get your vet to show you how. Remove after 10-15 minutes.

Tail-docking can be governed by animal welfare legislation in certain countries. Check what you can and cannot do by law in your locality. Some areas of the world require veterinary presence or permission for certain simple operations.

FLANK GNAWING

In my experience this is much less common than tailbiting (my records suggest about nine times less). It is quite often seen in hot humid climates, and in N.W. Europe in summer rather than winter.

Young pigs which have been inclined to the process called 'PINT' (Persistent Inguinal Nose Thrusting) while suckling the sow or when weaned, on other pen-mates, may start gnawing flanks later on in life – up to 18 weeks old in my experience. PINT (also called snout-rubbing) seems to originate when sucklers have missed out on a feed and try to massage the udder after milk let-down has ceased. They may then develop the action as a habit and go on to nose-thrust the flanks of their pen-mates in the nursery, which is rarely objected to, oddly enough. Licking the flank, usually the inner back leg area of a reclining weaner can also occur. Wheaver the action, the area eventually becomes sore in the attacked piglets and any resultant exudate encourages other (non-PINT-inclined) piglets to nibble at the spot, eventually causing a wound.

The writer has seemingly cured several bouts of PINT reported to him as a farm problem, in subsequent litters on the same farm, by fostering-off smaller pigs in the litter who have started to PINT (who may be the instigators) on to placid sows with a smaller litter and with plenty of milk to spare.

EAR CHEWING IN SMALL PIGS

Ear chewing or nibbling may have the same origin as tailbiting, but the writer has noticed it more often in suckling pigs where the ear could be regarded as a substitute nipple for small pigs forced off the suckle, and in weaners in hot conditions, where the back of a companion's ear can exude moisture which either smells or tastes good. In these cases the back of the ear, not the tip, is made sore.

It is essential, as with tailbiters and ear-chewed pigs, to hospitalise them immediately before the wound becomes serious/infected.

Good, alert stockmanship is essential, *e.g.* check functional teat availability.

BAR-GNAWING IN STALLED SOWS

This is pure boredom, some producers and all welfarists get very worried about it, which seems to be rather unnecessary. The sow is only trying to entertain itself, akin to watching television in our case. Thus, like head swaying, it is a relief of stress caused by the very act of stalling sows, which is now banned in several countries on welfare grounds. Both, to my mind, are a good thing which doesn't make me popular overseas!

If you are still allowed to stall your sows – don't worry about these odd behavioural relaxations/ releases; instead turn your attention to keeping them warm enough at night with ample water to drink at all times – two common faults seen when I've been called in to advise on bar-gnawing!

VULVA BITING

A most frustrating vice, and little understood.

Erosion of the sow's vulval area by biting from other group-housed sows. Often occurs in late pregnancy and in time can spread to the majority of sows in the group. In the worst (unattended) cases the whole vulva can be destroyed and lives lost.

CHECKLIST

✔   Check vulvas and for bloody noses daily, especially towards the end of pregnancy when the vulva swells and becomes noticeable.

✔   Remove a bitten sow to a separate pen.

✔   Initially there is usually one offender – if noticed, remove her.

✔   Have your sows enough floor space? 2.7m²/sow is advised (29ft²).

✔   The problem is more common now ESF systems are used. This could be due to sows getting impatient when waiting to feed, so a well-designed plan where resting sows can see the feeder(s) and where the imminent waiting area is not constricted by a wall or angle will help reduce its likelihood.

URINE DRINKING
UK and Australian experience

This does not seem to be associated with tailbiting/flank gnawing as cases appear quite independently of each other. It may be associated in pigs under 6 kg (13 lb) weight with pizzle and ear sucking, but it is much more common in older pigs (12-20 kg, 27-44 lb) where it was not often associated with the latter (<10%).

Likely causes

- Water deprivation. *i.e.* below 2.2 litres/day/pig (half a gallon).

- Difficulty in obtaining water, *i.e.* flow rates for 5 kg pigs less than 250 ml/min (0.55 gal/min) and for 12-35 kg (27-77 lb) pigs of 500 ml/min (1.1 gal/min).

- Hot weather, especially if drinkers are poorly sited or the pens narrow.

- Overcrowding.

- Water quality *i.e.* slime/balantidium.

- Poor drainage, especially if any of the above points are borderline or below.

- High relative humidity in covered pens, especially if the ventilation rate is well below recommended levels. In these cases it can occur in relatively colder conditions. However, cases (in Australia) have occurred in hot but very low RH conditions and cured at once when the pigs were misted.

All cases studies responded to the alteration of one or more of the above criteria.

The following cases may also be implicated from a study of the rather subjective literature on the subject:

- Too low/too high salt level in the diet (one authority has suggested urine drinking is a precursor to 'salt poisoning').

- Too coarse grinding, barley, wheat etc and general gut disturbance due to harsh feed texture. (In my experience this is more likely to result in aggression/tailbiting, ulcers and colitis).

REFERENCES

Aherne, F. (1997) News and Views. *Western Hog Journal*. Spring 1997. p 24.

Hillsborough, E., Beattie, V. (2000) Mushroom Compost May Cut Tailbiting. Report on Agr. Res. Inst of Northern Ireland trial. *Farmer's Weekly*, 31 March 2000, p 47.

Kyriazakis, I. (1996) Tailbiting Can be Caused by Extra Energy. *Pigs*, **12**, p 40.

Muirhead, M. (1989) Personal communication, but see also in Muirhead and Alexander, *Managing Pig Health*, 5M Enterprises, 466-467.

Noonon, G. et al. (1994) Behavioural Observations of Pigs Undergoing Taildocking. *Appl. An. Behaviour Science*, **39**, 203-213.

LEGS & LAMENESS 24

Lameness: Any condition in the foot or limb which causes deviation in the animal's normal gait.

INCIDENCE

After reproductive disorders, lameness is usually the second most common cause of premature culling in sows (Table 1).

Table 1 TYPICAL CULLING INCIDENCE GIVING LAMENESS AS A PRIMARY CAUSE IN SOWS (%)

USA		9 – 11
Canada		7 – 10
Denmark		8 – 10
Mexico		13
France		7 – 11
Australia		9 – 18
UK	Pre stall ban	10 – 12
	Post stall ban	8 – 10
Sweden	Pre stall ban	8 – 10
	Post stall ban	7 – 9

Source: *Data collected by the author within the countries listed*

Foot troubles in boars and weaners, and Joint Ill in sucklers are also commonly seen.

ECONOMIC COSTS

- This is difficult to quantify, but calculations suggest that if premature culling due to lameness on a typical herd incidence can be reduced by 25%, gross margin per sow may rise by 9%. I have seen plenty of cases where a reduction of 25% has indeed been achieved. The higher costs come from an increased need for replacement gilts, reduced farrowing index, more overlaying and a lower herd lifetime litter size due to culling before peak productivity.

• Foot lesions in weaners when corrected for subsequent batches has improved days to slaughter by 6 and MTF by 10.7 kg (23.6 lb).

• Cases of joint-ill (from scuffed knees in suckling piglets) were reduced by 80% after attention and improved daily gain (7 – 9.3 kg, 15.5 – 20.5 lb) by 8% over the whole herd.

• Splay incidence, typically 0.9% has reduced gross margin by 2% per herd for each 1% incidence.

• The value of meat lost to leg abscesses in the UK is valued at 10p/ finished pig.

In 1997 the loss in the UK attributed to lameness was estimated at £7 to £12 million per year (Smith, 1997). With an annual output of 13 million pigs at the time this was a reduction in income of between 0.8 to 1.3% per year.

CAUSES OF LAMENESS

Genetic

The general consensus is that genetics are a minor cause. This may be true in some countries which use bedding, but world-wide I feel sure that…

• Selection for faster growth has increased the problem.

• A hot, humid climate tends to aggravate leg weaknesses in such genetic lines.

• Certain lines which throw strong straight leg bones nevertheless have joint weaknesses/lack of springing.

• Certain lines with fine, 'aerofoil-shape leg bones (bred to minimise bone:flesh ratio and thus improve KO%) may be prone to leg problems.

• Uneven claws may be inherited and lead to foot problems.

I suspect – and this is a suspicion, not an accusation – that some pig breeding companies have rather let leg quality go by default. I have been told that as many as 20% to 30% of their boars and gilts need to be culled from leg weakness problems after performance tests and this is a point to ponder.

Congenital (congenital = evident from birth)

Splay leg is a typical and common example with male newborns being affected twice as much as females. There could be a genetic overlay as LR and Pietrain sows seem to throw more splays than LW. Pastern weakness may also be congenital, even if it causes trouble later.

Physical injury

Bad flooring is the commonest cause (foot erosion and damage) but slippery surfaces cause joint, ligament and muscle problems.
Group fighting, and clumsy mounting at service can lead to leg strain.

Infection

A very common cause – joint ill, infectious arthritis and a wide range of bacteria – often getting in through cuts and abrasions – clostridia, salmonellae, streptococci, brucellae, etc. But this is not a veterinary textbook – you are referred to the 'Further Reading Section' for some excellent practical advice on this subject by other authors.

Nutrition

Problems like bone rickets and biotin-deficient cracked cleys are now largely a thing of the past, due to better diet design and mineral/vitamin supplementation, but the value of *organic* zinc supplementation seems to be a recent and novel advance in this field in strengthening foot (keratin) tissue. It is not expensive.

SOME LEG WEAKNESS TERMINOLOGY

In my experience veterinarians and veterinary diagnostic laboratories are fond of quoting a range of technical terms. Table 2 may help to clarify things for you.

Table 2   SOME VETERINARY TERMS USED TO DESCRIBE LEG AND FOOT DISORDERS

Osteitis:	Bone inflammation, a localised protective response to damage or destruction of tissue
Osteomyelitis:	Bone inflammation due to septic ingress
Osteochrondrosis (OCD):	Abnormal cartilage growth
Osteochondritis	Inflammation of bone and cartilage
Osteomalacia:	Softening of the bone in adult animals
Osteoporosis:	Bone loss
Bursitis:	Inflammation of a bursa (a protective sac where friction occurs over bone)
Laminitis:	Hoof damage (mainly horses)
Arthritis	Inflammation of a joint

LOOKING FOR STRONG LEGS –
A CHECKLIST

As you will see from the section on *Choosing a Gilt,* early leg and foot examination of new or replacement stock is of paramount importance. Here are my own views taken from many years of selecting gilts.

The set of the bones

✓   Figure 1 indicates how the legs viewed from the side should relate to each other and to the ground. While there are differences between breeds – for example the Welsh and Canadian Duroc are at opposite ends of the angulation stance – the mid point illustrated in the Figure is the one to have in your mind's eye.

✓   From the front the forelegs should be parallel and those gilts and boars at 70-90 kg (154-198 lb) selection weight with small inner cleys (claws) rejected. About 1.2 cm (½ inch) is borderline. Above this difference I find such cleys will not stand up well to slatted floors and the trait seems quite heritable in the immediate progeny, e.g. 30-35% sire and dam. Whether uneven cleys are as important on straw-bedded or outside sows is debatable until more research appears.

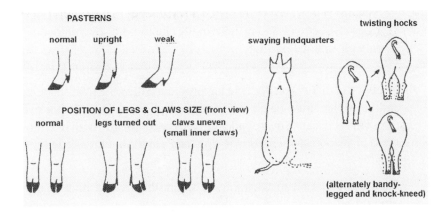

Figure 1 Examples of different clinical signs of leg weakness (Source: with acknowledgements to Dr I Johnson (Australian Pork Ltd) 1996)

✓   There is a tendency among pedigree breeders to praise 'fine bone'. While I understand the reason for this (said to be better KO%), it worries me. Such pigs display a pronounced aerofoil shape (Figure 2) particularly in the lower leg. I prefer a rather more oval, rounded shape particularly in hot humid overseas countries. It is no coincidence that one of the best selling breeds in the Far East has this shape of bone. Also that some indoor herds bedded on straw yards which I've known for over 10 years have fewer leg troubles than their neighbours in almost identical conditions which have had 'aerofoil cross-sections'.

In the author's opinion, when selecting against leg problems ….

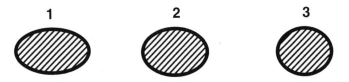

(1)   Aerofoil (ovoid). Fine bone, good KO% but tends not to stand up to hot climates and mud. Less good on straw.

(2)   Suggested compromise, neither too streamlined or too 'solid'. More of an oval shape.

(3)   Rounded, stronger, stands up well to soft floor conditions and hot weather but can give joint trouble and possibly higher bone to dressed meat ratio?

Figure 2 Cross-Sections, Pig leg bones

✓   Puzzlingly however, there seems no difference in these differing leg cross-sections among those of my clients with sows outdoors on well-drained pasture.

My advice is that if you do have leg problems, then the oval shape will stand up to the straw better than the aerofoil providing it is not over-done. This is because the rounder the cross-section the more likelihood there is of the poor, straight-down 'tippitoes' conformation seen in Figure 1.

Which brings us on to what causes leg problems and how to avoid them.

A SOW LAMENESS CHECKLIST
(PREDOMINANTLY FOR SLATTED SOWS)

Especially if the onset is sudden and/or visible, ***call the vet***.

For more gradually-occurring problems – those that have crept up on you I suggest the following:

✓   Get the animal up. Lameness will not be seen lying down.

✓   How does the sow get up/lie down? Stand? Move? Check *every* sow every day if possible for the first two actions, and as often as possible for the third, *i.e.* locomotion.

✓   Are the feet overgrown? If so seek advice on trimming.

✓   Are the cleys 'openly' cracked, allowing in secondary infection (bush feet)?

✓   Have you got the correct slat void areas? (Table 3).

Table 3   A GUIDE TO THE WIDTH OF SLAT AND GAP IN RELATION TO THE SIZE OF THE PIG

	Width of Slat (mm)		Gap size (mm)	
Farrowing Sows and Piglets				
(up to 30 kg, 66 lb)	50	(2 inch)	8 – 11	($^5/_{16}$ - ½ inch)
Finishers (up to 100 kg, 220 lb)	60 – 100	(2$^3/_8$-4 inch)	10 – 18	($^7/_{16}$-$^3/_4$ inch)
Sows, Boars and Finishers				
(over 100 kg, 220 lb)	80 – 100	(3¼-4 inch)	10 – 20	($^7/_{16}$-$^3/_4$ inch)

Source : MAFF (UK) 1996 (Revised to new draft Welfare Codes 2002, for discussion)

✓ Are the slatted edges pencil-shaped and unchipped?

✓ Is the floor well drained or slippery? Floor grip is important. Pressure-washing with a degreasing detergent helps materially.

✓ Are the stalls the correct length for the type of sow?

✓ Are the sows on new concrete? (Abrasive and chemically-caustic.)

✓ Any sharp edges to stall divisions, farrowing crate or flooring?

✓ Any awkward steps? Steps are OK except when animals are in a hurry.

✓ Do injuries occur when sows are moved? Check floor grip and fleeing space allowance. Provide hanging baffle boards if there is room.

✓ If lactating gilts are a problem, the farrowing crate floor space could be the area to examine. First parity gilts can have smaller, softer feet than 2nd parity sows.

✓ Is the slat surface sloped both ways towards the gap? If so *replace them* – the surface must be dead flat (or sloped no more than 2.0 mm, 1/16 in?) A common error in design done to assist drainage, but counter-productive in leg damage.

✓ Do not finish concrete surfaces for sows with a steel float. Consult the Cement Association's guidelines on surface screeds.

✓ Sow and gilt diets containing fat *must* have the floor pressure-washed with a degreasing detergent as routine. On 40% of farms with slippery concrete I found no degreasing detergent was used prior to clean-down. This is a very common error (*see Biosecurity Section*)

✓ If cleys are cracked, call the vet; there could be a variety of causes. But check the sow diet has contained Biotin at not less than 250 mg/kg for 6 months prior. This can be raised if horn problems are obvious.

✓ Do not grow gilts too fast. This can bring on lameness. 600 g/day (1.32 lb) up to puberty is adequate.

✓ Always try to exercise breeding stock where possible.

LAMENESS IN OUTDOOR SOWS

Soil contains a wide variety of pathogens of faecal origin which stone damage, especially from flints, can allow to enter. Mud accumulating between the cleys can harbour infection. Abnormal movement through restricted hut doorways and over badly designed creep restrainer boards causes bruising and strains.

Wet and waterlogged ground causes weakening of the horn tissue, especially among 'white-toe'd' breeds. The ability of pigs to move freely between huts, water and feeding areas without standing for long periods in puddled soil will reduce foot and leg problems substantially (by 60% on one client's farm). In my experience such waterlogged conditions are more vulnerable to either inertia/lack of conviction on the stockman's part, to lack of move-on ground and/or insufficient staff to accomplish it – than to climate!

The solutions to these dangers are obvious if sometimes difficult to carry out, but outdoor herds *must* be vaccinated against soil-borne Erysipelas as mandatory routine.

BABY PIGS : LAMENESS AND SWOLLEN JOINTS

Erosion of the sole and knees of recently-born piglets are the two most common problems. Once the outer tissues have been worn, cut or bruised, usually by 5 to 7 days old, a variety of organisms found in the farrowing pen can gain entry especially into the foot, and also joint cavities (septic arthritis or 'Joint Ill') and/or tendon sheaths over a joint (tenosynovitis) usually higher up the leg.

In my experience herd incidence is around 1% to 2%. While badly laid or eroded concrete and sharp perforated metals floors are the commonest and quite obvious causes, the condition is still puzzlingly common even when 'softer' more round-edged plastic slats are used.

Veterinarians suggest that this may be due to the infection gaining entry through the neonates' tonsils, from poor hygiene in the farrowing pen, clumsy teeth clipping and tail docking and even from the sow's udder during suckling, all compromised by a reduction in the piglet's immune defences for a variety of unclear reasons.

A LAMENESS CHECKLIST

✓ Catch infection early. Be observant. The signs of distress (shivering, lying in the sternum position while the others are supine, and a 'stary' coat, hairs raised) can well be apparent before obvious joint or foot damage or swelling – because the infection can gain entry from other sites and pass via the blood to the focus of infection, in this case the joint.

✓ Prompt treatment supervised by a veterinarian is essential.

✓ ***Concrete floors*** Examine, and if the aggregate is exposed (rounded pebbles exposed to 2 mm (1/16 in) or sharp-pointed grits only 1mm, 1/32 in, proud) rescreeding is necessary, preferably with one of the bitumastic epoxy-resin or latex/ rubber-chip products.

✓ Try the in-situ knuckle-test; clench your fist and rub your mid-joint knuckles with only modest pressure over a debris-cleared area where the piglet's knees and back feet are likely to scrabble when getting to a teat. If this process is noticeably painful, rescreeding is advisable.

✓ Bedding, *i.e.* shavings, chopped paper used sparingly will help, as will limewash, but because piglets will scrabble when suckling, if the concrete is worn, bedding may only be a palliative.

✓ Scuffed scabbed knees are often noticed but ***rear foot damage is less visible and a more common source of infection.***

✓ ***Perforated metal floors*** : Many have too much void area for a baby pig's foot and sharp edges – as in expanded metal, for example – are a nuisance. Table 3 suggests maxima/minima. Replace with a well-designed plastic or plasticised metal slat – there are plenty on the market.

Hygiene

✓ Improve post-farrowing pen hygiene if foot joint problems are over 3% and certainly up to 12-15% as I have seen on some farms.

✓ Ensure tooth clippers are sharp and clean. Wash between litters.

Do not cut tails and clip teeth with the same instrument. Consider grinding but follow directions carefully.

✓   Remove tails with a disinfected scalpel or very sharp scissors.

✓   Check up on your anaemia inspection routine.

✓   In bad cases spray the sow's udder with a teat antiseptic or apply dairy teat dip.

Seek veterinary guidance for all of these procedures. An annual check-up by the vet watching what is usually done is valuable.

SPLAY LEG

This condition appears to be failure of muscle fibres in the lumbar and especially ham areas to develop in sufficient numbers to provide adequate support at birth. As a result by 2 to 3 hours old the piglet cannot stand and subsides into a splayed and flattened stance.

The scientists are looking for the causes, but in the meantime up to 1.5% of our born-alives in the U.K. are starved out or crushed *even though their hind legs are taped.* That's too big a waste – about 30 pigs lost on a 100-sow farm each year.

Ten years ago I worked with farrowing attendants on four farms. We cut the loss by 90%!

Out of 16,500 born-alives, we had 322 splays – and we lost only 17!

We estimate, even with the taping that is currently recommended, we would have lost at least 150 of these splays.

SPLAY LEG – A CHECKLIST OF POSSIBLE CAUSES AND SOME SOLUTIONS

✓   Check on which *boars* may be throwing them, rather than sows?

✓   Males are more prone to it than females

✓   Predominantly LR or Welsh strains may benefit from LW blood

✓ Too low energy sow food in pregnancy *i.e.* only 30 MJ/DE day for a 250 kg (550 lb) sow at mating. Check with your nutritionist.

✓ Low Vitamin E/*organic* selenium levels, may be allied to high PUFA fatty acids. Check with your nutritionist.

✓ Mouldy feeds – fusarium mycotoxins in particular. Use a mould inhibitor *and* a mycotoxin absorbent in the sow's feed. (*See Mycotoxin Checklist section*)

✓ Slippery floors and all-metal slats may make it worse. Use temporary mats/carpet-pads/paper.

✓ Stressed sows throw more splays.

✓ Parvovirus infection may precipitate it, as can PRRS.

✓ Large litters – especially if farrowed a day early. Check your prostaglandin technique.

✓ Taping the legs helps piglets recover (*see below*)

✓ Correct massaging likewise, but is time-consuming (*see below*)

SINGLE-TAPING

There are two problems with taping (as distinct from double-taping, which I'll describe shortly.) First, having taped the piglet's back legs with ordinary surgical tape just above the pastern joint so that the legs are 2-2½ in. apart (6.5 cm), the stockman tends to plug it on a teat and go off on his rounds.

Not good enough! In fact, this in only half the job, as that piglet is now too stressed and too immobile to compete. So he loses his vital 40 cc of colostrum in that first hour.

In our procedure, we dry the piglet off, tape it up and then milk about 20 cc of colostrum into a clean, dry syringe, or have it ready in the fridge to be warmed. Any current sows colostrum will do. We dose the taped pig with that colostrum. And the dose needs to be repeated an hour later (within 1-1½ hours of birth, if possible). These problems do occur with overnight farrowings, but even then, it should be the first job on arrival in the farrowing room.

But even with the colostral feeding we are not finished. Next, all treated splay-legs are moved to a milky sow and allowed one full suckle while her litter is shut in the creep area. Each splay may need assistance to "settle" at a teat.

Lastly, the pigs are returned to their own litters and the tape is removed in two or three days, or earlier if locomotion is seen to be activated.

You will lose very few splays if this routine is followed. Just taping and hoping everything will be O.K is insufficient.

Refinement

There is a refinement to this technique:

DOUBLE-TAPING

In this refinement, the rear legs are taped normally, and then, usually with a little help from an assistant to hold the piglet while you do the taping, the legs are *gently* moved forward, under the piglet's trunk and slightly to one side (***see Figure 3***). Then a second band of tape is placed over the body, just forward of the pelvis, so as to hold the legs close to the body. The piglet is now well and truly trussed!

Figure 3 A double taped piglet.

This means you have to put it into a straw box, give it its dose of colostrum and leave it there with any similar unfortunates, for three to four hours. This is important, because it cannot move far and will certainly get crushed if left with the sow.

Why immobilize it so? Strange, but it seems that the attempts by the piglet to activate the rear leg muscles when the rear legs are in this position "kickstarts" the muscles into action. The body tape must be removed in three or four hours (not two or three days) and the piglet will achieve quite good locomotion at once. It is a quicker method of treatment, so the piglet returns to a good milk intake that much sooner.

MASSAGE

Here's another technique you may like to try. This practice is recommended and referenced by Philip Blackburn, an experienced pig vet, from observing a farmer do it successfully. The technique is explained in Table 4.

I've done quite a few splays this way and it certainly works, especially in the really bad cases which have front-leg trauma as well. But it takes time, perhaps eight minutes of massage in three or four two-minute sessions. However, it is a humane and positive technique compared to the cruder taping, which must stress them some. I recommend it for the real "stiffs" and for the farmer lucky enough only to come across the occasional splay.

Table 4   THE BLACKBURN MASSAGE TECHNIQUE

1.	Mark the piglet (you'll need to pick him out again).
2.	Sit or kneel down and place the piglet, head away from you, with his chest resting on your knee, which is inclined downwards.
3.	Hold the piglet with a hind leg in each hand, gripping the lower legs with your last three fingers, and the crutch with your index finger.
4.	Vigorously massage the lumbar muscles with your thumbs.
5.	Drop down either side of the tail to the ham muscles, dealing with the muscles over the pelvis as you go past it and using your index finger to massage the inside of the leg.
6.	After about two minutes, the muscles will relax and each leg can be flexed quite easily. What you have done is improve blood flow to the muscles and activate nerve responses.
7.	If the forelegs are affected, hold the upright piglet between your knees and give similar treatment to the front legs.
8.	Encourage the piglet to stand up and proceed with the colostrum-dosing as per normal.
9.	This may need to be repeated three or four times during the first day.

MYCOPLASMAL ARTHRITIS

An increasing cause of lameness particularly in gilts is caused by the organism *Mycoplasma hyosynoviae*. I mention this before referring you to a veterinarian because I find diagnosed cases are often associated with stressed, overcrowded gilts in dusty conditions.

Because there are so many other causes of lameness, the veterinarian needs to do a test to confirm mycoplasma is the cause. Once confirmed, strategic in-feed medication is effective but I find my own management checklist on how to avoid it has been useful.

A SUGGESTED PREVENTION CHECKLIST FOR MYCOPLASMAL ARTHRITIS

✓ Avoid 'backtracking' where older (immune) pigs transfer the organism to younger (susceptible) stock. Segregate by age and be careful not to allow movement around the farm where uncleaned passages used by older pigs – to the shipper's truck for example – are run across when moving weaners.

✓ Stress lowers immunity. Overcrowding, fighting, low temperatures (3°C, 37°F, or more below LCT), pronounced draughts (more than 3m/sec at LCT) and harsh flooring are particular stressors among young pigs and new intakes.

✓ I find MA is often associated with dust. This is because it seems to be a ***respiratory*** initial entry route, then the organism proceeds via the blood to the joints and tendons. Poorly ventilated (stuffy) kennels are a suspected source, I'm sure. Farmers are surprised that, of all things, dust could cause lameness!

✓ The susceptible period seems to be, in my experience, 4 to 6 weeks after arrival (gilts). *Ad lib* feed to boost their immunity at this time and consider a special replacement diet.

✓ Check out your housing and maximum ventilation rate at this time so as to minimise the effects of housing/re-housing stress.

REFERENCES

Blackburn, P (1988) 'Pig Farming' Health Supplement, Oct 1998

MAFF (UK) (1996) Action on Animal Welfare 'Lameness in Pigs' p5.

Smith W (1997) 'Lameness is Industry's Achilles Heel' 'Pig Industry' March 1997 p15

(UK) Cement & Concrete Assocation (1986 et seq) Farm Concrete leaflet Nº 6 Retexturing concrete. Nº 8 Repair of Concrete Floor Surfaces

MYCOTOXINS 25

Mycotoxin: A poisonous substance produced by a fungus.

THE PROBLEM (MYCOTOXICOSIS)

The fungus can be innocuous or even dead but still leaves mycotoxin residues behind, commonly in stored and mixed feed, and especially in mouldy grain. Processing the grain, while it may destroy the moulds, still leaves mycotoxin residues behind. These need only be present at a few parts per billion in the feed (*Table 1*) to cause problems in pigs – infertility, anoestrus, prolapse, false pregnancies and embryo mortality; poor growth and vomiting. They can pass through sow's milk and remain behind in any slaughtered carcase. Damp straw is another common source of moulds.

Table 1. WHAT LEVELS OF MYCOTOXINS ARE TOXIC?

Aflatoxin	200-500ppb
Fumonisin	5ppm non-ruminants; 100ppm ruminants
Deoxynivalenol	1-10ppm
T-2 toxin	100ppb
Zearalenone	200-300ppb

ppm = parts per million ppb = parts per billion

Mycotoxins may also compromise the pig's immune system.

PROBLEMS WITH DAMP GRAIN

Figure 1 shows the relationship of grain moisture content and humidity in the storage barn.

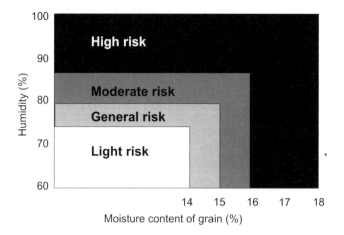

Figure 1 Likelihood of mould contamination in stored grains

BUT IS MYCOTOXICOSIS A SERIOUS PROBLEM ON PIG FARMS?

Probably. Mycotoxicosis is a growing disorder worldwide partly due to global warming, (El Niño, etc,) but badly maintained harvesting equipment and conveyors on farms and in mills can crack the grain pericarp that protects the grain and allow surface moulds, before drying takes place, to use the exposed starches and protein. Feed and grain bins with leaking roofs or poor ventilation can raise the moisture content over 14% (Figure 1). Even so moulds need less moisture to grow than bacteria. With the right conditions, they can increase tenfold over 3 days.

THE IMMUNOSUPPRESSIVE EFFECT OF SOME MYCOXINS

Immunosuppressive Mycotoxicosis (ISMT) is possibly much more of an insidious problem than farmers and their advisers recognise.
I have surveyed the opinions of the last 20 pig producers I've talked to in 4 countries and the responses were very similar . . .

Have you heard of mycotoxins?
Certainly.

Have you a mycotoxin problem?
Doesn't seem so.

How do you know this?
Our grain isn't particularly damp or mouldy. I haven't seen any
excessively swollen vulvas, and our nursery pigs don't vomit and they
eat well enough. No excessive cases of prolapse – a few, maybe.
So I don't see the symptoms you guys tell me to look for.

Do your pigs suffer from common disorders like E coli scours,
Swine Dysentery, Ileitis?
Good Lord yes, from time to time, but they're not mycotoxin
problems, are they ?

What's your daily gain?
Let's see now; we get from weaning to 100 kg (220 lbs) in about 160
days. Not brilliant, I admit, 585 g /day (1.3 lbs/day). Could be better,
I suppose. Here, are **you** saying that's due to mycotoxins – how do
you know ?

   . . . And so on. That little discussion just about sums up the situation
world-wide. Pig farmers and their advisors don't realise the insidious
nature of mycotoxins. If some are suppressing immunity to other
common diseases, then their effect is more serious than many people
realise.

Dr Devegowda, a world authority on mycotoxins, has said that 25% at
least of world grain is affected by moulds.

'WON'T GO AWAY PROBLEMS'

Some 25 years ago or more I was plagued, in my farm advisory work,
with what we called "Won't-Go-Away Diseases". Mostly they were
disease-based but some were production problems like sluggish growth.
Nothing we did seemed to help. Sure, things like E coli scours, Swine
Dysentery and the pneumonias responded to in-feed treatment, but this
cost a lot, usually knocking 20% off the profit picture. But at least it
was less costly than having the disease present, so it was done between
gritted teeth.

Eventually they went away and the medication was thankfully taken out. Back they came again! Repeat procedure, *ad nauseam.* "Won't-Go-Away Disorders". The nutritionist couldn't cure the process, neither seemingly could the vet or the ag. engineer – or myself for that matter. We were all stumped.

A SOLUTION

Then I met a Scottish pig vet called Sigurd Garden. He told me about mycotoxins; how the problems were more common after harvest, among farm-mixers, after a wet spell. He knew all about Won't-Go-Away problems, too.

"John, we're getting excellent results by just steam-cleaning their bulk bins. Not just the 'crud' at the top but doing the sides and bottom exit trunks. Trouble is, that it is such an awful job, you really can't get in there any sense – you need a special inside ladder and a boy to descend it as a grown man usually can't get his shoulders through the top hatch with the ladder present. And a special delivery pipe from the steam generator up-and-over. All too much hassle – and dangerous as well. We need a swing-away exit boot as well to drain off the detritus. But what it really needs is an entry door at the side (see Figure 2), then it all becomes simple enough. Then they'll do it as routine, twice a year, I suggest, but once in summer so it dries out quickly seems okay.

Can you write an article telling manufacturers to, for Pete's sake, put in a bulkhead door so bins can be properly sanitised?"

I did, three in fact, and that was in the 70's. The extra cost on a bulk bin would be 12% for the access door and about 10% more for a swing-away trunk rather than one which has to be laboriously unbolted and reconnected again.

Have the bin manufacturers taken any notice? Some have, but certainly not all. They need to.

WHAT'S THE SITUATION TODAY?

1. Almost certainly bulk bins and continuously-used feed hoppers are major source of mycotoxin build-up.

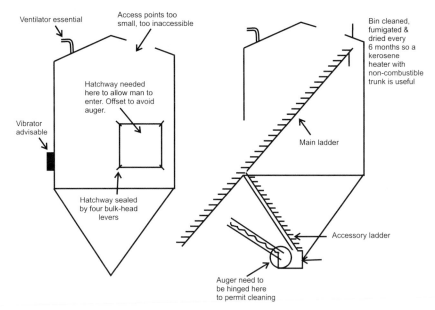

Figure 2 Bulk bin improvements/bulk bin hygiene

2. While bulk bins can periodically sanitised, at present this is far too laborious and can be dangerous.

3. Too much emphasis (in my opinion) has been placed on the direct problems from mycotoxins, and nowhere near enough on the way that a build-up of quite small quantities of a variety of them can depress the animals' immunity to a wide range of quite common bacteria and viruses which otherwise wouldn't present much of a problem.

    These are the "Won't-Go-Away Diseases".

4. The indirect immunity problem could be an enormous one. I have no real idea how big, but the indirect effects of slower growth and a higher disease incidence must, I guess, be reducing income / raising costs by 20% on those pig farms (25% at least) who have (unknown to them) the problem of ISMT

5. Then there are the direct effects. Table 2 gives case history results which are significant in themselves.

6. What about slower growth? Table 3 gives 3 case histories where growth was improved by 18% directly after steam-cleaning the bins. In terms of labour cost, this gave an REO (Return to Extra Outlay Ratio) of over 30:1 across 6 months, when the steam cleaning was done again.

Table 2   CASE HISTORIES OF SUCCESSFULLY REDUCING OR REMOVING ENTIRELY THE FOLLOWING PROBLEMS AFTER STEAM-CLEANING BULK BINS TWICE YEARLY (SPRING AND AFTER HARVEST)

	Positive result	*Negative or inconclusive*
1. Swine Dysentery	3	1
2. Abortions in Gilts	2	–
3. Prolapses	2	1 (*Other measures tried*)
4. Mummies	1	–
5. Returns to service	2	–
6. Respiratory infection (non-specific)	3	1

Source: Author's records

Table 3   CASE HISTORIES OF BEFORE-AND-AFTER GROWTH RATES ON 3 FARMS WHERE BULK BINS WERE STEAM-CLEANED

	Weight range kg (lb)	*ADG before, g/day (lb) (Av. 12 months)*		*ADG after, g/day (lb) (Av. 6 months)*	
Farm 1	6 - 90 (13-198)	572 (1.25)	Little disease	652	(1.43)
Farm 2	30 - 90 (66-198)	607 (1.33)	Disease present (SD then colitis/ileitis)	781	(1.72)
Farm 3	25 - 86 (55-190)	616 (1.35)	Little disease	697	(1.53)

Source: Client's records

CHECKLIST: WHAT TO LOOK FOR ...

...ON THE FARM

✓   If you have a 'Won't-go-away-disease' picture, suspect moulds.

✓   Have you black/grey detritus on the upper surfaces of bulk bins?

✓   Any above-normal bridging problems?

✓   Deposits in unscoured corners of conveyors?

✓   In *ad-libitum* feed troughs? (Last year I noticed stale, fermenting/putrefying food residues in 38% of the self feed hoppers for group-fed dry sows)

✓   In the unscoured areas of wet mixing tanks?

✓   Have you mixed a wettish batch of grain in with a clean sample recently?

✓   Check your grain store for damp, discoloured patches.

✓   Any detectable heating in feed or grain?

✓   Know the moisture content of your bought-in grain, <12%.

✓   Are the grain storage areas well ventilated?

✓   Check straw and wet bedding. Have shavings got damp?

...WITH THE PIGS

Direct effects: · Poor growth

·   Vomiting, especially in baby pigs

·   Inappetance

·   A variety of reproductive signs – returns to service, anoestrus, swollen vulvas, rectal prolapse in young stock, embryonic death, abortions. Consult your veterinarian as there are plenty of other causes for these disorders.

The mycotoxins most commonly involved on pig farms are:-

Mould	*Mycotoxin*
Fusarium spp	Zearalenone; Deoxyvalenol (DON), T2, Fumonisin.
Aspergillus spp	Ochratoxin, Aflatoxins

'Indirect'effects: **(Immunosuppression)**

Aflatoxins, T2 and DON have been suspected but others may be involved.

Knowledge is still sparse, but I suspect a variety of common bacterial ailments may be involved.

Also, is a mycotoxin one of the 'trigger' factors which turns PCV-2 circovirus from a harmless virus into one causing PMWS? And an aggravating factor in PRRS?

CHECKLIST: WHAT TO DO IF YOU SUSPECT MYCOTOXICOSIS

✓   Check and sanitise all feed troughs.

✓   Remove any suspect grain/feed. If you have to use if off, dilute not less than 1:10 with clean feed for slaughter pigs (not breeding stock); preferably dispose of it completely.

✓   Take a sample. Because of the very small amounts likely to be present, 10 samples must be taken from equi-spatial sites in the bulk bin or grain store.

✓   Each sample must contain 1 kg (2 lb) of material. Mix thoroughly together.

✓   Take 4 x 1 kg samples from the primary mix, bag separately in *paper* bags (so as to avoid moisture formation), cool to 5°C and transport to drying facilities as soon as possible, *i.e.* to the analytical service. Some people use two laboratories in view of the low levels to be detected.

✓   Keep one sample yourself; store at 5°C (41°F) or less.

✓   When the results arrive, liaise with your veterinarian.

CHECKLIST: ROUTINE MEASURES TO TAKE TO PREVENT MOULD GROWTH

✓   *Dry all grain* to <12% moisture and store well-ventilated. Moisture is the single most important factor determining mould growth, especially if humidity is above 80-85%. Carbon dioxide production is a measure of mould activity and Figure 3 shows the dramatic reduction once 12% moisture is achieved. Above 16%, and even a modern mould inhibitor is blunted in its effect.

✓   *Sanitise all storage bins* and troughs. Bulk bins should preferably be steam-cleaned in spring and late summer to allow thorough drying before re-use, otherwise use a kerosene blower drier.

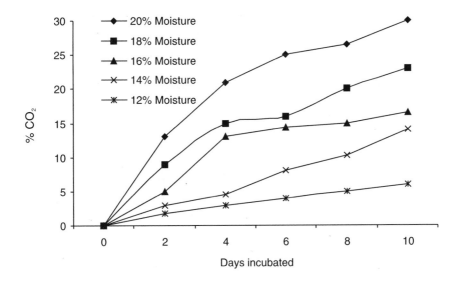

Figure 3 The effect of moisture on mould growth

✓   *Be extremely careful* about entering bulk bins due to residual gases. Have a second person present and prepare an extrication procedure.

✓   *Use a Modern Mould Inhibitor*. In the past the simple and relatively cheap proprionic acid was used in stored grain or feed. As well as being corrosive to metal surfaces, propionic acid soon loses its punch. Today science can do much better. By using a combination of similar organic acids the activity is lengthened, and by buffering them the corrosion aspect is much reduced. (Table 4)

Table 4 COMPARISON BETWEEN PROPRIONIC ACID AND A NEW COMBINED MOULD INHIBITOR (MOLD-ZAP)

Treatment	Relative active agent content	% loss after 6 days	Corrosive effect
Propionic Acid	100	67	100
Mould Inhibitor Complex	70	3	4

Source: Alltech (1998)

The advantage of a combination of organic acids is that they compensate each other for the minor disadvantages of each acid when used alone. Also, dose-response curves have been well-established so that the correct levels of each acid are present to achieve maximum effect, minimal cost, widen the range of moulds treated, and improve safety and ease of handling.

✓ *Use a Mould Absorbent* even with drying and the use of a mould inhibitor. This is because some mycotoxins will get through just because they are dangerous at such tiny quantities. The problem here is that continued ingestion by the animal of extremely small levels of some mycotoxins, *e.g.* 15 to 20 parts per *billion* *, can cause direct problems in the animal, but even worse, may get into the human food chain where maximum permitted levels have now been established – and they are extremely low. The European Union, by the way, has the lowest permitted mycotoxin ceilings in the world in food products.

*I am told this is equivalent to 7 to 10 grains of sand in a 25 kg (56 lb) bag.

THE DRAWBACKS OF CLAYS

Up to now certain clays, like bentonite, have been used quite effectively as nutritional binders for the major mycotoxins. Trouble is, a lot is needed (a minimum of 4 kg/tonne and up to 10 kg/tonne for some of them, 9-22 lb/ton) and they can bind other useful nutrients. They pass straight through the pig and so form a hard sediment in the slurry pits, difficult to dislodge. Even at the lowest effective usage level of 4 kg/t a 100 sow farrow-to-finish unit will accumulate 2.2 tonnes of clay binder in the slurry pits every year !

GLUCOMANNANS

What is eminently preferable is to use natural nutrients in the feed – forms of sugar – to do a much better job. These 'lock-up' a much wider range of mycotoxins (clays mainly affect aflatoxin) at an inclusion rate at least 8 times lower than with bentonite. Table 5 shows the broad spectrum attack possible with this type of naturally sourced sugar derived from yeast.

There is no insoluble waste in the slurry, and research from several institutes in India, Canada, USA and Argentina confirm better pig performance, higher immune levels and it is easier to keep below the permitted maximum mycotoxin thresholds in milk and meat.

HUGE SURFACE AREA AVAILABLE TO 'MOP-UP' MYCOTOXINS

Mycosorb (Alltech) is used at between 0.5 to 2 kg/tonne of feed (1-4.5 lb/ton) according to the likely severity of mould presence. 0.5 kg (or 1 lb) of the commercial product Mycosorb has an absorptive surface area equivalent to 1 hectare or 2.5 acres!

So when added to feed a typical pig would eat sufficient to provide $8m^2$ (almost 8½ sq yds) of absorptive area each day.

Table 5   GLUCOMANNAN'S BINDING CAPACITY

Aflatoxins (B1 + B2 + G1 + G2)	85%
Zearalanone	66%
DON	13%
Ochratoxin	12%
Citrina	18%
T2	33%
Fumonisin	67%

Source : Trenholm (1997)

COST BENEFITS

The cost of the commercial product (Mycosorb, Alltech) is about £2.00/ tonne of feed, maybe £1.30 when the clays are replaced. Preliminary estimates from the research which is measurable econometrically suggest MTF (Meat Per Tonne of Feed) could be increased between 9-14 kg (20-31 lb) in pigs, giving an REO (Return on Extra Outlay) of 5 to 7:1. What part lowering disease plays in this is uncertain – the figures all come from improved physical performance.

SUMMARY AND CONCLUSIONS

- Mycotoxin effects on performance and disease incidence are underestimated by pig producers, and possibly by their advisors.

- Much better cleaning of bulk bins and troughs is advisable.

- After this four countermeasures are available – which should address the problem.

- The in-feed precautions are new and sophisticated; the mode of action is well understood and thoroughly tested and published by recognised and independent research establishments in several parts of the world. The resultant products are freely available at a reasonable cost.

- The econometrics are encouraging and farmers should consider making provision in their supplemental feed costs for the products/measures described.

IMMUNITY 26

The condition of being immune, or non-susceptible to the invasive or pathogenic effects of micro-organisms, viruses and cancers. The mechanisms of immunity invoke the body's ability to detect and combat substances within it which it interprets as foreign to its wellbeing. When such substances enter the body, automatic complex chemical reactions are commenced to defend the body's cells and tissues.

When I was a farm student many years ago, we had very few vaccines and no in-feed antibiotics. As a result we tried to know all about what we then called, in our ignorance, 'natural resistance'. As a possible result disease levels were surprisingly rather less on pig farms than they are today.

A CHECKLIST OF 40 YEARS AGO. HOW TO MAINTAIN 'NATURAL RESISTANCE'

✓ We knew for example that our farm could be too dirty – and too clean; thus the solution was to try to get the balance right.

✓ We knew that we needed a good proportion of mature sows in the herd – that way disease was less prevalent.

✓ We knew that we needed good strong weaners – not to wean too soon or at too light a weight. We ignored the new fashion for 21 day weaning and weaned only when all the piglets were well over 11 lbs or 5 kg – more like 24-26 days. (Small pigs we back-fostered until they were 5 kg.) So we weaned by *weight*, not by *date*.

✓ While we weren't all-in/all-out, we knew that meticulous cleaning of the farrowing and weaner follow-on pens was essential.

✓ We quarantined all new stock and practised our own method of feedback to in-pig sows (afterbirth and a six month roll-over

395

mixture of minced-up piglet guts kept frozen in the fridge. Today, however, this may not necessarily be the right thing to do).

✓ We didn't serve our gilts too early – in those days over 115 kg (254 lb) seemed adequate, but not today – they need to be even heavier and grown less quickly to 125 kg or heavier (275 lb).

✓ Stalls had yet to arrive. We kept sows as far as possible in groups on bedding and could afford plenty of space for them.

✓ We didn't use the vet very often as a result. (There were no pig specialist veterinarians in our area.)

Now I'm not saying the modern pig breeder needs to follow that advice to the letter, as some of it was misguided or expensive and parts of it dangerous under today's conditions. But re-reading my students' notes of the 1950's one thing stands out clearly compared with today's average producer, **we knew – instinctively – about immunity.** We had to!

IMPROVE YOUR KNOWLEDGE ABOUT IMMUNITY

We have to raise our awareness of how to stimulate natural immunity in all our pigs, but especially the sow and the baby pig up to about 20 kg (44 lb). Why ?

First : Because several 'new' viruses appearing on our farms are at present shrugging off existing preventive vaccines – and for many others of them there aren't any, yet. But even if one new virus is protected against – another seems to appear.

Second : The old-favourite drugs – especially the in-feed ones – which are useful to control the secondary infections which the viruses allow in, are being increasingly constrained by bureaucrats and the buying public mainly on grounds of antibiotic resistance.

Third : We are stressing our pigs more in the race for productivity at all costs. Stress neutralises or inhibits immunity.

Fourth : We have been shown how much food energy and other nutrients the pig uses to rebuild his damaged natural immunity (Table 1) or to set up the necessary defences should the challenge be high.

Table 1 BY HAVING TO COPE WITH A HIGH DISEASE CHALLENGE, GENETICALLY IMPROVED† PIGS 6.3-27.2 kg (14-60 lb) EAT LESS, GROW SLOWER AND HAVE A POORER QUALITY CARCASS.

| | *Immune stimulus required* | | |
	Low	*High*	*Difference**
VFI, kg/day (1b/day)	0.97 (2.1)	0.86 (1.9)	12.8% more
ADG, g (lb)	677 (1.5)	477 (1.1)	42% more
FCR	1.44	1.81	25% better
Protein gain (g/day)	105	65†	62% more
Fat gain (g/day)	68	63	8% more

* In favour of low immunity needs
† The leaner the genotype the more the protein gain is damaged
Note: Both sets of pigs could be considered "healthy". A high disease challenge is described typically as a 'pig sick' building and a low disease challenge environment as 'all-in/all-out/multisite' scenario, properly disinfected.
Source: Stahly et al (1995)

This subject of immunity is so important to lowering our costs and in defending our profits that farmers need to understand it thoroughly. Failure to do so, and not act on what such understanding reveals must mean the costs of keeping their pigs healthy will rise substantially.

ACTION PLAN TO IMPROVE YOUR PIG HERD IMMUNITY

Here is my advice on what every pig farmer needs to do to establish a better immune status in his herd. Time is short and you need to act now. Do it at once – the viruses won't wait.

1. *Study the subject.* Go to every meeting you can on the subject of immunity. Read and file articles. Talk to your veterinarian about your own circumstances. Within one year you must be as knowledgeable on the subject of immunity as you are today, for example, on mating procedure at which you are expert.

2. *Contact a specialist pig veterinarian.* Compared to the 1960's a good local pig vet is often available. Many of us have them on our doorsteps. So use them – it is one big advantage Europeans possess over the 'low-cost' pig producers in the Far East, for example, where pig veterinarians are scarce or a long way away.

Show him (or her) round the farm, give him time to think (and maybe do a few tests) and then have a 'what-to-do session' with him.

His or her action plan may or may not involve remodelling expense – it very much depends on a lot of things, including how you both decide on your present and future exposure to the 'new' virus diseases. *My experience is that the remodelling needed is often – maybe usually – far less costly than the theorists have proposed in print.* (Table 3, page 401, line 4) So don't panic. For example, some disease-breaking ideas need not be onerous, but they will involve an altered and meticulously-followed routine. If you understand how immunity functions you will convince/discipline yourself to do – and spend – what is necessary.

3. ***Adopt a more disciplined approach.*** We are very much on a tightrope situation with regard to the present virus diseases and the 'killer-secondaries' they let in. It is very easy to fall off a tightrope, but if you are trained to it and become practised and never lose concentration, then it is relatively safe. But you and your staff have to do exactly what you are trained to do in disease control, with no deviation or omissions. This is particularly true of cleaning and disinfection.

   Your veterinarian is the keystone in the disciplinary structure. You must allow him, *ie* pay for him, to ***disease-profile your herd*** on a regular (6 to 8 week) basis and set up what the Americans – now well versed in this – call a 'protocol' – a clearly set-out programme of pre-vaccination, medication and management which may change month to month according to how the disease challenges rise and fall in your herd.

4. ***With your vet – and with the help of other advisers – ag. engineer, nutritionist, geneticist, general consultant – there is a need to analyse what is stressing the pigs, and reduce it.*** Stress lowers immunity to disease. We know this with our human ailments in our relatively comfortable life at home. Do a stress audit. There are so many stress-inducing things we do to pigs which lowers their immunity these days, from 15% overstocking, to culling too early, to allowing in mycotoxin poisons, and a whole group of other stressors. Identify and ameliorate.

HELPING THE PIG BY LOWERING THE IMMUNE CHALLENGE – A CHECKLIST

How can a producer avoid high levels of chronic immune stimulation and hence maximise productivity at least cost?

✓ Reduce the need for immune stimulation from other pigs. Older pigs are a major source of disease challenge to younger pigs, so segregate by age.

✓ Adopt an all-in, all-out policy wherever possible.

✓ Thoroughly clean and disinfect weaner, grower and finisher accommodation between every batch. This includes correct pre-cleaning with detergents (not just plain water), fogging enclosed air spaces and sanitising the water system.

✓ Reduce dust levels in pig houses. Dust particles are virus 'taxis', and inflame the problem.

✓ Where continuous production has to be practised, institute short production breaks either by selling young pigs or following the 'partial depopulation' idea. Utilise these breaks to clean and disinfect thoroughly.

✓ Adopt tight on-farm biosecurity, especially from vehicles delivering supplies and removing stock. Biosecurity involves at least 30 other things apart from showering-in/showering-out which is what many people think 'biosecurity' means. Study the subject in depth; many of you need to catch up with the latest advice and transfer to the latest products. (*See Biosecurity section.*)

✓ Avoid stressing the pigs. Do a stress audit. (*See Stress section*)

✓ Don't overcrowd/overstock.

✓ Have plenty of 2nd to 5th litter sows in the herd.

✓ Follow a new-stock induction programme agreed with your veterinarian. This is *not* the same as (also essential) quarantine, which is preliminary total isolation. Induction is planned progressive merging, *not* isolation.

✓   Be careful about vaccinating 'as routine'. The ideal situation is for a pig veterinarian to disease-profile your herd and advise on what natural immune stimuli are needed, backed up if needs be by specific vaccination. However, there is already suspicion that some American 'needle-happy' farms are overloading their pigs' ability to acquire a robust immune defence. Table 2 gives a typical American vaccination protocol for a breeding herd.

Table 2   SUGGESTED BREEDING HERD IMMUNISATION SCHEDULE (USA 2001)

Time/Age	Immunisations/Treatments
Gilts/Sows	
6½ months	Leptospirosis, erysipelas, parvovirus, PRV. Feed fresh anure from boars/sows; repeat one week later.
7 ½ months	Repeat vaccinations
6 weeks before farrowing	E coli bacterin, AR, TGE, rotavirus, PRV
2 weeks before farrowing	E coli bacterin, clostridium toxoid, mycoplasma, rotavirus, TGE, AR
3-5 weeks after farrowing	Leptospirosis, parvovirus, erysipelas, PRV
Boars	
First 30 days in isolation	Blood test for brucellosis, lepto, parvovirus, APP, TGE, PRV
Every 30 days in isolation	Erysipelas, leptospirosis, parvovirus
Every 6 months	Revaccinate for PRV, leptospirosis, erysipelas, parvovirus
Pigs	
Day 1	Clostridium antitoxin
Day 3-7	AR, TGE
Day 7	Mycoplasma
Week 3-4	Revaccinate for AR, mycoplasma
Weaning +20 days	Erysipelas, APP
10-12 weeks	PRV; revaccinate for erysipelas, APP

Source: Pork Industry Handbook, PIH-68

TGE = Transmissible gastroenteritis, AR = atrophic rhinitis (bordetella/pasturella), leptospirosis = 6-strain leptospirosis, PRV = pseudorabies virus (called Aujeszky's elsewhere), APP = Actinobacillus (Haemophilus) pleuropneumonia

Quite a workload! And quite a load on the pig's response system! My advice is to see which of these or others is definitely necessary or just advisable from your own veterinarian's experience of your conditions. So consult him – often.

✓ Adequate colostral intake is vital. Being there at farrowing is a good idea. Do not cross-foster until at least 5 good suckles are achieved. If this proves a problem, there are artifical colostrum products *eg* 'Resus' developed from the foremilk of cows which are hyperimmune to ten common E coli pathogens of piglets. *Back*-fostering *may* not be the thing to do these days; discuss it with your veterinarian. Also talk with him over piglet-swapping between sows - but only up to 24 hours from birth.

DISEASE PROFILING

Getting a pig specialist veterinary practice to disease-profile your herd is a worthwhile investment. We did this at our Dean's Grove Farm in the 1980s and got superb performance. Recent figures from clients in the USA confirm the value of the idea as Table 3 shows.

Table 3 BEFORE-AND-AFTER RESULTS FROM USING A PIG SPECIALIST VETERINARIAN TO DISEASE-PROFILE 3 FARMS, WITH EXTRA VACCINATION & RE-MODELLING EXPENSES COSTED IN. (US$ PER SOW)

	Before			*After*		
Farm	*A*	*B*	*C*	*A*	*B*	*C*
Estimated cost of disease per year*	284	186	300	80	96	109
Cost of veterinarian	8	3	12	30	27	31
Cost of vaccines & medication†	26	18	30	18	20	21
Cost of remodelling (over 7 years)	–	–	–	27	45	33
Total Disease Costs (US$)	318	207	342	155	188	194
Difference (Improvement %)	–	–	–	51%	9%	43%

* Disease costs *estimated* from items like the effect of post weaning scour and check to growth on potential performance; respiratory disorders, ileitis, abortions, infectious infertility, *etc*.
† Note that the cost of planned preventive medication was *lower* than for reactive curative medicine.
Source : Clients' records and one veterinary practice

WHAT DOES INADEQUATE IMMUNITY COST?

There must be a hundred thousand answers to this! It is impossible to quantify in general terms any more than there is an answer to "How much will I save if I don't get disease?" or "How much will I lose if I don't fertilise my fields?"

Immediate detectable losses

No reader needs reminding that the cost of inadequate immunity lets in very serious diseases some of which, especially the viruses causing PRRS, PMWS/PDNS; Swine Fever (Hog Cholera), Swine 'Flu, Coronaviruses, etc are deadly to profits. The damage these pathogens have done in eroding my clients profits – even when the pig price was good – have varied from 40% to 100%, sometimes lasting as long as 18 months, and sometimes with carry-over losses extending to six months after the disease had seemed to have gone. To these losses must be added the costs of vaccinations and vet/med attention. Such routine preventive costs alone can be 8% of production costs, excluding labour, with protective in-feed medication adding another 1.5%.

Insidious/hidden costs

What many producers fail to realise is the penalty which the pig's body puts on performance when its immune system has to respond to a high degree of challenge. From time to time useful research has appeared to quantify this and I illustrate two examples, both from the pioneering work, on the interaction between immune demand and nutrition which Iowa State University carried out in the mid to late 1990s.

Table 1 (page 397) summarised what can happen when a young growing pig has to activate its immune system to a high degree compared to one which has no need to do this to anything like the same extent. Notice how protein gain – the primary objective of any of us as meat producers – is severely reduced, and what is more, we are the very people in farming leading the trend to purchase high lean-gain genetics!

I have attempted to put a cost to this fall-off in nursery performance in Table 4. It is important that we try to quantify what this can cost the wean-to-finish producer because this underachievement *from perfectly healthy-looking growers* (the activated immune system has seen to that) is far higher in lost performance than the cost of providing a low immunity challenge environment. This is where many producers are falling down today by not realising that...

- It is cheaper to provide a low challenge environment in the first place than to make the pigs protect themselves by (over) stimulating their immune system.

- And it is cheaper than loading the feed with protective drugs – often needing more over a period of time to achieve the same effect.

Table 4   HOW FAILING TO MATCH DIETARY QUALITY TO THE CURRENT DISEASE STATUS CAN AFFECT ECONOMIC PERFORMANCE IN MTF, PPTE AND REO TERMS*

Immune activation	*Lysine needed per day 7-102 kg (15.4-225 lb)*		*Advantages from altering diet density (average + 2.81g/lysine/day) for high health pigs*		
	(g)	*Extra feed needed (kg)*	*MTF (kg)*	*PPTE*	*REO*
High ('Low health' pigs)	5.7 to 16.1	+25.26 (55.7 lb)	286 (631 lb)	–	–
Low ('High health' pigs)	7.9 to 19.3	–	321** (708 lb)	+£33.25/t ** (or +24%)	5.3:1

*   The New Terminology.......
   MTF   =   Saleable Meat per Tonne of Feed
   PPTE =   Price Per Tonne Equivalent – a figure relating the MTF
               improvement to an equivalent reduction in feed/cost/tonne.
   REO   =   Return on Extra Outlay. Extra outlay in this case is £6.25 to
               provide the better diet, thus REO is £33.25÷£6.25=5.32:1
** MTF includes a 0.91% improvement in yield. Base calculations from Williams (1995) and Stahly (1996)

What this table shows
Failing to match dietary quality to immune status can be equivalent to a 24% price rise in the cost/tonne of all feed from 7 to 102 kg (15.4-225 lb)

Not matching the diet to the current immune status can be equivalent to your paying a quarter as much again for all the food you need from weaning to slaughter (Table 4). That pays for a lot of disinfection, better housing, and veterinary monitoring/guidance.

THE SITUATION IN BREEDING SOWS

Iowa State workers have also shown that continuous activation of the immune response during an 18 day lactation reduced sow feed intake by 0.5 to 1 kg day (1.1 to 2.2 lb). This resulted in a reduction of litter weight gain of 0.32 kg/day (11.26 oz/day) probably from poorer quality milk (Table 5). This, at slaughter could itself cost 9 kg MTF (approx 20 lb) or a 6.5% increase in the cost/tonne of all grower feed from 7 to 25 kg (15.4-225 lb).

This is why any attempt to keep the sow more comfortable, cleaner and to reduce the strain on her system in lactation is so cost-effective, not only to herself – but to her progeny right through to slaughter.

Table 5   IMPACT OF IMMUNE SYSTEM ACTIVATION ON LACTATING SOW AND LITTER PERFORMANCE

	Immune system activation			
	Low		High	
Sow traits				
Feed intake – kg/day (lb/day)	5.36	(11.80)	4.80	(10.60)
Body weight change – kg/day (lb/day)	0.74	(1.63)	0.69	(1.50)
Backfat change – mm/day (in/month)	0.19	(0.22)	0.24	(0.28)
Litter traits				
Number of pigs weaned	12.6		12.6	
Litter weight gain – kg/day (lb/day)	2.60	(5.73)	2.28	(5.02)
Estimated weaning weight – kg/pig (lb/day)	5.53	(12.20)	4.93	(10.90)
Milk & milk component yield				
Immunoglobulin G (mg/ml)	4.3		5.4	
Immunoglobulin A (mg/ml)	12.4		17.8	
Yield – kg/day (lb/day)	11.5	(25.40)	10.1	(22.30)
Energy (Mcal/day)	14.4		12.7	
Protein – g/day (lb/day)	683	(1.50)	612	(1.34)
Fat – g/day (lb/day)	726	(1.59)	675	(1.48)

Source: Sauber *et al.*, 1999

And that if a higher degree of challenge is imposed in lactation, a better lactation diet is needed.

Standards of biosecurity and cleanliness to reduce the need for the pig to activate a high immune barrier are given in the Biosecurity section.

IMMUNITY IN THE YOUNG SOW

The breeding herd is exposed to debilitating disease some 5 to 7 times longer than the grower/finisher, and the young sow therefore needs a good solid immune barrier as soon as possible and for as long as possible (Figure 1).

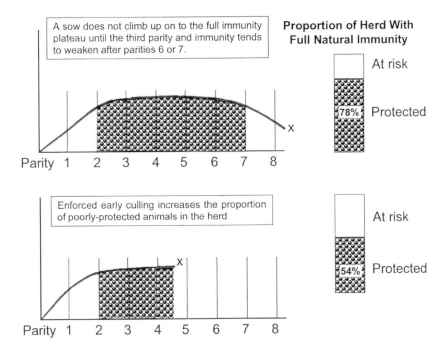

Figure 1 The danger of a short herd life. If the immune status of the gilt/young sow needs to be high to protect her future productivity - so the feed they need has to be of high quality to sustain this extra immune demand.

To my mind there are several important reasons why the immune status of our breeding animals is often too low: -

---

A CHECKLIST: STIMULATING GRADUAL AND NATURAL IMMUNITY IN THE BREEDING HERD.

✓   We don't give the gilt a long enough acclimatization period from entry on to the premises to fully merging her with the herd. 5 weeks is minimal, 6 to 7 weeks may be necessary to combat these "new" viruses (PDNS, PMWS, PRRS, circo- & coronaviruses etc).

✓   As to a minimal length of time, ask your vet, who should be monitoring the disease profile of your herd, and also knows the prevalence of the viruses in your area and can liaise with the vendor's veterinarian if you are buying-in replacements.

✓   We are going about the challenge protocols in the acclimatisation period too casually, often using the same old techniques (afterbirth, fence-line culls etc) when a specific planned and varied programme is needed as the pathogen population changes, including vaccination.

✓   Ask the vet again to advise on what challenge procedures to adopt, for how long and when in the induction period. Normally this will be in the first 14 days; the next month being a 'rest & recuperation' period.

✓   We are growing gilts too fast between purchase or selection at 90-95 kg (198-209 lbs) and first service at 125-135 kg (276-298 lbs). The range is dependent on lean gain genotype. Slow them down! Let immunity acquisition catch up with the modern gilt's precocity. Let her hormone system catch up with her ability to grow fast. She looks like a sexy 21 year old woman but with the hormone and immune development of a 14 year old schoolgirl – or should I say 12 years old in modern 21st century society!

✓   Consult a nutritionist about a Gilt Developer Diet to grow them no more than 650 g/day (1.43 lbs) or 700 g (1.54 lbs) if you must (gilts grow very fast these days and a few take some holding back). This needs to be high in certain nutrients but fed under control.

✓   Feed and manage the pregnant gilt and first-litter sow differently to your standard, established sows. She is a totally different,

developing animal, and quite apart from her nutritional needs,
failure to do so could compromise her subsequent immune status.

IN SUMMARY

Bone up on this whole subject – developing the gilt to service, the latest
flushing technique, first pregnancy feeding and management, helping
her cope with that strange, stressful first farrowing, and using a special
first litter sow lactator feed. Examples are given in the Further Reading
Section, particularly in Close & Cole 'Nutrition of Sows & Boars' 2000,
the cutting edge textbook on pig breeding nutrition and very clearly
written/easily understood.

IMMUNITY IS COMPROMISED BY TOO-RAPID SOW TURNOVER

During speaking tours across the world I am worried about my audiences'
knowledge gap between what is known and published about the gilts
physiological needs from initial selection at 5 months – to the end of
that first litter 8 months later – and the listener's acceptance of what is
needed.

There seems to be a disturbing trend towards faster and faster sow
replacement rate. I don't think the breeding companies are entirely to
blame as some say; it must be the unacceptably-high young sow culling
rates mostly due to early-in-life infertility.

40% TO 45% REPLACEMENT RATE A NONSENSE!

It is precisely because we are being forced to cull so many sows
prematurely for reproductive failure at the end of the second or third
parity that we have ended up with herds which are only partially
immunized to disease – in some cases under 50% protected (Figure 1,
page 405). Moreover, these young sows may be unable to produce
sufficient of the correct antibodies to protect their piglets from eventually
succumbing to new diseases like PDNS/PMWS.

Replacement percentages should be in the mid to slightly upper 30s – my clients who have a good long herd life (averaging 4.5 to 4.8 parities) seem to have fewer disease problems, with vet/med costs half to two-thirds lower than most. Look again at Figure 1 – the key to their success lies there.

BUT WHAT IS THE COST?

They all have longer induction periods under veterinary control and while the cost of this is substantial in extra housing and feed needed, raising the cost of the first litter by about 15% (range 12.8% to 17.1% from my clients' records) the payback from a higher and longer herd life seems to far outweigh the time, trouble and money invested early on into the gilt and that first parity sow.

Table 6 is from a selection of my UK clients. It shows that if the extra cost of getting a gilt properly prepared and looked after is, say 15%, the longer productive life likely to result from this early investment actually makes the gilt 50% cheaper per pig sold, so the REO is 50÷15 or 3.33:1 on this basis – more than a 3:1 return for the extra cost and hassle involved.

Table 6   WHY SOW LONGEVITY IS IMPORTANT – YOUR GILT INVESTMENT IS HALVED

Cost of getting sow to first litter	Sow lasting 3 plus litters (40 pigs) (per pig sold)	Sow lasting 6 litters (70 pigs) (per pig sold)
£268	£6.70	£3.82
Plus empty-day lag in replacement female at 4th parity		
25 days at £2.50/day ÷ 40 piglets =	£1.56	Nil
Total per pig	£8.26	£3.82
		(around 50% less)

UK Costings (2001)

WHAT HAPPENS WHEN A VIRUS STRIKES

Simplifying a complex procedure, readers will forgive me if I draw an analogy with a modern battle, but it is so important that immunity is understood.

When a virus invades, a firefight commences. A variety of soldiers are called up to deal with the pathogen enemy, which has already invaded healthy cells in the animal's tissue, reproduced inside them and emerged to take over other cell territory.

The reconnaissance

As the virus bursts forth from its bridgehead, ***Helper T cells***, already on watch in the body, identify them to HQ as ***antigens*** or foreign invaders. This is an alarm alert.

Mobilisation

1. However ***natural killer cells*** are also an on-watch force which do not need the antigen alarm. They immediately go ahead, seeking out and killing some virus and cancer cells. But they soon need to be reinforced.

2. Responding to the antigen alarm from the Helper T cells, ***Macrophages*** (white blood cells) are a rapid reaction force who in peaceful times live in the bone marrow 'barracks'. (These bone marrow barracks in wartime – *i.e.* when the disease organism invades – become training and replacement camps.) Macrophages deal with some, not all, of the invaders, especially bacteria, fungi, cells invaded by viruses and cancer cells. Thus they identify the enemy and liaise with the Helper T cells to mobilise the assistance of B cells.

   B cells are the heavily armed troops. Their armament being ***antibodies***. Specific antibodies for specific antigens – in the same way that an army uses different weapons to deal with different challenges – anti-aircraft, anti-tank, mines, machine guns, etc. It can take 14 days for full mobilisation to happen (*i.e.* with vaccination) but under attack the rapid reaction troops hold the line as a much more solid defensive build-up develops. So during mobilisation, as well as the macrophage defence troops, one type of Helper T cell goes for the invaders, destroying virus-infected cells, while another variety organises the correct armament (the right antibodies) needed by the B cell heavy-duty troops.

The battle commences

After the build-up in response to attack by disease (during which time all the troops involved need all the help you - the civilian population/ government edicts - can give them by reducing stress, keeping things clean, not overstocking and managing warmth and ventilation well) while the heavy duty B cells start tearing into the antigen invaders. Each B cell 'regiment' recognises and reacts to only one specific antigen by destroying the body cells harbouring the enemy antigen (virus) or neutralising the virus itself.

Battle over

Another form of Helper T cells have been in reserve, called ***Suppressor T cells***. These detect that the battle has been won and stand most of the troops down. Without it the various troops, by now in full fighting mood, could begin attacking healthy body cells, too – not just those occupied by the enemy. If you remember, the heavily and specifically armed B cells recognise and react to only one form of virus enemy, and ***Memory B cells*** and some T cells stay in the body waiting for any re-invasion of the antigen 'enemy' *i.e.* if there is re-exposure to the same antigen, or invader.

DON'T CONFUSE ANTIGEN WITH ANTIBODY

The Antigen is the invader, and there are many different types: ***unwanted***, *i.e.* viruses, bacteria, fungi, cancers etc, as well as the various families within each group. And ***planned***, as in a vaccine.

A vaccine antigen is a 'teaser' invader to alert, stimulate and mobilise just sufficient defensive troops to fight off an unwanted invader should it materialise.

The Antibody. These are protein structures (principally IgA, IgM, IgE and IgG) which fight the foreign invading agents. In our analogy, the weapons and ammunition the troops can call on.

CAN NUTRITION HELP WITH IMMUNITY?

The pioneer work of Stahly, Williams, Cook, Sauber, Zimmerman and others has already been mentioned, *e.g.* on page 397. It is important to discuss the nutrient density of your diets with a nutritionist who understands the work of these pioneer researchers. Feed design may be affected both by appetite (intake) and immune status differences, and a nutritionist experienced in designing diets to match – as far as present knowledge allows – both these variables is a useful ally for the future.

We immediately run into a problem. Not insurmountable even if it presents, at first sight, practical difficulties.

The problem lies with diet design. Farmers, always pragmatic realists, ask "Fine, but these pigs I'm looking at, how do I know – and more important how do you know – where they are on the immunity activation ladder?"

Good question ! An absolutely vital question, in fact, because getting it wrong could raise, in UK terms, at least 15% of the cost of producing a pig from imprecise nutrition alone.

3 OPTIONS

As I see it these are three possible solutions.

1. **Serology**. Here one uses the vet to blood-sample the herd to try to establish a disease profile. Snags are that even the cutting edge of serology cannot identify certain diseases – so what happens if the current challenge happens to be mostly from one of these ? Secondly, it is expensive and could be time-consuming. Serology, as knowledge advances, could help more in future. But what can we do now?

2. **Challenge or test feeding**. This concept takes 50 typical growers, feeds them on what the nutritionist calls a 'non-limiting diet' and periodically, say every 14 days, monitors growth, FCR and lean gain (by using a deep muscle scanner). In this way, along with carcase data at slaughter, the nutritionist has a good idea of the grow/finish herd's lean accretion curve and can design a farm-specific diet or diets to satisfy it. Done twice a year as routine, or if the disease picture changes markedly (Table 7).

    Snags:– The scanner is expensive so it is really the province of a feed manufacturer who can organise it and whose clients have a computerised wet feeding facility. Why wet feeding? Because with this equipment any variety of diet can be made on-farm from only two (or three) basic formulae. This cuts down a custom mix inventory drastically; in fact one feed compounder known to me has, across two years, reduced his normal pig diet list by 50% despite increasing his custom mix clientele by 60% – and his pig business by 300%! He also dispensed with several feed reps as he had no need for selling on price – the predicted lean gain curve dictated the price (Table 8).

Is this the shape of the future? Could be.

Table 7   CHALLENGE OR TEST FEEDING CONCEPT TO MATCH NUTRIENT-INTAKE TO IMMUNE STATUS

Method

1.	50 representative pigs 25 kg – 105 kg (55-232 lb)
2.	Fed non-limiting diet
3.	Weighed every 14 days
4.	Ultrasonic test every 14 days
5.	Results sent to nutritionist
6.	Lysine accretion curve calculated
7.	Least cost diet designed to match it – Farm Specific Diet (FSD)
8.	Done x 2 per year, Summer/Winter or if the disease picture alters abruptly.

Table 8   FINANCIAL & PHYSICAL PERFORMANCE BENEFITS FROM USING LEAN GAIN FEEDING VS CONVENTIONAL FORMULATION METHODS

	Conventionally-formulated grower/finisher for all clients		Diets designed specifically for the lean gain and appetite potentials of the genotype used		
Physical Performance					
Deadweight FCR	2.97		2.87		3.36% better
Av. daily lwt gain – g (lb)	786	(1.73)	846	(1.86)	7.63% better
Av. daily saleable carcase					
gain* – g (lb)	581	(1.27)	660	(1.45)	13.6% better
P2 Backfat – mm (in)	12.1	(0.47)	11.6	(0.45)	
Financial Performance					
Av. cost/tonne of feed	100		107		7% more
Margin over feed cost	100		116		16% more
Nett return	100		121		21% more

* assuming pigs started with 80% saleable carcase wt at 25 kg (55 lb)
  Source : Farm Trials (1996 - 1998)

3. **Measuring growth rate**. At present – and it is early days yet – there could be a possible correlation (linkage) between growth rate and immune activation. Measuring growth rate accurately is something the producer can do if he sets his mind to it, so the idea looks workable, is farmer-friendly – and doesn't cost much!

Snags? Of course, there are other things besides immune stimulation which can easily affect daily gain. Temperature, stress, feeder management, overcrowding, water, wet/dry feeding and so on. What we need is confirmation from research that this potentially simple and workable guideline is indeed a viable option.

MATCHING DIETS TO IMMUNE RESPONSE – A NUTRITIONIST'S HEADACHE

Most commercial nutritionists at the present time view the subject as a "nightmare", to quote a leading European formulator. When pigs encounter pathogenic challenge, cytokines (a type of protein chemical messenger) are released which reprogram the animal's metabolism to divert nutrients away from growth, especially lean growth, in order to ensure the immune process is prioritised. Cytokines alter nutrient intake and utilisation which – first headache for the nutritionist – need to be compensated for in order to lessen the damage to productivity.

At the same time – second headache – metabolic changes are occurring which both increase and decrease nutrient requirements. Fever places demands on energy and while the consequences of fever – reduced activity and more sleep – lessens it; a reduction in growth rate lowers the demand still further. On top of this, appetite reduces when immune response is high, even if the animal feels healthy enough.

I quote Paul Toplis, a leading European commercial pig nutritionist responsible for making sense of it all:–

"Appetite changes can be unpredictable. For example, if a healthy growing pig with an appetite of 1.5 kg a day requires 15g of lysine then the diet specification for lysine should be set at 1.0% (10g/kg). Now if this pig encounters an immune challenge its lysine requirement might fall to 14, 13 or perhaps 12g per day and the feed intake might fall to 1.4, 1.2 or 1.0 kg per day, giving the nutritionist nine possible diet specifications to work with."

Toplis (1999)

BUT DOES IT MATTER ?

An understandable question from the producer. Yes, it could well do. Taking the variables Toplis quotes and the reduction in performance quoted in the Iowa State results, even for the less extreme differences, under current UK prices at September 2000 if you *underachieve* immune demand intake this could reduce saleable meat sold per tonne of feed fed by slaughter by 11.6 kg, and also incur 12 days longer to slaughter in overhead costs. But if you *overachieve* the immune demand this might cause you to pay an unnecessary 6% more for your food as the pig won't use all of it, and will just excrete the overage – another extra cost in more slurry disposal.

CAN FEED ADDITIVES HELP WITH IMMUNITY ?

Zinc

Zinc supplementation has long been recommended by the medical profession, particularly for the older human patient, to help bolster their immune defences. In animals it is known that zinc plays a critical role in both reproduction and immune-competence, but unfortunately there are no clear guidelines as to the optimum requirement for the latter. The levels are likely to be considerably higher than the requirement for growth. In terms of the immunocompetence of the animal, zinc has a positive effect on both the immune response to pathogens and the prevention of disease by maintaining healthy epithelial tissue (epithelial = tissues involving many varieties of cells, in zinc's case those deterring or delaying invasion by pathogens).

So if the zinc needs are higher to assist immunocompetence – how much higher? I don't think we know yet, not fully. What could be an important lead ('breakthrough' is a too dramatic term to use yet) is the way proteinated or 'bioplexed' trace elements are better used by the animal. Let me try to explain it in layman's language at the risk of over-simplifying a complex metabolic pathway – academics please bear with me!

The Bioplex Concept

In the case of zinc, the mineral is linked to an amino-acid, in this case methionine, which 'tows' it through the point of absorption (for methionine) in the intestine, which also happens to be where zinc is absorbed. Result, more zinc is absorbed so less is needed in the diet. Such linkages care called bioplexes. Therefore, less is excreted as unused by the pig and pasture contamination and watercourse pollution due to small but prolonged soil build-up is reduced substantially (amino acid-linked trace minerals are also called 'proteinates').

Is the zinc, when more of the bioplexed form is absorbed, better used for both production and immune status? It seems so, although rather more evidence has accumulated to date on the productivity side than on the immunocompetence area, which is not surprising as it is much more difficult to measure.

The experts still seem to be undecided whether all zinc should be derived from the bioplexed form, or whether just some of it, or even a good proportion of it. Meanwhile follow their advice, which at the time of writing seems to be moving towards total replacement away from conventional inorganic sources.

What we are certain of is that more **nutrients** are needed when disease starts to challenge. For example in poultry, some 1% of all nutritional needs are used to maintain a normal immune level, while this rises to 7% when disease activates the bird's immune defences. American work hints that this could be even more in the case of the pig, especially high lean gain genotypes. Anyway, using a bioplexed form of zinc either in whole or in part seems to be a good idea. It looks – at the dose rates advised – as if it won't harm anything i.e. through over-availability, and it could do a lot of good. REOs of between 5:1 and 22:1 for either bioplexed, organic zinc, iron or copper have been obtained as the inclusion costs/tonne are low.

Oligosaccharides

Oligosaccharides are simple sugars derived from brewing by-products (fructo-oligosaccharides or FOS) or yeast manufacture (mannan-oligosaccharides or MOS).

Both these additives have come into prominence now that antibiotic growth promoters are being increasingly banned. Being natural and safe by-products (sugars), they provide a useful and cost-effective alternative.

Originally they were thought to work mainly by competitive exclusion of gut pathogens on the gut wall. They do, but in the case of MOS, other more complex mechanisms seem to be at work.

Again at the risk of oversimplification, the latest evidence so far seems to be that Biomos – the bestselling source of MOS – is getting results with some disorders which are surprisingly good and could be unlikely to do so solely to its proven 'capturing' effect of pathogenic bacteria and holding them fast until gut peristalsis (peristalsis = wavelike movement on down the gut) removes them out of harm's way to the outside of the digestive tract in the faeces. Something else may be at work – could this be a strengthening of immunocompetence?

MOS appears to enhance immune function in a variety of ways ...

•   Oregon State University reported a 25% increase in secretory IgA.

•   Researchers have found that MOS enhances macrophage response.

•   Other workers find that in germ-free pigs MOS influences both humoral and cellular (B cells and T cells) immune systems, although the levels measured seem to be widely different. (For definitions of these technical terms, see Table 9.)

So... Biomos seems to facilitate the complex interactions of all these disease-fighting substances. This is called immunomodulation (the effect on immune response). Much still remains to be discovered, but already research suggests Biomos helps resist infection from E coli, campylobacter and salmonella, so the beneficial effects of this useful alternative to antibiotic growth enhancers can now, with confidence from a growing number of such trials, be added to the original benefits from the inhibiting effect of 'pathogen-capture' under which banner Biomos was first announced more than 5 years ago.

Table 9   SO YOU DON'T GET LOST ... SOME IMMUNOLOGICAL TERMS

Humoral immunity = B cells, lymphocytes
   Memory cells which remain behind after an infection, recognize the reappearance of the pathogen and quickly call up the correct defences again.

Cellular immunity = T cells
   These stand guard against pathogen challenge, are limited to body cells in various tissues susceptible to pathogen ingress.

Systemic or mucosal immunity
Local humoral or cellular antibodies ideally present when body surfaces are exposed to the outside – nose, throat, gut, outer reproductive tract.

Active immunity
After exposure to infection, stimulated antibodies remain in the sow which are transferred to the offspring via colostrum for a while in the form of antibodies IgA, IgG, IgE, IgM etc.
The dam is **active** in passing on the immunity.

Passive immunity
The piglets accept the antibodies (i.e. are **passive**) and this lasts as long as the maternal antibodies survive. As no memory cells (lymphocytes) are provided or formed so the immunity is not permanent.

Acquired immunity
After a pig recovers from disease or vaccination it develops acquired immunity.

Antigens
Foreign material which triggers the body's defence mechanism – pathogens or vaccines.

Antibodies
Protein structures (IgA, IgM, etc) which fight antigens and unless overwhelmed, prevent disease

Phagocytes
Cells which ingest and so destroy pathogens.

Macrophages (white bloodcells)
Large immobile cells, usually originating in bone marrow, which become actively mobile when stimulated by inflammation, immune reactors and microbial products.

Cytokines
Messenger proteins which control macrophages and lymphocytes.

Immunosuppression
When an immune system is not working properly because of dirty conditions, overcrowding, a poor diet, pre-existing disease, stress and mycotoxin presence, etc. some of the newer viruses seem to carry their own immunosuppressive capabilities, which is giving them a head start at present.

Titre (titer)
A numerical measure or test of a pig's immunity. An antibody titer measures how much antibody is in the pig's blood. Expressed as 1 followed by a number. For example if one volume of blood was diluted with 64 volumes of saline solution and antibodies were still detectable the titer is 1:64. The higher the number after 1 the more antibodies are present and stronger the immunity present.

Serology

The expression of antigen:antibody reactions by laboratory test.

Inflammation

A localised protective response caused by injury, destruction of tissues or injected poisons (eg insect bites) to block off, or destroy or dilute the injurious agent and protect the affected tissue.

REFERENCES

Close W H & Cole D J A *'Nutrition of Sows & Boars'* Nottingham University Press (2000) p 359.

Molitor T (1992) 'Immunization' Nat. Hog Farmer Blueprint Series, Spring 1992 p 28.

Sauber T E, T S Stahly and B J Nonnecke,(1999) *J Anim. Sci* **77**: 1985-1993.

Stahly T S and D R Cook (1996) ISU Swine Research Report, Iowa State University Ames, ASL-R 1373 pp 38-41.

Stahly T S, S G Swenson, D R Zimmerman and N H Williams (1994a) ISU Swine Research Report, Iowa State University Ames, AS-629, pp 3-5.

Toplis P (1999) 'Interactions Between Health & Nutrition' Procs JSR Genetics 10th Annual Technical Conference p2

Williams N H and T S Stahly (1994) ISU Swine Research Report, Iowa State University Ames, AS-633, pp 31-34.

Williams N H, T S Stahly and D R Zimmerman (1994) *J Anim. Sci.* **72** (Suppl 2):57.

STRESS AND STRESSORS 27

Stress : The total of all the biological reactions of an organism (in our case pigs) to any adverse stimuli. These can be physical, mental or emotional – those which disturb the smooth functioning (stability) of the pig's metabolism.

Stressor : Any individual factor or action (collectively called stimuli) which disturbs this stability. There is a long list of known stressors, and there may be more to be discovered due to physiological interreactions.

It is a reflection of how little we know about stress that this section is not one of the longest in this book. It should be, as I am convinced that stress is certainly one of the most important subjects for the pig producer to be aware of, not only affecting the welfare of his pigs (and ourselves as their guardians) but his profits, too.

Especially profits – stress is a big drain on profit. It is almost certainly one of the major influences on it and I suspect the extent is poorly recognised by producers.

SO WHAT DOES STRESS COST ?

There is almost no information available. I only wish I had collected before-and-after evidence on stress on the many farms I have visited (as I have done for more easy-to-measure subjects) where the results of stress alleviation might have been measured. Producers who have been persuaded to follow my stress audit checklist have certainly noticed improved performance, and in Table 1 I give a very approximate *estimate* of what this might be – from 30 years of the audit's use.

419

Table 1   CONJECTURAL FIGURES POSTULATED FOR WHAT STRESSORS OF ALL VARIETIES MAY COST A COMPETENT BREEDER/FEEDER PIG PRODUCER IN, SAY, THE TOP THIRD OF HIS NATION'S PERFORMANCE TABLES*

Breeding unit	3 fewer pigs weaned/sow year (Comprised of 6 more annual empty days; 2% more pre-weaning mortality, 8% greater sow replacement rate, and 10% fewer born alives)
	Plus 12% more vet/med. costs, 2% more housing costs and 1% more labour costs
Growing Finishing unit	Food conversion worse by 0.15
	Daily gain 7-100 kg (15-221 lb) lower by 30g (1.06oz)
	M.T.F. lower by 19 kg (42 lb)

Plus general overheads increased by 3%

* This is purely the author's assessment of improved performance when the producer was persuaded to concentrate on stress relief alone as a permanent part of his daily management routine. Even so it should be viewed with caution until research confirms these suggestions.

At farm level there are three areas we need to consider:

1.   *Anticipation*   What stressors are we likely to see in and around the pigs?

2.   *Observation*   What responses are the animals making to these adverse stimuli?

3.   *Action*   What action can we or must we or should we carry out within the bounds of cost and feasibility so as to alleviate the stress?

HOW STRESS WORKS

Stress is characterised by two quite different groups of reactions. Acute Stress or Fright. Chronic Stress or Anxiety.

Fright (fear, nervousness, alarm) is quite easy to recognise. Anxiety, (worry, low-level frustration, discomfort) is much more difficult.

The pig's body nevertheless has to make readjustments, if it can, to both types of stimuli. The stimulus is processed in the brain which tells the body how it should respond either by sending direct instructions through the autonomic nervous system, or ANS (autonomic = not subject to voluntary control) or indirectly through the neuroendocrine system – NES (neuroendocrine = hormones activating nerve response).

ANS : In the case of an ANS response to acute stress the response is automatic – for example the hormone adrenalin is generated which increases heart-rate and respiration rate and puts digestion on hold so the animal can at once flee or fight. That's why your heart starts thumping immediately after a fright and you don't feel like eating (or even sex!) for a short while. This is all done automatically.

NES : With low-level stress, if the animal's brain decides it cannot do much about the stimulus (such as to get away from it or combat it) then it goes into NES mode. Here the brain calls up other hormones which help the animal cope with more long-term stressors. Different hormones to those activated under ANS are released which dull down the effects of other hormone-producing glands such as the pituitary – a sort of 'centre-half' organ which controls several other hormone players in the team. These are the thyroid (growth and metabolism), the pancreas (digestion) and the sex organs (breeding efficiency, especially at ovulation and implantation). The idea is to slow things down so that the body can cope and is given less of a load to carry while the stress persists.

HOW STRESS AFFECTS DISEASE

However, the NES process has a dark side to it. It also stimulates the production of corticosteroid hormones such as cortisol which reduce the number of white blood cells called lymphocytes. These produce antibodies against disease – thus the animal's immune shield is weakened, and if the low level stress goes on long enough its defences against disease may be severely affected.

So in reducing disease, how __long__ you stress your pigs without attention can be as important as how __much__ you stress them.

LEAN GROWTH AND STRESS

Cortisol can also have a inhibitory effect on protein formation; a sort of protective, calming effect on a highly-demanding part of the pig's routine physiology. If you look at all areas where protein metabolism is economically critical – for example the young pig, the grower, the developing litter, the sow's fleshing down through lactation – you can soon see where low-level stress can plunder profits.

So continuous low level stress can be quite a drag on the animal's potential performance – quite apart from letting disease gain a foothold.

WHAT ARE THE STRESSORS ?

Stress responses vary widely between individuals – we see this in humans. Look about you, we all know phlegmatic and conversely, excitable people! Responses in pigs – like humans, are affected by age and experience (conditioning/adaptability) and presumably by genetics too – the halothane gene is an example of this.

Table 2 gives a list of the main stressors I have come across. I divide these up into Acute (ANS) and Chronic (NES) stressors and subdivide them into natural occurrences and those which we impose on the pigs through poor planning, lack of observant stockmanship and under-investment, signified by an asterisk.

We see in this review that it is the effect of low level NES stress which is so often under the stockperson's control and which is a main problem affecting potential performance.

THE STRESS AUDIT

A periodic stress audit is a very useful exercise, and is probably needed every 6 months on many units. The purpose is three-fold.

1. *Observation.* To identify, from the body language of the pigs before you, their responses to the likely stressors they are encountering.

2. *Welfare.* To check that the conditions you have imposed on the pigs are suitable, that they haven't slipped, and that the equipment and management impositions on the animals are up to standard and within acceptable boundaries.

3. *Action.* To decide on, and provide instructions for, rectification of any shortfall *i.e.* areas found wanting or approaching borderline conditions.

Table 2 A CHECKLIST OF NATURAL STRESSORS UNDER TODAY'S CONDITIONS

		ANS	*NES*
The small pig	Birth	✓	
	Pathogens		✓✳
	Establishing itself/competition	✓✳	✓✳
	Temperature/cold		✓✳
	Thirst		✓✳
	Weaning/re-establishing itself	✓	
The farrowed sow	Parturition	✓	
	Lactation/water availability		✓✳
	Temperature/hot	✓✳	
	Comfort		✓✳
	Weaning	✓	✓
	Parasites		✓✳
The pregnant sow	Ovulation Helped by the		✓✳
	Implantation 'Feel Good Factor'		✓✳
	Confinement/comfort/boredom		✓✳
	Competition (groups)	✓	✓✳
	Gutfill/fibre		✓✳
	Temperature/cold		✓✳
	Parasites		✓✳
	Legs / floors		✓
The gilt	Onset of puberty		✓
	Competition/bullying	✓✳	
	Poor light		✓✳
	Space		✓✳
The boar	Temperature / hot and cold		✓✳
	Lack of Exercise		✓✳
	Frustration		✓
	Gutfill		✓✳
	Boredom		✓✳
The grower/finisher	Temperature/variation/diurnal		✓✳
	Space		✓✳
	Pathogens	✓✳	
	Food and water access		✓✳
	Sleep adequacy		✓
	Boredom		✓✳
	Transport/lairage/handling	✓✳	

✳=Stressor under your control

Table 2 can be used as a basis for a stress audit.

A FEW EXPERIENCES FROM 30 YEARS OF STRESS AUDITING

- *Do not wear white overalls*. Wear green or dark blue. Keep quiet, move slowly.

- *Do it quietly!* Observe the pigs under their normal behaviour patterns. Open nursery doors an inch, and listen before switching on the light and/or entering. Listen to the breathing, for restlessness, wheezing, 'snittering' (light, irritant sneezing). The same in the farrowing house and in the grower houses.

- *Observe them unawares.* Still with the door slightly open, switch on the light if needed and look for the resting pattern, huddling (piling) and where they are lying relative to air placement. Do not enter until you have looked at as many areas as possible before the pigs disturb themselves. A good time to detect problems is last thing at night.

- *Observe them stirred up.* Enter the room and immediately move quietly along its length breathing the atmosphere for gases and checking temperature. At the same time look for stiffness/lameness/reluctance to move and listen for laboured respiration/coughing.

- *In nurseries and farrowing rooms*, look for piglets which are lying awkwardly/lifting their undersides in the semi-sternum posture. This is advance warning of digestive upsets.

- *In lidded kennels etc*, try to get a quiet peep inside but ensure all occupants are eventually 'banged-out'. Scrutinise those last to leave carefully.

- *Use the farrowing crate to palpate the sow's udders*, feeling for unusual conditions/discomfort.

- *Check hoppers and troughs*, for stale food, contamination, cleanliness. Check evidence of wastage. Are pigs 'nosing' the food in the troughs? A sign of nutritional dissatisfaction from hunger to unpalatability.

- *Get as many stalled sows* to stand as you can. Condition score those that do and check legs/sores/rubbing (see Lameness Checklist). Check water adequacy *ie* colour of urine. Check for discharges.

- *For group-housed sows*, get them to move past you and study gait and alertness. Check their water supply.

- ***It is a good idea*** to do an audit with a neighbour present occasionally and you do it for him likewise.

There are many more tips than this of course, embodied in Table 2, but the above will give you a flavour of what a stress audit does.

MEASURING AND MONITORING VITAL SIGNS

A stress audit cannot be carried out to its full potential unless you are able to check the physical constraints which you have imposed on the animals in your care. These are:-

- **Temperature.** (*Common weakness*). Thermometers should be properly positioned as close to the pigs height without being damaged. At least two are needed per house located at a 'neutral' spot on the floor plan *i.e.* out of extremes of ventilation pattern. While auto-recording is best as this gives day/night variations, individual max/min thermometers are adequate as long as they are:–

Clean; consulted frequently; calibrated to no more than a 1°C error against a BSI-rated instrument.

- **Check air movement.** (*Major weakness*) : Part of a stress audit involves checking that fan speeds, placement and operation are as designed/ intended. ***A ventilation engineer should do this at least once a year***. It is remarkable how often a specialist will pick up individual major errors causing a 0.2 worsening of FCR for growing pigs in the error zone – which easily pays for his visit.

Things you can do between such visits are to have smoke tubes/phials to 'see' air movement, and to wet the back of your hand or bared arm to detect cold draughts. This idea is particularly useful to discover cold downfall draughts on to the backs of sleeping pigs close up to a wall especially at night. This can be simply and cheaply counteracted by nailing a 3 to 4 cm (1½") triangular wooden batten to deflect the air falling down close to the wall surface into the rising current of warm air rising from the sleeping pigs and so 'lose' the draughts by natural means. There is no substitute for knowing how air moves in a piggery and the importance of correct air placement (*see Ventilation section*).

- **Check stocking density** *(Common weakness)* : Check, check, check that you are within the safe spatial guidelines *(see Stocking Density section)*. Stocking density quickly gets out of control, and the stress audit draws this to the attention for every pen. Fully half the farms I visit are infringing stocking density recommendations somewhere or other – and as you can see from the Stocking Density section it does matter!

- **Attend to water** *(Common weakness)* : Again, the audit must measure that not only water flow rate is up to standard, but the ***ease of access*** is adequate.

  Examples are: – only one drinker in a pen; no use of height-adjustable fixings; bite drinkers in a farrowing crate, not troughs/bowls; drinkers set too low; no separate, additional water-only points in a wet-feed system or wet/dry troughed pen; siting a drinker in a corner; dry sows in yards watered from bite drinkers, not a trough.   All are common errors which raise stress.

- **Monitor flooring and bedding** *(Major weakness)* : I have always been a 'bedding' man, certainly for sows and young pigs. However 80% of the farms I visit outside Sweden and Britain use little or no bedding for economic or logistical reasons. This means that correct floor design is paramount. You need to check that many in-contact areas are not too small for tiny feet in the interests of cleanliness. If so, provide a temporary solid comfort-board for tender-footed weaners so that they can at least get on to a solid area as a respite.

  Maintenance of floor quality (slipperiness, gaps/holes, roughness) must all be checked during the stress audit – reference to the Lameness Section will provide more information.

  Deep-bedded (straw) yards are bound to increase in future. Keeping them 'sweet' and free from dead, coagulated, over-fouled areas in the yards, particularly corners and pen-fronts in hot weather, needs physical aeration which is one of the hardest physical tasks I've ever done! Nevertheless there are mechanical devices to do this, free-moving aerators suspended from a monorail for sawdust can take the effort out and so encourage more frequent attention which is essential – and save on bedding as well.

- **Review conditions at implantation** *(Major weakness)*: Far too many sows are stressed during the 7-28 day period post-service. They need rest and quiet, freedom from aggression or discouragement from aggressing others. They need adequate gutfill even though the nutritionist cautions against too much feed despite some of them being out of condition. Skilful use of supplementary hay, edible straw and especially dried sugar beet pulp can make a huge difference to a feeling of well-being.

They need to be kept out of draughts, warm (18°C; 65°F) but the air needs to be free of gases, *e.g.* the marker gas ammonia present at under 12-15 ppm. This can easily be measured by an appropriate chemical discolouration tube.

- **Take time**. Finally, I find it is essential to take time over a stress audit. Just watching pig behaviour, especially when feeding, when stirred up, when being moved and when at rest will guide you. *Pigs talk to you all the time in a variety of physical ways* but often we don't grant them the value of 'listening' to them because we are so busy with routine.

- Talking to them? It helps, I'm sure, establish empathy.

- Playing music? The familiarity of background sounds must reassure animals when quarters are changed and pigs are mixed.

- Sticking to a time routine? All animals habituate and following their expectations as to when things ought to happen should lower stress.

- Lighting – periods of distinct light and darkness must help sleep patterns – as it does with us.

- Giving them toys? Sure, why not. They must get bored stiff.

   Of course it is dangerous to become anthropomorphic and assume pigs respond to what we as humans may prefer, but I find *good* stockpeople do use anthropomorphism more than we care to admit.

   It helps to lower stress – on both sides.

BIOSECURITY 28

Biosecurity involves everything which needs to be done to protect a farm, its livestock and its workers from disease.

Scope

Biosecurity is therefore a massive subject, encompassing not only hygiene (cleaning and disinfection), but also the protection of premises from disease ingress from other animals, (farm, domestic and verminous), birds, humans, insects, as well as from the weather, transport, air movement, waste food and liquids, the water supply, drainage, rivers and carcase disposal. Finally vaccination and farm location and layout as well as animal movement are all involved..

This section concentrates mainly on the measures needed to prevent infectious diseases getting into the farm; to deter them from increasing should they gain a foothold; and to eliminate those present wherever possible.

TARGETS

The ideal is to reduce the level of pathogens to that which is low enough for the animals own defence mechanisms to cope with those remaining. No farm can ever be sterile, so the target is to obtain a working balance between disease challenge and effective defences against disease.

The level of natural immunity sufficient for the animal to do this varies due to the age and condition of the animal, to the degree of exposure to the organism responsible, and to beneficial or detrimental conditions under which the animal lives.

Some clues as to how 'clean' things should be have appeared from research and general pig farm experience. For example Waddilove, a pig veterinarian specialising in this area, suggests the targets in Table 1.

Table 1   TYPICAL TOTAL VIABLE COUNTS (TVC) OF BACTERIA AFTER PIGS HAVE BEEN REMOVED

State of house	TVC/sq cm
Immediately after pigs out	50,000,000
After plain washing	20,000,000
Hot wash and heavy duty detergent	100.000 (650,000 per sq in)
Target after disinfection	1,000 (6500 per sq in)

Source: Waddilove (1999)

DOES THE ORDINARY FARMER ACHIEVE THESE LEVELS?

I suspect – rarely! On the few occasions that I have seen where swabs have been taken even after a clean-up 'blitz', the TVC before disinfection has been 3 million or more and in places many were at over a million *after* disinfection with one or two over 5 million! If the bacteria weren't being reduced sufficiently, then the viruses, which are more difficult to kill, were even less likely to have been controlled.

Ask most serious pig producers (as I have) and they say "But we clean and disinfect pretty well", "We have a set routine and stick to it", "Yes, we are now AIAO" (All-in/All-Out), "Don't disinfectants deal with everything?" and so on.

When I ask a few questions on what they actually do, great gaping holes appear between what we are told by the experts is necessary and what producers are still doing on their farms. (Table 2)

Table 2   SURVEY REVEALED NINE OMISSIONS OR ERRORS

I've surveyed 105 pig farmers across 18 months. Here are the results.

- Three quarters of them used no detergent in their pressure wash-down.
- 68% of these did not use a *hot* pressure wash, even so.
- Four-fifths of those who used a detergent did not use a farm-specific detergent.
- Only 2% troubled to monitor the effectiveness of their cleansing and disinfection procedure.
- Of the 9% who were sampled with swabs after disinfection, none at all had

Table 2 (contd)

levels of viable bacteria remaining at or below the target level of 1,000 viable bacteria/cm^2 (6,500 per sq inch).

- 50% of those swabbed had over 5 million/cm^2 viable bacteria from at least one swab.
- 80% did not sanitise the water.
- 40% did not *regularly* combat vermin.
- 90% "only fogged after a disease storm", not as routine. No one fogged their loft space.

If this is representative of pig farmers' attitudes, are they complacent? I don't think so. Complacency implies that farmers know what is correct but it is not done for reasons of time, or labour, or money or a failure to monitor things diligently.

To my mind more likely reasons are ….

- Failure to realise how much disease costs you in performance. Probably 0.3:1 food conversion from 7 – 100 kg (15-221 lb) and 4 fewer pigs sold per sow per year in the breeding herd. Yes, that much! And these are probably minima, I'm told.

- Failure to recognise that subclinical disease – the continuous effect of rumbling, low-level, largely invisible disease – *possibly costs you more*, over a period of say 2 years, than the outbreaks of clearly visible clinical disease we all worry about and take action on when they happen.

- Failure to realise that modern pathogens are tougher, more resilient and more virulent than ever before and so need uprated detergents and disinfectants to combat them.

- Failing to clean properly before disinfection. Ever painted the outside of a house? I'm sure you have. Experience from previous disappointments tells you that it is the *preparation of the surfaces* which lead to a long-lasting effect, not so much the care in application or number of coats of paint subsequently applied. In the same way, pre-cleaning before disinfection has gone by default. We don't clean adequately so we end up not disinfecting properly, with disease being the disappointment.

We now have better virucidal (virus-killing) disinfectants, but they tend to be neutralised by organic matter and fat deposits.

These new viruses have stronger protective biofilms around them – they've changed so as to be better survivors. The 'old' disinfectants, like the phenols and quats, aren't so good at getting through this protection. Newer oxidative disinfectants (peracetic acid and also peroxygen) are more effective at the job, as well as being more biofriendly – one spin-off bonus is that pigs can even breathe one of them in at advised dilutions (fogging). One can also be used, with care, in the pig's drinking water.

However, the new virucides do tend to be weakened by organic deposits on surfaces. In addition, fat and grease on the piggery surfaces can make the job of any disinfectant harder. The nutritionists are using more fats in lactating sow, baby pig and nursery diets these days, so there are protective grease layers all over the in-contact surfaces.

You must remove these barriers to get a good kill of pathogens from the disinfective process – down to about 100,000 TVC (Total Viable Count) per cm^2 ***before*** disinfection (650,000 per sq. inch). To think that you can start with 50 million/cm^2 and only reduce this to 20 million/cm^2 if you just use a cold pressure wash before applying your disinfectant wash , you can see the problem you are giving any disinfectant! Waddilove moreover, puts it very succinctly.

> *"To understand the problem, think about the material you are trying to remove. It is often dried on, strongly adherent and greasy. Now think about washing up your own dinner plates after leaving them overnight after a greasy meal. With cold water this is nearly impossible, with hot water it is difficult, but add a detergent and it is much easier. So why, when they pressure wash, do most farmers use just water (often cold) and don't use a detergent?"*
>
> Waddilove (1999)

PRE-CLEANING PARAMOUNT

Over the past ten years, as research into better disinfectants received prominence in the media, the importance of pre-cleaning has been sidelined.

A PRE-CLEANING CHECKLIST

✓ Disconnect the electricity supply.

✓ Remove all moveable equipment.

✓ Open all inaccessible areas – fan trunking etc.

✓ Physically remove as much organic matter as possible from all in-contact surfaces.

✓ Flush out slatted storage pigs and gullies.

✓ Use an approved degreasing detergent to loosen up the dirty surface

✓ Allow time (a minimum of 20-30 minutes) for it to soak in, but longer is better – half a day if possible.

✓ Pressure wash the complete building at 500 psi (max) with hot water (to remove all traces of grease). The temperature must be 70°C or higher (160°F).

✓ Inspect for thoroughness.

Next, make sure you are using a satisfactory detergent.

HOW TO CHOOSE A GOOD DETERGENT: A CHECKLIST

It must be farm approved. What does this entail?

✓ Capable of working well on all surfaces found on a pig farm. Unlike urban factories, there are many kinds of farm surfaces. Several of them are semi-porous (*e.g.* concrete, plastic and some metals). This variability makes it more important to use a product specifically designed for on-farm use. A heavy-duty formula is essential, stronger than those used in a catering establishment, for example. In my survey 18% of the farms used a well-known catering detergent 'because it was cheaper'.

✓ Contamination in crevices and other poorly accessible places is more easily removed with a heavy-duty formula.

✓   Slats are more thoroughly cleaned. The build-up of dung on the surface facing between the slats is more easily dislodged. This is especially important with enteric organisms such as *E. Coli* and *Serpulina hyodysenteriae* (Swine Dysentery) and *Lawsonia intracellularis* (Ileitis).

✓   Good degreasing is vital. Just because a surface looks clean it does not mean it is clean of all pathogens. The presence of a greasy layer on the surface increases protection of micro-organisms by long chain fatty-acid molecules. A heavy-duty alkaline formula helps remove this protection. This is important as the newer, essential and better virucides don't work so well with fat protecting the organisms.

✓   Vital if time is limited – a heavy-duty detergent works quicker and faster.

✓   It mustn't interfere with the subsequent disinfectant's activity. This highlights the importance of using a fully integrated programme, such as the Antec Pig Biosecurity Programme, when the products are specially chosen to be compatible, or in some cases help each other.

✓   Ideally it should be applied through existing equipment with minimal modifications.

✓   Foaming can be helpful. This increases the contact time and allows operatives to see where it has been applied. The foaming decreases the amount of water needed in the soaking and pressure-washing phases of cleaning. Reducing water reduces costs and problems with excess run-off to dispose of.

✓   It does not leave residues that can make the floor slippery and harbour micro-organisms. Especially, it should not leave cumulative residues.

✓   It should work in hard water situations.

✓   It should be non-toxic to pigs and operatives.

THE VALUE OF PROPER CLEANING DOWN / DETERGENT USE

Table 3 from Australia shows the performance improvement of growing/finishing pigs from proper precleaning.

Table 3 PRE-CLEANING ITSELF BOOSTS PERFORMANCE

Class of pigs	Cleaned buildings before disinfection		Uncleaned buildings before disinfection		% increase
Weaners	572	(1.26)	500	(1.10)	14.4
Growers	736	(1.62)	692	(1.52)	6.3
Finishers	671	(1.48)	621	(1.37)	8.1
Birth to market	569	(1.25)	530	(1.17)	8.2

The buildings were AIAO (All-in; All-out) All weights in gms/day (lb/day)

Source: Cargill & Benhazi (1998)

ECONOMICS

A 39g/day (1.37 oz) improved gain from 6 to 90 kg (13-198 kg) results in an improvement in saleable meat per tonne of feed used (MTF) of 24 kg (53 lb). This is equivalent to a 15% reduction in cost/tonne of feed at a modest deadweight pig price. The extra cost of cleaning and the special biocide detergent used is equivalent to 5% of the cost of one tonne of feed. Therefore a payback or REO (Return to Extra Outlay) of 3:1 is achieved, even in times of very low pig returns.

DISINFECTION

We should now have a clean and exposed surface with a TVC of 100,000/cm^2 or less bacteria. Such a bacterial threshold should also bring down viruses to a controllable level. The *surfaces* are now ready for disinfection, but we also have water tanks, lines and drinkers harbouring pathogens, and pockets of air, such as in lofts, which need attention to prevent recontamination. The problem with the older disinfectants has been that ...

- They are poorly effective against some of the newer viruses unless used at impracticable and costly concentrations.

- Due to toxicity they cannot be used in water lines or as space foggers.

- There is a wide range of correct dilution rates and coverage areas to deal with certain pathogens. Stockpeople risk getting them wrong and owners order up the wrong disinfectant basing their decision on price.

- Some are toxic or irritant. See Table 5.

A DISINFECTION CHECKLIST

✓   There is a wide range of disinfectants on the market.

✓   You must choose one which is approved for the disease spectrum you and your veterinarian are likely to encounter.

✓   So either take advice from your veterinarian , or only buy from a well-known primary manufacturer who can advise you on which one to choose.

✓   Equally important as the choice of disinfectant is to follow the approved dilution rate which will differ according to the disease situation. Some manufacturers have a simple colour dipstick (*e.g.* Antec) so that you can *quickly and easily* check that the product is correctly diluted for the purpose in mind.

✓   Also important are the instructions as to cover rate. Most people use a 'chisel' pressure washer at 200 psi or a spray nozzle. Managers should check the usage rate of the disinfectant purchased against the surface area which should have been covered in, say a month's use. In the examples reported to me during the recent serious Foot and Mouth outbreak in Britain, I suspected from cans used and discarded that the cleaning contractors had not fully followed the instructions either as to dilution or coverage rate.

✓   Check the time rate recommended for the disinfectant to act fully.

✓   Some disinfectants take longer to work in cold weather (Table 4). You may need a higher concentration, so check with the manufacturer in cold weather conditions

✓   If in doubt or you are unwilling to go into all these careful details, just use a peracetic acid disinfectant, the best known is Virkon or Virkon S. These are powerful oxidising agents and can rapidly kill most viruses as well as all bacteria; especially if you clean down well.

Table 4 IS YOUR DISINFECTANT EFFECTIVE IN WINTER?

Dilution	Temperature		*Able to stop bacterial growth*	
			Formaldehyde	*Modern formulated product at specified dilution*
1%	20°C	68°F	Yes	Yes
2%	10°C	50°F	No	Yes
3%	4°C	39°F	No	Yes
4%	0°C	32°F	No	Yes

Table 5 SOME PROPERTIES OF DISINFECTANTS

Toxic	*Irritant*	*Corrosive*	*Taint*
Chlorines	Chlorines	Phenols	Chlorines
Phenols	Phenols		Formalin
Some iodophors	Formalin		Iodophors
Formalin			
Good speed of kill	*Long persistence*	*Good virus control*	*Chloroxynol phenols*
Peracetic acid	Phenols	Peracetic acid	Chlorines
Peroxygen		and peroxygen	Iodophors
		are best followed	
		by iodophors	

	*Action in presence of organic matter**	
Best	*Moderate*	*Poorest*
Phenols	Peracetic Acid	Chlorine
Iodophors	Peroxygen	
Formalin	Quaternaries	

* To be safe, always preclean with a detergent

AFTER DISINFECTION

'Resting' the Building

The writer has found that the common practice of resting the building before reoccupation (3 to 4 days winter; 2 days in summer) is generally not needed if the surfaces are quite dry before stock are put in.

An industrial kerosene space heater is very useful to achieve this in quite a short time, but check slatted areas and crevices in particular when the hot air currents can be deflected.

Again, as with cleaning down, check out the surfaces and, if possible, do a periodic swab test on close-contact areas. Your veterinarian can arrange this. The target threshold is 1000 TVC/cm^2 or 6500/sq in.

SANITISING THE WATER LINES

The drinking water system is a potent source of viral reinfection (PRRS, PMWS especially) as well as digestive-disturbing bacteria like Balantidium coli. While QAC (Quaternary Ammonium Compounds) have been used for this, any organic matter – like slime – can inactivate them. Much better to use a peracetic acid like Virkon S.

CHECKLIST FOR SANITIZING WATER LINES

✓   Make certain that the product you intend to use is recommended for this purpose.

✓   Check whether or not pigs can be present during sanitation. *The dilution rate will be different for in-situ pigs*.

✓   Ensure the water system is not blocked, or if any drinker is leaking badly.

✓   Read the product instructions carefully.

✓   Dose the header tank. At terminal disinfection, (pigs absent) the strength can be much greater than if the water is santized with pigs present.

✓   For terminal disinfection know the volume of each header tank and the volume of the pipe-run it serves. Leave for at least 30 minutes, then allow to drain.

✓   For sanitation with pigs present, there is no such intense activation period, of course. Frequency of treatment depends on disease level (as does the dilution rate). See advice from your veterinarian or from the manufacturer.

✓ For sanitation with pigs present, know the volume of each header tank and use a ***dilution test strip*** as a double-check.

✓ Routine water treatment can depend on climate *eg* systems in the tropics may need more frequent attention, especially the header tanks.

✓ Do not mix water medication products with the water sanitizer.

✓ Keep your header tanks covered against dust, insects etc.

✓ Where pressure-washing the pens, ***attend to the underside of any drinker tongues***. The water sanitizer will insufficiently reach such an inaccessible surface which continually comes so close to a pig's mouth. Feel for any slipperiness/slime with your finger.

Remember: Water systems are often overlooked as a source of disease spread. Incorporate water sanitation into your routine.

AIR FOGGING

Dust and microglobules of water are pathogen 'taxis', carrying hostile organisms directly into the pigs mouth and lungs as well as acting as breeding surfaces especially in the upper reaches of a piggery, like ceilings and lofts. These microparticles can enter the smallest cracks, and changes in air pressure from outside or in ensures the movement and release of disease organisms *even after the structures/rooms below have been carefully cleaned and disinfected*.

The disadvantages of previous fogging technique

Routine fogging with 500 ml liquid Formalin 40% poured on to 200g of potassium permanganate per 28m^3 (989 ft^3) of air space which is closed up for 12 hours and not used for 8 hours after opening up again was developed by the poultry industry. This formula is effective and cheap but has serious drawbacks.

·It is laborious · It is dangerous · The surfaces need to be wetted first and openings sealed   · A second person is advisable as a standby · Livestock cannot be present · Protective clothing ***and a face mask*** is essential   · Stockpeople dislike it   · The pot. permang., if bought in quantity, doesn't store well as it absorbs moisture from the atmosphere and then cakes.

A major breakthrough

When oxygenating disinfectants appeared and could be used through an electric misting device most of these disadvantages were overcome. At the correct dilution rate, Virkon S, for example can be used to fog the atmosphere not only at terminal disinfection but *when the pigs are present*. It is thought that this in situ application as a breathable mist could well be beneficial where respiratory infection is present, even on a daily basis for a time. Seek veterinary advice on this.

With products of this sophistication, virtually all the disadvantages of the formalin method disappear, *but it is much more expensive*. Do your sums to explore this. I give some guidelines below.

So should we fog as routine?

There is little concrete evidence of how easily or how much several pig viruses can spread via the internal aerosol (air droplet) route, though it is expected to be very likely. Routine fogging is likely to be a worthwhile method of killing airborne bacteria and viruses, and to sanitize the more inaccessible surfaces in a building where they can lodge.

At the time of writing fogging costs up to £13/sow/year in Britain. A lot! This cost pays for the products used for all surfaces and air space to be sanitized *each time* terminal disinfection is done (x11 times a year, farrowing house, x5 times nursery and x3 times finishing house) and also includes the cost of the misting equipment and all the labour involved.

As some viruses are such predators of profit these days, incurring costs varying from £500 per sow per year for Classical Swine Fever through Aujeszky's (£168) TGE (£124) and EP (£100) and £75 for Infertility Viruses, the spending of about £13 per sow/year is put in perspective. In the light of these potential costs, £1/month on fogging seems to me to be a worthwhile investment now that viruses are attacking us so remorselessly.

LIMEWASH

Another old and well-tried technique. After solid floors have been cleaned and disinfected the in-contact surfaces are brushed with a mixture of hydrated lime to produce a thickness of thin salad cream.

CHECKLIST - BUYING A DISINFECTANT

These notes can be used as a checklist for *any* disinfectant under contemplation before purchase.

Why are peracetic acid and peroxygen disinfectants such a step forward?

✓ **Effectiveness**. The longest list yet of many hundreds of bacteria and viruses killed in test trials including virtually all the critical bacterial, viral and fungal families known to cause disease in pigs.

✓ **Stability**. Very stable which ensures long-term killing power.

✓ **Biofriendly**. Eventually breaks down into water, oxygen and carbon dioxide so do not remain a threat to the environment.

✓ **Organic challenge**. Work quite well in the presence of organic material.

✓ **Low temperatures**. Many disinfectants have to be increased in concentration to work well in winter conditions. Peracetic acid and peroxygen keep on killing in cold weather at their advised quite high dilution rates.

✓ **Note: Comparing the cost of disinfectants**: Always compare disinfectants on their power relative to the *approved dilution rate* and recommended *surface cover rates*. What may seem an expensive price per drum can often provide a cheaper cost per square metre treated.

✓ **Sewage and earthworms**. Even with strict European standards, they do not pose a threat to either sewage treatment or earthworms in soil, if used as directed.

✓ **Corrosion**. They are non-corrosive to metal, rubber or plastic.

✓ **Animal safe**. Can be used for fogging and water treatment purposes with the pigs present.

✓ **Operator safe**. Maximum Exposure Levels can be up to 40 times more than other disinfectants – for example Glutaraldehyde.

Cement paint and phenolic disinfectant are another combination. Particularly popular in solid floored farrowing pens of the old-fashioned type, where even the crate rails were coated using a soft household brush. After 48 hours drying a hard but flesh-friendly antiseptic skin was left on treated farrowing pen surfaces. The technique is still popular in hot/dry countries where water is short and the atmosphere dry and warm.

However, wherever possible producers should use the modern technique described in this section as hydrated lime is not a match (even with some phenol disinfectant added at 30 ml to each 45 litres (10 gallons) of limewash) for the modern cleaning and disinfectant agents.

WHY ARE MODERN PERACETIC ACID/ PEROXYGEN DISINFECTANTS SO EFFECTIVE AGAINST VIRUSES IN PARTICULAR?

A virus has a protective **outer shield** consisting of …
- a fat layer
- a membrane
- a peplomer, or protein structure.

Inside this is a **capsid** or **inner shell** which envelopes the DNA-containing nucleus. This capsid consists of …
- capsomers; protein structures to hold and sustain the nucleus
- a membrane
- the nucleus itself.

The disinfectant has to break into both the shields and destroy or dysfunction the nucleus.

It has four weapons to do this …

1. **A surfactant** to dissolve the outer fat layer and help attack the protein of the peplomer and capsomer layers.

2. **Organic acids** which also attack these protein layers. Being organic, these are non-corrosive.

3. **An oxidising agent** which cuts through the now-damaged protein structures and goes for the nucleus – even in cold conditions.

4. **A buffering agent** (to increase acidity) which improves the biocidal effect, and also reduces the neutralising effect of hard water and organic material.

Thus these new sophisticated disinfectants are a composite of carefully-chosen chemicals each specifically damaging a part of the virus structure. Viruses are not only small organisms and thus difficult to target, but many are very tough and resistant (*i.e.* virulent).

HOW THEY WORK. A SIMPLE SUMMARY

The surfactant *locates* the virus and *opens up* its outer surfaces to allow the nucleus to be *attacked* by the other two ingredients – the organic acids and the oxidizers – and the buffering agent *fends off the defensive action* of organic matter etc.

Simpler, cheaper disinfectants only complete some of the processes, and at a slower rate.

DO YOU HAVE TO USE PERACETIC ACID/ PEROXYGEN DISINFECTANTS FOR EVERYTHING?

Not necessarily, although there is a tendency – laudable I think – to just use one or at the most two approved products for simplicity's sake and to avoid mistakes in mixing and application. There is also the adage '*Better safe than sorry!*'.

A generally agreed list of what to use/where could be:

Virus infections. Peracetic acid, peroxygen or iodophors.

Foot dips. Iodine based, or if frequent replenishment is not likely – peracetic acid.

Fogging. Peracetic acid.

Hands. Quaternaries and soaps.

Water sanitation. Peracetic acid.

Concrete surfaces. Phenols. For very rough, broken *outdoor* surfaces, use an oil-based phenol.

Loading ramps. Peracetic acid or peroxygen (because you have no knowledge of what notifiable or transmissable diseases may be being brought on to your farm, so as broad a spectrum of control as possible is wise).

Delivery and collection vehicles. Peracetic acid as above, plus the advantages of minimal corrosion.

COSTS AND PAYBACKS OF MODERN HYGIENE PRACTICE

Of course these better detergents and disinfectants cost more, even at the higher dilution levels their efficiency allows. Added to this are the extra tasks of things like pre-soaking time, hot pressure washing, fogging, water sanitation and thorough drying before re-occupation.

IS IT ALL WORTH IT?

We need to examine typical results, both on-farm and research, and put some economic benefits to them. Then cost out the extra investment needed to update the countermeasures to satisfy the most up-to-date advice available.

Table 6 is based on 17 comparative trials (11 farm and 6 research) where on 16 of them the pigs were not suffering from any particular or obvious disease. This is important, as good biosecurity should lessen the incidence of clinical disease outbreaks, where the benefit would be much greater – possibly up to 5 times as much.

THE COSTS OF DOING IT PROPERLY

So if these are the possible benefits, what does a meticulous hygiene system cost?

I make a stab at it in Table 7 and go on to set this against what the trial work to date suggests you will gain in increased income/overall saved costs.

Table 6 THE BENEFITS FROM BETTER BIOSECURITY TECHNIQUES

Reference	Trial type, basic details	Calculated value of finishers (Against controls or former practice)
Cargill & Benhazi (1998)	Cleaning AIAO buildings before disinfection with a detergent	+ £3.12/pig
Overton (1995)	Salmonella outbreak controlled	+ £3.86/pig
Jajubowski *et al.* (1998)	Using a peracetic acid disinfectant instead of NaOH	+ £8.80/pig
Sala *et al.* (1998)	Full Antec programme v. iodine	+ £2.10/pig
Sala *et al.* (1998)	Full Antec programme batch disinfection v. terminal disinfection only	+ £5.66/pig
NCASHP Denmark (*nd*)	Partial v. total biosecurity programme	+ £7.77/pig
Antec Trial (G&M, 1999)	Change to AIAO and updated disinfectant, result after 3rd batch	+ £7.15/pig
Gadd (1994-1998)	Average of 10 clients uprated to full biosecurity protocols	+ £5.63/pig
	Average : all results	£ 5.51/pig

Assumptions:

Weights ranged from 6–90 to 30-100 kg (13-199 and 66-221 lb)
Food in last 14-21 days, range 2.2 to 2.25 kg/day (4.9-5 lb/day)
Finisher feed price £130/t (about twice the value of a finished pig in Europe today)
KO% standardised at 73%

The additional cost of proper biosecurity is divided between the extra cost of the modern materials over what you use now (*i.e.* **virucides** rather than just bactericides); the costs of extra tasks now considered important (*i.e.* **hot** pressure washing using a **heavy duty** detergent; **fogging** the airspace and **sanitising** the drinking water) and the cost of the **extra** labour all this entails. Table 7 is an attempt to draw all this together for the first time. I have not seen this done before.

If from the trials cited in Table 6 the expected benefit from a complete biosecurity protocol is £5.50/pig then the REO (Return on Extra Outlay) is £5.50 ÷ £0.44 *or 12.5:1* (Table 7). When a good growth enhancer typically obtains 6:1 REO at best, this puts proper biosecurity into its true perspective – a very good bargain indeed.

Table 7   PROPER BIOSECURITY – THE COST PICTURE

Figures based on the surface areas/labour and materials needed to produce 100 finished pigs on a farrow-to-finish farm, including all breeding unit surface areas needed for 4.33 sows

Expenditure on materials	*Correct biosecurity protocol*	*What you do now*	*Extra Cost*
Cleaning materials	1.07p – 1.3p/m^2	No detergent used	1.07p – 1.3p/m^2
Disinfection	0.79p-1.7p/m^2	0.61p/m^2	0.18-1.09p/m^2
Airspace fogging	38p per 100 pigs	Rarely done	0.38p/pig
Water sanitation	£5.43 per 100 pigs	Rarely done	5.43p/pig

Labour (& Cleaning Equipment)

Cleaning	No difference	–
Hot pressure wash	Cost of steam jenny over 8 years £1.60/100 pigs	1.6p/pig
Disinfection	No difference	–
Airspace fogging	£1.30 per 100 pigs	0.13p/pig
Water sanitation	65p per 100 pigs	0.65p/pig

Total extra cost per pig is as follows:
At 15m^2 total surface area needed to be sanitised per finished pig (including the breeding farm) and taking the most powerful/costly products available at the more concentrated dilution rates at 2.39p/m^2 the extra cost per pig is 15 x 2.39 = 35.85p. Add to this the remaining extra costs/pig listed above of 8.19p, the total extra cost over what is done now is 44p/pig.

Note: UK costings and metric measurements are used here because (see below) we are only comparing costs against likely return.

CONCLUSIONS FROM THIS ECONOMETRIC EXERCISE

• Far from being expensive, proper biosecurity offers a very handsome payback.

- Farmers often baulk at the extra labour cost of doing these extra tasks. In fact it is only 5.4% more. Several producers and I who measured it carefully and obtained this average figure between us were really surprised how low it was.

- Farmers also complain about the higher cost of the correct, more powerful products per drum/bag, but it is still under 36p/pig (or 0.6% of production cost), and 6p of that is in the water sanitation area. Sure, it is over three times more expensive than using just a simple disinfectant alone – but as Table 7 suggests it should give you *over a 12 times return* on the extra cost.

- Farmers say that they just haven't got the labour to devote to the extra time needed. Then maybe they aren't correctly prioritising the work of their labour pool? A 12:1 return is just as important as most of the other vital labour tasks in potential value – farrowing, breeding, weaning.

In fact, moving up a gear to correct biosecurity is one of the most important routes you can take to more profit. I strongly recommend all pig producers reading this to reassess, with their veterinarian, exactly how much trouble they are presently taking over biosecurity, the adequacy of the products they currently use, and think it all through carefully.

You could be losing £11 on every £1 you are saving. Enough said?

VEHICLES AND TRANSPORT

Recent serious outbreaks of CSF and FMD in Britain have emphasised how dangerous a disease vector are the vehicles which visit your farm regularly.

At the time of writing the full recommendations on vehicle disinfection and the advised precautions to take from the various UK enquiries into the spread of FMD have yet to be made public. However I give below my own advance conclusions and recommendations as follows:

A FUTURE CHECKLIST FOR FARM BIOSECURITY?

✓   Farms should have *perimeter* unloading and loading facilities for pigs, feed and general goods. No delivery/collection vehicle, including salesmen's cars or even the vet's vehicle, should be permitted past the farm boundary, or alternatively only allowed into a biosecure area with separate access on to hard standing with cleaning facilities attached.

✓   No pig collection or delivery vehicle should be accepted without a form/certificate issued by the buying organisation /vendor signed by the buyer/vendor's biosecurity supervisor that the vehicle has been properly cleaned and disinfected – inside, under and out – before calling at the farm. These organisations to train and appoint a biosecurity clearance officer as routine by law.

✓   In addition, each vehicle must have available a written load record which you can examine.

✓   No driver is allowed to assist in loading/unloading pigs.

✓   Anyone driving a transporter must not be allowed *by law* to keep pigs on any other premises.

✓   All abattoirs, markets, feed mills and commercial pig-breeding companies must invest in satisfactory tunnel-type disinfecting bays with a supervisor attached to ensure thoroughness and avoid 'dodging'. Legislation may be needed to enforce this.

✓   Farm pig loading bays are essential, with wash-down facilities on tap. Drainage should be away from the farm.

✓   Casualties must be incinerated on site. In certain circumstances they can be left under licensed arrangements for collection *eg* rendering etc at an off-site venue, suitably protected from exposure and degradation, with the site disinfected after removal.

✓   Wheel dips on routes to and from the farm are a wise precaution but must be adequate *i.e.* with automatic or manual spraying of wheel arches or undersides of all vehicles passing through. Such devices are now on the market.

✓   Place clear signs at perimeter exit/entrances, and provide your suppliers of goods and services with specific instructions on biosecurity-related measures you will expect them to follow.

✓ For bulk feed deliveries it is preferable to have your own off-loading blower hose, as many truck-mounted hoses are dragged across farmyards and rarely disinfected for fear of contaminating feed and for delaying deliveries.

Remember – all vehicles are a potentially serious disease risk. After the world's worst outbreak of FMD we in Britain realise this only too well. Think about minimising that risk where you are concerned.

YOU AND YOUR STAFF - A PERSONAL DISEASE CHECKLIST

✓ Zoonoses are dangerous. You can easily catch some pig diseases.

✓ So wash your hands before you leave a main building. This is for your protection as much as the pigs. (The Japanese do this quite commonly and have a portable washbowl and soap/towel on a metal stand by the door.)

✓ Use a foot dip, replenished regularly, outside each main building.

✓ Wear one set of boots for outside (office and stores) premises, another for inside. The same with coveralls.

✓ If visiting a farm (like myself) leave your car off the farm boundary, *walk*, and carry your things with you! It's a pain, but I do it.

✓ Do not wear other people's boots.

✓ Shower and also wash clothes which have been on a piggery immediately after getting home.

✓ Wear disposable gloves when taking blood, castrating or doing post-mortems. Cover all cuts/grazes.

✓ Shower-in/shower-out (in my opinion) is probably over-rated, but continue until it is proved so, if ever. The quality of these facilities is generally far too low, anyway.

✓ Dust and mould spores can be very dangerous. Wear a proper, approved filter helmet when home-mixing or dealing with poor quality straw. Dust is much more dangerous to you than to the pigs. You live 15-150 times longer!

✓   After you have dosed a litter of young pigs against scour, your hands, and somewhere on your clothing may harbour as many as 10 billion (ten thousand million) of the organisms responsible. Only about 0.02% (or 2 million) bacteria ingested by you are enough to give you diarrhoea, especially if your current immune system is shaky.

So after treatment and flushing out the scoured areas in the pen with water, *wash your hands and change your overalls* before handling further piglets or eating rest-room food. Have special waterproof aprons to facilitate this. Stockpeople spread a lot of disease internally on a pig unit.

✓   Carbon monoxide from slurry pits stirred up before emptying is lethal – it nearly cost me my life, when I was a stockman. You cannot smell it. It knocks you out in a few seconds and kills you in a few more. Fortunately I fell down unconscious outside the piggery door, not inside it, otherwise this book wouldn't have been written.

✓   Likewise never enter a bulk bin. Cleaning must be done from the side through an inserted bulkhead door, with the top inspection hatch open.

DOING A BIOSECURITY AUDIT: AN EXTENDED CHECKLIST/ QUESTIONNAIRE

This list, while long, is not exhaustive. Nevertheless one or two items may make you or your suppliers sit up a bit!

✓   **Who visits the farm?** What precautions have you taken to keep them out/sanitise them on entry and exit? This includes your veterinarian and his vehicle, bless him!

✓   **How do you dissuade bird vectors?** Netting over entry points, covering feed hoppers.

✓   **How do you control flies?** Have you a fly-control person?

✓ **How do you control rodents?** Have you a rodent 'king' in the same way?

✓ **How do you take in replacement stock?** Does your vet liaise with the vendor? Do you quarantine? Far away enough? For long enough? Inspect them frequently enough? Have a disposal procedure agreed in advance for suspects/failures? Buy from one source? Never buy from markets?

✓ **Have you vehicle sanitising facilities**, both for you and for visiting vehicles if allowed entry? Hard standings for visitors' cars?

✓ **Have you footbaths outside each main building?** Replenish correctly and often enough.

✓ **Do you encourage/insist on handwashing/hand protection** with all surgical, farrowing and dosing tasks?

✓ **How clean** are your toilets/staff-rooms?

✓ **Have you put up correct guidelines in the staff-room** for inspection techniques/needle use?

✓ **And for AI procedure likewise?**

✓ **Have you made it clear** to all your suppliers that you insist that they follow a protocol when delivering to your farm? Or at least discuss feasibility with them.

✓ And the same for your haulage-out requirements?

✓ **Have you asked them both** how, when and how often they sanitise their vehicles? Check up on this once a year with a visit.

✓ **Have you a locked perimeter fence and call bell/telephone?**

✓ **Do you have bulk bins on the perimeter?** Do you use your own input hose?

✓ **Do you sanitise your bulk bins?** It is scandalous that some bin manufacturers don't build in a bulkhead door on one side to get a ladder in. This should be mandatory by law.

✓ **If wet feeding** do you sanitise the tanks, line, troughs, periodically?

✓ **Do you load pigs out at a separate, specific place on the perimeter?**

✓ **Do you wash and sanitise this area after every batch is shipped?**

✓   **Do the washings drain away from the farm?**

✓   **Have you asked your vet** for a drug use record and guidance on usage/storage?

✓   **Do you sanitize your water header tanks and lines?**

✓   **Do you fog your buildings?** Have you discussed this with your vet?

✓   **Do you do water tests/bacterial swabs periodically?**

✓   **Have you asked your feed supplier what salmonella tests he has done?**

✓   **Have you trained/made aware to your staff** that handling diseased animals needs *extra* hygiene and self protection/ disinfection beyond their ordinary routine?

✓   **Have your staff got girl/boy friends on another farm?**

✓   **Do you ever borrow/share gear from/with other livestock farmers?** If so, how do you sanitise it before entry to the farm?

✓   **If outdoors, have you studied how to minimise disease spread** *e.g.* using the box of matches, field rotation/air and sunshine? Take advice.

✓   **Have you got chickens/sheep or goats among the pig area?** Ban them!

✓   **How clean and tidy are the spaces between the houses?** The chaos I often see encourages vermin, the puddles encourage flies, spilt feed other predators – all of which bring in disease. *Make time* to have a blitz twice a year.

✓   **Have you a method of checking periodically that:-**

   •   you are using the right detergent
      in the right way

   •   you are using the right disinfectant
      in the right way

   } Dilution and cover rates are correct

✓   Use dilution strips and have a check on what you've used from sales dockets covering the required total area every 3 or 4 months. The survey revealed that 3 farms were under-using the products by 35%!

   Get the salesman to monitor this for you. At 35% less sales it is worth his while keeping you straight!

✓ **Do you sanitise the slurry pits?** It is scandalous that many floor manufacturers/house designers don't allow for a hinged access trapdoor in every pen to allow a lance to get in there as a minimum. We may need future legislation to make them do so.

✓ **Do you cover manure heaps against rains/leaching?** Where does the run-off go?

✓ **Do you inspect ad-lib hoppers for stale feed?**

✓ And wastage?

✓ **Finally**:

Biosecurity also involves protective vaccination. You need to review your needs with your veterinarian twice a year. Pathogens change that quickly.

REFERENCES

Waddilove, J (1998) 'Disinfection on the Cheap?' Antec International, Monograph

Ibid (1999) 'Get it Clean Before You Disinfect' Antec International, Monograph

NCPBHP Denmark (1996) National Committee for Pig Breeding Health & Production Report p 37

Cargill, C & Benhazi, T (1998) Proc. IPVS Conf. 3, 15.

Sala et al (1998) Proc. IPVS Conf. 2 110 et seq.

Antec International (1999) G&M Field trial report.

Gadd (2000) 'The Revolution in Biosecurity' Vétoquinol Trans Canada Swine Seminars, March 10-15 2000.

LIQUID FEEDING - PROBLEMS AND GUIDELINES

29

Liquid feeding of pigs uses a liquid vehicle – primarily water, skim milk or whey, but any suitable liquid by-product will do – to carry solid nutrients (usually in meal form) in suspension, or alternatively some co-products in solution, to the point of consumption, which is usually a trough.

Liquid feeding, also known as pipeline feeding, should not be confused with wet/dry feeding (sometimes called, erroneously, 'single-space feeding'), where water is available from a displacement or nose-press valve operated by the pig itself to dampen or liquify meal, and sometimes pellets, into a shallow pan to a consistency of its choice.

Both methods of feeding are excellent systems when operated properly, with wet/dry feeding particularly popular at present for the younger grower, and fully liquid feeding for all classes of pigs, including breeding stock.

MANY ADVANTAGES

This said, the author believes that the liquid/pipeline system has many advantages over the wet/dry system as detailed below.

However I must 'declare an interest' as the politicians say. I have grown up with pipeline feeding, and was fortunate to be among the pioneers of the system in the 1950s and 60s, taking an active part in its development across 40 years of farm advisory work.

This experience leads me to forecast that liquid feeding will eventually become the predominant method of feeding all swine across the world.

THE ADVANTAGES OF LIQUID FEEDING FOR PIGS

1.   ***Pigs do better.***   Most users report significant improvements in performance comparing liquid feeding to dry feeding. Something the American hog industry, currently one of the lowest users of the concept, could ponder? A summary of the latest position is in Table 1.

Table 1   EVALUATION OF PERFORMANCE OF PIGS ON LIQUID OR DRY FEEDING REGIMES: CLASSIFIED RESULTS FROM PAPERS PUBLISHED SINCE 1956

Growth rate	Feed:gain ratio	Carcass quality	N° of reports (Braude, 1972)	N° of reports (Geary, 1997)
+	+	+	5	2
+	+	0	7	4
+	+	N	11	1
+	+	-	1	1
+	0	N	2	1
+	N	N		
+	0	0	3	
0	0	0	5	
0	0	N	4	
0	+	0	1	
0	0	+	1	
0	-	N	1	1
-	-	0	1	
-	-	N	2	

+   represents experiments where liquid feeding was better than dry feeding
-   represents experiments where dry feeding was better than liquid feeding
0   represents no difference between liquid and dry feeding
N   represents no information available for that performance measure
Source: Chesworth (2001)

2.   ***Pigs waste less feed.***   Wasted feed can be direct (down the slats; trodden in; dust) or indirect (wrong nutritive ratios, etc). Wasted food from pipeline troughs is not significantly different to that from wet/dry feed hoppers (*Table 3*). Feed waste is literally money down the drain. Most dry feeders waste 6%; some as much as 15%

Table 2 EXAMPLES OF BOTH BEFORE-AND-AFTER / SIDE-BY-SIDE TRIALS ON 3 FARMS ADJUSTED TO 5,000 FINISHERS PRODUCED PER YEAR (TONNES/YEAR)

	With wet feed	Without wet feed	Saving on wet feed
Farm 1	838	770	68 tonnes
Farm 2	900	803	97 tonnes
Farm 3	984	942	42 tonnes

Source : Clients' records

Table 3 COMPARISON OF ONE FARM WHERE WET/DRY AND FULLY WET FEED SYSTEMS ARE RUN IN PARALLEL. (WASTAGE COLLECTED FROM GRILLES UNDER EACH FEED STATION. AD-LIB FEEDING IN BOTH TREATMENTS.)

	Fully wet N = 20 pens	Wet/dry N = 20 pens
Food collected as waste (dry matter basis)	2.1%	2.0%

Source : Clients' records

3. *Sows fed wet eat more.* During the dry stage this is not important, but for lactating sows (Table 4) especially in hot weather or in the tropics (Table 5) this can be of critical importance.

Table 4 THE EFFECT OF WET AND DRY FEEDING ON LACTATION FEED INTAKE IN SOWS IN A TEMPERATE ZONE (UK)

		Wet (& water)	Dry (& water)
Average food eaten/day sows, kg (lb)		6.2 (13.7)	5.2 (11.5)
Piglet mortality		10.6%	12.3%
Farrowing index		2.26	2.12
% pigs served by 5 days or	Farm A	62	48
less after weaning	Farm B	66	51

Source : Clients' records

Table 5 SOW TRIALS ON WET V. DRY FEEDING OF SOWS IN THE TROPICS

	Farm 1		Farm 2 (average of 2 trials)	
	Wet	*Dry*	*Wet*	*Dry*
Farrowings	136	130	161	85
Av daytime temperature, °C (°F)	28 (82)	28 (82)	30 (86)	30 (86)
Feed eaten in lactation, kg/day (lb/day)	6 (13.2)	5 (11)	6.2 (13.7)	5.5 (12.1)
% sows served by day 5 after weaning	64	46	60	50
Weaner weight per sow per year, at 21 days, kg (lb)	148 (326)	126 (278)	120 (265)	94 (207)
	+17%		+28%	

Source : Clients' records, Thailand, (1993)

4. Because critical nutrient intake is higher, sows lose less condition.

Table 6 THE EFFECT OF WET AND DRY FEEDING ON SOW CONDITION AND PERFORMANCE

	Sows fed wet	*Sows fed conventionally*
Average condition score	2.6	2.5
Pigs per sow per year	21.4	19.1
Weaner weight per sow per year, kg (lb)	147.7 (326) +17%	126.1 (278)
Weaner weight per crate space per year, kg (lb)	773 (1705) +11%	696 (1535)

Source : Clients' records

5. Wet feeding reduces dust substantially. Many researchers report figures similar to those of the early work (Table 7).

Table 7 A SURVEY OF DUST CONCENTRATIONS IN PIGGERIES

	Dust concentrations (mg/m^3)
Feeding meal	14-79
Feeding pellets	5.1-23
Wet feeding	0.5-14

Source : Cermack (1978)

Carpenter (1986) reported that airborne micro-organisms are 3 times higher in the pens of dry-fed pigs, and Robertson (1991) found that 45% of dust particles were in excess of 10mg/m^3 (the operational exposure limit for UK COSHH Regulations) during home milling-mixing operations. With liquid feeding the feed mixing is done in a tank.

6. ***Piggery workers are healthier.*** A medical report I found in my files reveals the type of illnesses encountered on the pig units of that time where wet feeding was perhaps only available on 6% of the finishing pig units in Britain; and less than 0.5% on breeding farms. It could be interesting to update the table, and see how much better things are today, now that in some countries liquid feeding is used on 50% or more pig farms.

Table 8 PIGMEN REPORTING SYMPTOMS

1.	Cough	58%
2.	Phlegm	39%
3.	Chest tightness	26%
4.	Throat irritation	39%
5.	Nose irritation	39%
6.	Eye irritation	25%
7.	Fatigue	35%
8.	Muscle pain	22%
9.	Joint pain	23%

N°s 1 - 6 are directly attributable to or aggravated by dust in piggeries.
Source : Watson (1978)

7. ***Better use of labour*** World-wide, producers complain about the difficulty of getting and keeping good replacement labour. Moving food around has always been a heavy, onerous task (Table 9) and liquid feeding removes this completely. Hydraulics do the work!

Table 9 THE AMOUNT OF FOOD HANDLED BY PIG PRODUCERS IN A YEAR FOR EVERY 100 SOWS

	*Feed handled per year (tonnes)**
Breeding herd	142
Weaner herd	127
Finishing herd	271
Total	398

* Farrow to finish – 100 sows, 20 gilts, 5 boars, 22 pigs/sow/yr sold

Table 10 EFFECTS OF DRY AND WET FEEDING* ON LABOUR ISSUES. (AVERAGE OF THREE FARMS CORRECTED TO 5,000 FINISHERS AT ANY ONE TIME)

	Manual delivery		Wet fed by pipeline
	Individual pens	*Ad lib groups*	
Man hours per week (feeding)	50	20	5
Labour cost per finished pig**	£3.07 ($4.60)	£2.38 ($3.57)	£2.01 ($3.01)
Hours lost to sickness/non attendance (per year	270	212	89
Staff turnover across 5 years (%)	64	58	10

* Farmer with 5,000 finishers to feed twice daily in five piggeries.
** Including all other tasks
Source: Clients' records (1989-94)

8. ***Reduced slurry volume***. Logically one would have thought there was more slurry, not less from liquid feeding. I find this is often not the case (Table 11).

Table 11 THE EFFECTS OF WET AND DRY FEEDING ON SLURRY VOLUME. (SLURRY REMOVED FROM TWO DRY SOW HOUSES IN WINTER [SEPT-MAY].)

	Wet-fed	*Dry-fed*	*Wet-fed*	*Dry-fed*
Per sow per week, litres (gals)	126 (28)	148 (33)	117 (26)	115 (25)
Tanker loads/herd	4	5	2	2

Source: Gadd (unpublished)

9. ***Easier to recruit good labour***. Surveys of stockpeople have shown that a progressive farm is more likely to attract good employees (Table 12).

Table 12 RANKING OF REASONS FOR ACCEPTING OR QUITTING EMPLOYMENT ON PIG FARMS IN THE UK. (ATTITUDINAL SURVEYS OF SKILLED STOCKPEOPLE.)

Accepting		*Quitting*	
Modern attitude*	11	Work is hard, dirty, repetitive	12
Convenient location	10	Nobody listens to me	10
Automation	10	No future/lack of time off	8
Wages/money	10	Not keen on pigs	6
Benefits	8	Don't like co-workers	3
Working hours	7	Need a change anyway	1
Need the job	3		
	59		40

Source: Staffing Agency (1988)   Source: Gadd, Survey (1990)
*Includes being able to use new technology, *e.g.* computerised wet feeding

10. *Less chance of salmonella*. From Denmark, the processors Steff-Houlberg and Danish Crown showed there was less risk of salmonella on wet feed (Table 13), which rather puts paid to the objection that wet feeding breeds pathogens. Of course it *does* if the facilities are filthy, but as the Danish work suggests, under decent conditions the risk (of salmonellosis) seems less. An interesting finding, backing up the view that with a one-hour soak before feeding, the acidity rises sufficiently to penalise salmonellae (and possibly other pathogens though only salmonella was measured).

Note that this research shows the risk of having over 33% samples positive for salmonella in meat was five times greater for units feeding dry feed than for units feeding wet feed.

Table 13 THE EXTENT OF SALMONELLA IN WET- AND DRY-FED HERDS

	Wet feed	Dry feed
Over 33% positive	4 (0.85%)	92 (4.2%)
Under 33% positive	466	2189
Total	470	2281

Source: Steff-Houlberg (1998)

11. *Wet-fed pigs are more contented*. Everybody likes quiet, happy pigs. Bishop Burton Agricultural College in Britain did some interesting student work on growing pigs a few years ago as shown in Table 14.

Table 14 PROPORTION OF TIME PIGS SPEND IN VARIOUS ACTIVITIES 20-50 kg (44-110 lb) DURING WET AND DRY FEEDING

	Wet-fed (%)	Dry pellets (%)
Asleep/dozing	53	45
Feeding/drinking	7	12
Social activity	35	32
Fighting/playing	5	11

Cambac Research drew this out most elegantly with group-housed sows, who settled down dramatically faster after being wet-fed (Figure 1)

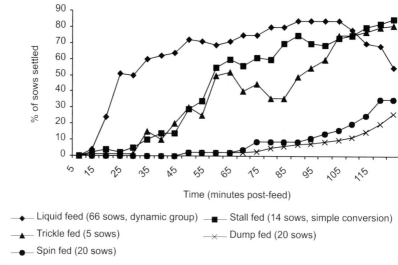

Figure 1 The effect of differing sow feeding systems on post-feeding behaviour

12. ***Quick and accurate medication.*** As little as 250g or ml/tonne (8 oz/ ton) gives rapid and even distribution. According to Taylor (1976), with wet feeding, there's a "50% less mixing charge than from a compounder."

FUTURE REASONS FOR USING PIPELINE FEEDING

Up to now I have listed some of the evidence why large numbers of pig producers have already switched to wet feeding. But the future of pig nutrition is even more exciting and is changing quickly.

Pipeline feeding is superbly poised to accommodate these developments because, as distinct from dry, or even wet/dry feeding .....

*   It is remarkably flexible and adaptable.

*   The computer technology needed is already here.

*   The equipment is ready and waiting as are distributorships and spares/ service facilities in many countries.

*   The know-how / track-record of companies in this field is considerable, *e.g.* Big Dutchman have over 5000 installations world-wide.

AHEAD OF ITS TIME

So, for once, we have a proven delivery system in place which can accommodate these new nutritional advances rather than being designed in a hurry to render them feasible on-farm.

What are the future advances ?

1. *'Challenge' or 'Test' Feeding* *(See also Immunity Section)*

Solves the problem of wide differences in protein accretion curves *among pigs of the same genotype* on different farms caused particularly by differences in disease thresholds and also by variations in environment between individual farms. A small sample of growing/finishing pigs are tested regularly on a non-nutritionally limiting diet and carefully monitored. The results are computer-modelled and a Farm Specific Diet (FSD) is least-cost formulated for the whole herd based, for example on the lean accretion curve revealed by the test results.

<u>Value</u> : Between 20-40 kg (44-88 lb) more saleable meat/tonne feed is suggested. Feed cost/tonne rises by 6% to 8% but gross margin increases by 10%-13%; nett margin by as much as 20%

Wet feeding best suits this concept

2. *Blend Feeding*

The obvious problem created by FSD is the multiplicity of diets demanded of the feed manufacturer. Initially some compromises can be made, for example by having a range of diets with nutrient densities most closely fitting the commonest protein growth curves, or supplying different basic diets for high/low disease status.

In future however full FSD will be available – one for each farm, reviewed regularly – and to avoid multiplicity of formulations, all can be made from only *two* diets delivered to the farm and placed in separate bins. One is of high nutrient density, one of low nutrient density, with a small premix hopper in between.

By blending varying amounts of the two primary feeds into a wet mixer, *every farm variant can be made on-farm* and the product range the feed compounder needs to carry is drastically reduced. Formulation and blending is all computer-controlled.

Wet feeding can accommodate this process in the most economical way.

3. Multi-pen feeding

In the short term future (5-10 years) most piggeries will continue to contain up to 9 different weight bands of growing/finishing pigs in one house. (Beyond this, multi-site production will adopt the poultry concept of batch rearing similar weights of pigs all placed in one house).

Only wet feeding by computer can easily, economically and accurately feed 9 weight categories of pigs with up to 14 different diets

4. Multiphase feeding

At present we **Step-feed** (3 diets and only 3 nutrient ratios for a starter, grower, finisher) which is very inefficient. **Phase feeding** (about 5 early steps, 3 later steps) is better, but not ideal. **Multiphase Feeding**, where there are 30-50 nutritive-ratio changes across a pig's growing life, gives slightly improved performance over Phase Feeding but marked reductions in N +P pollution.

Multiphase Feeding and FSD (under research) is likely to give much improved performance as well as modest improvements in pollutants, but considerably less slurry **volume** as less water is needed to metabolise more efficient protein use.

Only pipeline feeding can cope with this degree of sophistication, again effortlessly and accurately.

5. Choice feeding

Up to now all these developments involve the **nutritionist** deciding when to change dietary allowances and nutrient values. Choice Feeding allows the **growing pig** itself to make the dietary changes – quite accurately, it seems where the all-important protein intake is concerned.

While choice feeding for older pigs is still under research, a development of this – Menu Feeding – has worked well with nursery pigs.

6. Menu feeding

Provides two feeds of slightly differing nutrient densities to be on offer at any one time all through the nursery period of 6 to 25 or 30 kg (13-55 or 66 lb) by which time a total of 6 diets have been on offer. By a process of 'leap-frogging' one diet is changed every 7 to 9 days. The diets also vary in flavours to further stimulate appetite as young pigs may quickly get bored with one added flavour throughout. Dramatic increases in FI (24%) and DLWG (23%) have been obtained in the nursery period. However improvements in the all-important FCR are usually modest (1%-2%).

The real benefit occurs at the end of the finishing period even if the pigs are fed conventionally from 25/30 kg (55/66 lb) to slaughter. Menu Feeding early on can give between 7 to 21 days quicker to slaughter.

And an increase of up to 20-40 kg (44-88 lb)saleable lean/tonne of food.

While the different diets can be added dry, by hand or auto-control, it is easier and cheaper to do this by pipeline to avoid hassle and mistakes.

7. *Wet feeding nursery pigs (gruel feed)*

Still under development, but the signs are favourable. It is well established that the suckling pigs' intestinal surface, once weaned, is far less damaged if transferred to a thick wet gruel than to dry or even wet/dry feed. (Figure 2)

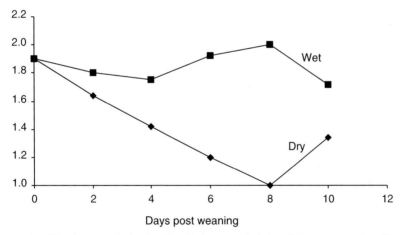

Figure 2 Villus/crypt ratio in the distal jejunum of piglets fed a wet or a dry diet

So the weaners get away faster and reach slaughter sooner.

A pipeline system is needed to achieve the degree of wetness needed.

8.   *Inoculation by probiotic bacteria, with or without enzyme addition.*

Inoculation of wet feed for weaners with fermentative bacteria markedly raises acid levels in the stomach thus reducing susceptible pathogens to harmless levels. The feed is held soaking with the inoculum for several hours before feeding, thus both physical softening and enzyme formation helps predigest the food in the relatively poorly-developed digestive tract of the weaner, itself under severe stress trying to cope with solid feed rather than milk. (Figure 3)

Only a pipeline feeder can do this

This concept is fully described under the FLF section at the end of this chapter.

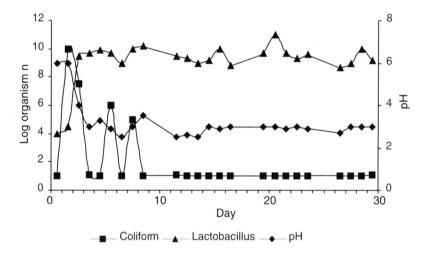

Figure 3 Effect of *Lactobacillus* spp. growth on pH and *E. coli* numbers in liquid feed. Source: Brooks (1997)

What the graph shows: The satisfactory drop in pH (*i.e. rise in acidity*) of the digestive tract after 4 days. This makes it difficult for hostile coliforms to survive beyond day 10 while the acid conditions create favourable conditions for many beneficial bacteria *e.g.* lactobacilli.

Phytase is rendered more active and thus more phosphorus is released when it is included in wet feed (Figure 4).

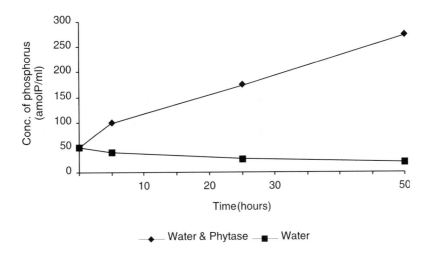

Figure 4 Effect of steeping soyabean in water or water-phytase. Source: Brooks (1997)

9.   *Nutritional Biotechnology*

We are seeing this area grow. It will grow more in future. Often, with nutrients like organic selenium (0.3 ppm) and organic chromium (200 ppb) only tiny quantities are involved – but they do a lot of good: far greater than their inclusion levels suggest.

Only pipeline mixers can cope with very small amounts accurately without recourse to unnecessarily bulked-up (thus more expensive) carriers.

10. *Liquid feeds*

Crystalline supplementary amino-acids cost a lot to dry into a free-flowing powder-form. To add them wet, as made, is much cheaper.

Future enzyme technology will allow us to use 'wet' foods, presently too indigestible for pigs, like grass, grass silage, brassica tops, banana leaves, potato haulms and possibly even forest foliage.

Several factory by-products from the petro-chemical industry, the canning / sugar / confectionery industries can be used for pigs, to add to those from the dairy industry in use now.

Only a pipeline system can use all these materials and more

So there are 10 future reasons (some ready for use now; others for trial) why wet feeding is essential to deliver the technology to come.

But back to the present. Any system has snags – what are they?

THE SNAGS AND DRAWBACKS TO PIPELINE FEEDING: HOW TO AVOID THE PITFALLS

About 20% of the problems encountered have been at the installation stage, and another 20% have been due to home-installed circuits where the producer himself, 'good' with plumbing, welding (both metal and plastic) and machinery, nevertheless got things wrong, sometimes invoking rectification which cost more than any savings he enjoyed from not employing an experienced installation team.

Table 15 summarises my own experiences from dealing with liquid feeding problems.

Table 15 ANALYSIS OF 62 COMPLAINTS/CALL-OUTS OVER THE PAST 25 YEARS INVOLVING THE INSTALLATION OF LIQUID FEED SYSTEMS

Overfat pigs	13
Blockage	16
Food wastage	7
Dirty conditions/wrong machinery	9
Respiratory problems	2
Poorer performance	8
Computer origin	6
Trough fouling / lying in troughs	3
Young pigs / feeding messiness	2
Food delivery (valves)	2

Some are duplicated *i.e.* more than one problem was encountered on the farm.

Note: The vast majority of these problems were sorted out, some very quickly. There were 4 complaints about frost which did not need a visit to rectify. Problems with FLF (see end of this section) are not included as this concept is still under development.

Many problems have occurred with wet feeding because insufficient 'homework' was done before adopting the system. As with any radically new concept in pig production, ***do your research first***. Take time to visit, compare, enquire and argue/negotiate. The pig industry has at least 40 years experience of basic liquid feeding of pigs, and apart from a few imponderables about FLF, *the answers are all there*.

A START-UP CHECKLIST

✓   *Choose a reliable manufacturer with a proven track record.* If your research throws up hearsay problems, question him assiduously on a "what if" basis and mull over his replies, especially valuable if he can direct you to customers who know about or have encountered the problem in particular.

✓   *Be careful about new gimmicks.* Ask for evidence that the new idea/development/cost saving has been thoroughly tested under farm conditions and where, then go and see it in action. Ask yourself 'do I really need the modification/update?'. My maxim has always been… 'Simplest and most rugged is best'. Having said all this – I am impressed with the track records of most of the firms involved in liquid feeding today.

3.   *Make sure the installation team is qualified and experienced.* The best equipment, poorly installed, will give trouble. If the manufacturer has his own installation division, this is excellent. If he subcontracts out – even under his own supervision, question him closely about the evidence of their experience and do a telephone check of customers who have used the subcontractors *recently*. Be especially vigilant if local electricians/plumbers are employed as sub-contractors.

✓   *Ensure advice, spare parts and prompt service are available.* You and your pigs will come to depend entirely on the system, so a prompt rectification/service is essential. *Author's note*: Especially over the long Christmas/New Year break! I once had to feed 10,000 finishing pigs by hand for 4 days (and nights!) over Christmas because a spare couldn't be located. Never again!

✓   *What facilities exist for initial training of your stockpeople?* Will the supervisor stay on for 2 or 3 feeds to ensure you/they get it right? How good is the instruction manual? Does it have a 'what-if' section? Insist you get taken through it – make notes. Is your supervisor available on call/mobile subsequently? He knows your installation in detail, and can spot errors at once.

WHAT ARE THE SNAGS OF WET FEEDING?

These fall into two categories: True Snags and Perceived Snags. Perceived or apparent snags are those the newcomer worries about (justifiably) but can avoid with know-how and foresight. True snags have to be dealt with and absorbed.

True snags

- *Capital cost*. Installing a pipeline is expensive in capital needs, especially in converting an existing dry/wet-dry farm to it. Experience from 137 farms over 30 years suggests it raises housing costs, normally 9% to 11% per pig, to 11% to 13% across a ten-year amortization period.

  Also, updating the ventilation system (see below) will put about another 1.5% on top of this.

  For this you should get:

  - A 0.1% improvement in FCR at least on every pig sold. Often more.

  - Or another 20 kg (44 lb) more saleable meat/tonne of finisher food.

  This latter is a convincing sales point in liquid feeding's favour, and helps put the somewhat daunting extra capital cost in perspective, like this:-

  You know how much income you should get per kg (lb) of saleable (*i.e.* dcw) meat. Next, a good pipeline installation should last 10-15 years (I've known early models, *e.g.* Taymix, last 27 years). Now calculate how many tonnes of feed the circuit will handle over 10 to 15 years and relate the value of 20 kg (44 lb) – or even 10 kg (22 lb) – more meat sales per tonne of feed handled, to the total capital cost. If this figure doesn't convince you, nothing will – so I'll leave you to do the sums!

  Other economic benefits can be added to this:

  - vet/med costs down by up to 33%

  - a happier, healthier workforce staying longer and easier to recruit (Table 12)

  - between 4% to 6% lower labour costs overall or much better *use* of labour (Table 10)

- 5% to 9% less annual housing maintenance costs (fans, structure)

- a minimal-cost entry into home-mixing, itself worth up to 18% cheaper feed-cost/tonne (25% with by-products).

  From this equation you can see it easily pays for itself. Farm data collected in the 1990s show an REO of between 2:1 and 6:1 (average 4.1:1). The main difficulty is often finding the capital in the first place among other pressing needs.

- *Ventilation*: Wet feeding is a wet process! Ventilation will often need to be reviewed by an agricultural engineer to remove the excess humidity, especially in winter. Don't forget to check on this.

  Sadly, the max/min ventilation rates in liquid feeding growing/finishing buildings, coupled with the need for accurate air placement, is often adrift. In my experience *all* the complaints of poor or disappointing performance can be laid at this particular door, and I find them operating years after the original conversion to liquid feeding. It is a ventilation problem NOT a wet feeding one.

- *Bloat*. Wet fed (older) pigs tend to bloat, especially with whey. Anti-bloat formulations can be obtained but it is still a problem. You will lose 1% more finishing pigs due to this, and I have never managed to get it much lower. You can ameliorate its effect, but rarely erase it entirely.

- *Young pigs – less than 25 kg (55 lb).* Quite a few problems here – initial inappetance (odd, but it can happen), messy eating/wastage, lack of dry matter intake, oedema, and bed-wetting are all problems particular to young pigs. All these can be eventually overcome by trial and error alterations to management and improving nursery living conditions, but the move to FLF (Fermented Liquid Feeding) (see later) seems to be overcoming a lot of them.

Perceived or apparent snags

All the following snags are commonly encountered and made much of by the adherents of dry feeding. In fact they are rare in a well-installed pipeline system.

- *Frost* is rarely a problem. Design the layout to accommodate frost and wind chill however acute. The degree of insulation the Swedes and some experienced Canadians use is surprising, but totally effective. If you are really worried, then drain the system after use *as long as the circuit does not have sagging pipelines*. If so, keep them charged full but insulate well.

- *Pipefouling* is rare if the equipment is used daily. Use water flushing and a dump tank to keep the circuit clean. Anyway sanitisation (pigs present or absent) is possible. To avoid mixing tank contamination of its upper, unscoured surfaces, use the whirlspray device periodically.

- *Blockage* can happen, but very rarely. Design the problem out at installation, so if a block does occur you can easily remove it. For example, a rat could get in.

- *Never pump downhill*. Pumping down is a logical idea so that gravity saves a little electrical energy or increases pressure at a distant pen. Trouble is, unless the circuit is flushed out and left with air (frost) or plain water, the feed particles flocculate down to the foot of a vertical or acute slope, leaving the supernatant liquid as a cushion protecting the solid plug below. But by pumping *up* a vertical, the turbulent liquid/solid mixture eats into the plug from below dispersing it into the liquid above and freeing any overnight blockage due to settling.

- *Do not have loops (sags) in horizontal runs*. A good installer will design this out.

- *Install an Oxford Union at each right angle bend*. Then all straight runs can be rodded out if needed.

- *Never seal a pipe underground*. Place in a channel covered with metal plates.

- *Overfat pigs* often occur on changeover as the diets aren't altered to allow for the overeating due to improved palatability, which can occur on *ad lib*. Consult your nutritionist. The same phenomenon occurs when changing from dry feed to wet/dry feed, and the solution is the same – adjust the diet.

- *Water deficiency.* Always have a drinker available. Liquid feeding is not a substitute for water, merely a physical method of moving bulky feed. Indeed, whey and skim 'carriers' are thirst-making in themselves, and whey concentrate is high in salt, as are some other edible industrial by-products. Supplementary water is absolutely essential.

- *Ironing out the snags pays hands down.* Finally, here are the consistent results across 3 years from 3 farms who had needed a good deal of attention and rectification in their early stages, having not got several of the perceived snags ironed-out before converting to liquid. They were on three different manufacturers' systems.

Table 16 FEED CONSUMPTION AND FCRS ON FARMS BEFORE AND AFTER WET FEEDING (AV 30-88 KG) 1994-1998.

	Feed eaten kg kg (lb)				FCR		Yield of extra saleable meat Per tonne feed kg (lb)	
	Before		*After*	*Improvement**	*Before*	*After*		
Farm A	168	(370)	154 (340)	8.3%	2.89	2.61	+29.3	(65)
Farm B	180	(397)	161 (355)	10.6%	3.05	2.72	+31.9	(70)
Farm C	197	(434)	188 (415)	4.6%	2.81	2.69	+10.9	(24)
Average				**7.8%**			**21.0**	**(46)**

*The average of 7.8% wastage accords well with Dr Mike Baxter's work which suggested most producers waste 6% of their dry feed

REFERENCES

Brooks P H, Geary T et al (1996); Procs. PVS (*Pig Journal* pps 43-67)

Carpenter, G A; *J Agric Eng Res* 1986 **33** 227-241

Cermack, J P (1978); *Farm Buildings Progress* **51** 11-15

Chesworth, K; Procs. Australian Pig Science Assoc. Conference, Adelaide Nov 2001

Gadd, J; 'Pipeline Feeding' in '*The Pig Pen*' Vol 4 Nº4 Jan-March 1998.

Robertson, J F; Dust in farm mill & mix plants (1991) *Farm Buildings Progress* **106** 14

FERMENTED LIQUID FEEDING (FLF) 30

The FLF concept takes the successful concept of 'porridge' feeding of newly weaned pigs one step further. As well as using specialised 'non-antigenic' feed raw materials designed to cushion the shock of the weaner's gut surface in responding to solid feed rather than sow's milk, these diet formulae are fermented for a while. This is done under as controlled on-farm conditions as is feasible by adding a bacterial starter inoculant, and possibly some enzymes, organic acids and gentle heat (around 21ºC, 70ºF) to a central tank using modified wet-mixing equipment.

Pioneered in the 1990's by research nutritionists Drs Peter Brooks, Tina Geary and others, this intriguing development in young pig feeding tends to be shrouded in mystery for many pig producers.

The post-weaning check to growth, caused by the time it takes a 4.0-4.5 kg (9-10 lb) 17-day weaner (USA) or a 6.5 kg (14 lb) 24-day weaner (Europe) to recover its growth rate and readjust its digestive competence after weaning, has always led to a severe check to eventual performance at slaughter (Table 1).

Table 1   LOSING TYPICAL GROWTH IMPULSION AT WEANING – A SERIOUS REDUCTION IN THE AMOUNT OF LEAN MEAT SOLD FOR EACH TONNE OF FEED REQUIRED

Estimated post weaning check* to growth	Days to 94 kg (207 lb)	Daily gain –g (lb)	% First graders	Carcase lean	Saleable lean meat per tonne food
5.8 kg 12 days (12 – 75 lb)	156	567 (1.24 lb)	72	52.3%	166 kg (366 lb)
5.8 kg 3 days	142	621 (1.36 lb)	86	53.1%	182 kg (401 lb)

* Defined as the time it takes for the weaner to recover full growth potential
Source: Based on data from various feed manufacturers

Anything which shortens this 'stall-out', as the North Americans call it, is to be welcomed. Nutritionists now know how to choose the right materials in a 'Link Feed' so as to minimise gut surface damage after weaning. Certain ingredients are banned, others are favoured, others are pre-treated. FLF takes this knowledge onwards in setting up a beneficial fermentation in the feed to further help digestion at a very tricky time.

BUT DOES FLF WORK?

Yes and no! At least in my experience. *Yes*, because under controlled research conditions, and on at least half the farms now using the concept the performance results are fine (Table 2).

Table 2　STARTER PIG PERFORMANCE ON VARIOUS FEEDING TECHNIQUES

	Daily Gain %		Feed:Gain %	
	Average	*Range*	*Average*	*Range*
Liquid vs Dry (10)*	+12.3	−7.5 to 34.2	−4.1	−32.6 to 10.1
Fermented vs Dry (4)	+22.3	9.2 to 43.8	−10.9	−44.3 to 5.8
Fermented vs Liquid (3)	+13.4	5.7 to 22.9	−1.4	−4.8 to 0.6

Liquid = liquid feeding of non-fermented feeds
Dry = conventional dry feed
Fermented = liquid feeding of fermented feeds
* (number of trials)
Comment: The results are positive but nevertheless variable
Source: Jensen & Mikkelsen 1998: Scholten *et al.* 1999

No, because rather too many farms are initially disappointed. It takes them some while to get the lack of improved performance up to speed and a few (15%?) never did manage it sufficiently and have abandoned FLF in favour of non-fermented but nevertheless special post-weaning diets. This leaves somewhere around 80% of producers who are happy enough with the improved performances of FLF in the nursery stages as shown in Table 2. While the FCR improvements versus dry pellet feeding over the 5-25/30 kg (11-55/66 lb) range are worth pocketing (reducing the cost of the expensive post-weaning feed by about 9%) a major advantage could be the 22% increased kick-start to growth against conventional pellet feeding (however good the pellets are) or 13% better against liquid (porridge consistency but non-fermented) ingredients. Producers experienced in the concept comfortably get FCRs of 1.05 (7-12 kg, 15-26 lb) and 1.65 (12-35 kg, 27-77 lb).

ECONOMETRICS

This degree of greater laying down of lean-gain early on in the 'Acceleration Phase' of lean growth – a happy term coined by Prof Whittemore all those years ago – usually means another 7 to 12 days sooner to 100 kg (220 lb). This alone provides an extra 25 to 45 kg (55-99 lb) of MTF (Saleable Meat per Tonne of Food fed). With a deadweight price of say £1/kg in the UK as I write, this is equivalent to an 18-33% reduction in the cost/tonne paid on all the feed fed from 30-100 kg (66-220 lb). And this doesn't include the savings in overheads, worth about a further £3 to £5/tonne. One FLF installation in the nursery should handle 1000 weaners comfortably or, with a turnaround of 7 batches a year over a 7 year lifetime it could feed about 50,000 weaners of which most will become finished, shipped pigs. All these could enjoy an 18% to 33% reduction in their feed bill, so the economics are excellent and the payback swift. Even halving these figures gives an attractive and competitive return against other investments.

So where is the system falling down, and what could need more research?

AVOIDING PROBLEMS WITH FLF A CHECKLIST

✓ Check the water to feed ratio with your nutritionist. While for older pigs a water:feed ratio of 3.25 to 3.5: 1 improves energy digestibility and protein substantially, for young pigs the dry matter must achieve at least 200g/kg (3.2 oz/lb) of intake, which comes out at less than 3.0:1 water:feed ratio.

✓ Do not add antibiotics to the feed, they will 'take out' the bacterial starter inoculant. Seek another medication route.

✓ You must provide supplementary water.

✓   Feed troughs must not be empty for long, and feed allowance carefully supervised. Liaise closely with the equipment supplier, and cross-check his instructions with your nutritionist to ensure optimal daily intake is achieved.

✓   Trough length in relation to piglets to be fed seems important. Do not depart from the advice you are given.

✓   Continually check probe depth as advised by the manufacturer.

✓   Do *not* use supplementary probiotic products. As Professor Peter Brooks says "The P word is *banned!*" A simple starter inoculant is usually all that is needed.

✓   Always have at least one third of the volume sitting in the system ready for the next feed. This is needed to keep the fermentation 'fire' going.

✓   About eight hours fermentation seems to be adequate. Does your capacity allow this over the one third 'reserve' advice?

✓   If you are using liquid whey as a co-product, seek nutritional advice as you could overdo sugars (keep to under12%).

✓   About 50-60 weaners/group is okay. Above this seems to be more difficult to manage.

✓   Fermented wheat, sometimes fermented separately beforehand, is popular - but don't over do it (Figure 1).

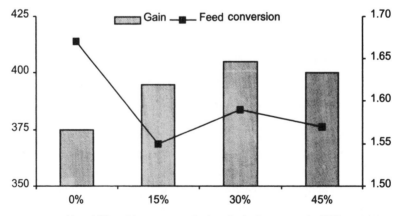

No additional improvement when inclusion exceeds 15%
Starting body weight 10kg; fixed feeding

Figure 1 Fermented wheat in weaner pig diets: effects on growth performance

If you avoid these potential pitfalls you should enjoy the benefits listed in Table 3 by a leading researcher.

Table 3 BENEFITS OF LIQUID FEEDING & ESPECIALLY FERMENTATION

- Enhanced nutrient availability
- Improved digestive capacity:
    - Lower stomach pH ('kills' harmful microbes; initiates protein digestion)
    - Better mucosal structure (increased effective gut surface for nutrient absorption – in small intestine)
- Improved gut health:
    - Fewer toxin producing coliforms & more beneficial lactobacilli
- Reduced maintenance energy requirements:
    - Reduced physical activity of pigs; pigs spend more time sleeping.

(De Lange, 2001)

Two questions at the time of writing

- How to cope with ingredient variability?

- Why do young nursery pigs seem to sleep more on FLF in comparison to non-fermented liquid feed? Several people remark on this phenomenon – it seems to be quite noticeable. Would it be worth extrapolating whatever causes it to older/more stress-susceptible pigs?

CONCLUSIONS

- Econometrically FLF seems to be a very promising investment.

- There is good track-record guidance available, sufficient to get positive results.

- But some producers are letting some of it go by default and get disillusioned.

- There are still research refinements to be made. At the time of writing, I give below some suggestions which will help refine the process (Table 4 *et seq*).

AREAS THE RESEARCHER MAY NEED TO INVESTIGATE FURTHER.

As fermentation can produce negative as well as positive reactions (Table 4) we still seem to need better information on the storage and fermentation areas of: -

Table 4   NUTRIENT CHANGES DURING STORAGE/FERMENTATION

- Nutrient losses:
  - About 25% of synthetic lysine (during 24 hour fermentation of complete feed; Pedersen 1999)
  - Other amino acids and vitamins?
- Amines (alteration of amino acids; fermentation of protein) can have toxic effects.
  - may explain variable animal response (especially when feeding fermented diets, rather than fermented ingredients)
- Conversion of starch, sugars, fibre to:
  - Lactic acid (acetic acid) – positive
  - Alcohol & CO_2 – negative; represents nutrient losses
  - (the changes vary & may be manipulated)

- Time and temperature; inoculants and enzyme addition; pH and buffering reactions.

- Ingredient type, for example adding about 15% of a fermented relatively low protein:high carbohydrate ingredient like ground wheat rather than fermenting it up with an inoculant after it is added to the tank. But for how long can we keep on fermenting each ingredient before it loses its potency?

REFERENCES

Brooks PH, Geary TM, Morgan DT, Campbell A (1995) New Developments in Liquid Feeding. *Pig Veterinary Society Journal* **36** 43-64.

Brooks PH, J D Beal and S Niven 2001. Liquid feeding of pigs: potential for reducing environmental impacts and for improving productivity and food safety. *Recent Advances in Animal Nutrition in Australia* **13**: 49-63.

De Lange CFM, BJ Marty, SH Birkett, P Morel and B Szkotnicki. 2001. Application of pig growth models in commercial pork production. *Can. J. Anim. Sci.* **81**: 1-8.

Jenssen, BB and LL Mikkelsen. 1998. Liquid feeding diets to pigs. In: *Recent Advances in Animal Nutrition*. (Eds PC Garnsworthy and J Wiseman), Nottingham University Press, Nottingham. Pp 107-126.

Scholten, RHJ, CMC van der Peet-Schwering, MWA Verstegen, LA de Hartog, JW Schrama and PC Vesseur. 1999. Fermented co-products and fermented compound diets for pigs: a review. *Animal Feed Science and Technology* **82**: 1-19.

PROBLEMS WITH VENTILATION 31

The process or act of supplying an enclosed space with a continual supply of fresh air.

WHY DO WE VENTILATE?

Table 1 lists the primary reasons why we ventilate a building or part of it.

Table 1   WHY WE VENTILATE

- Provide oxygen
- Control temperature
- Remove
  - moisture
  - excess heat
  - gases
  - dust
  - bacteria
  - viruses
  - spores
  - chemicals
- Keep stockmen
  - healthier
  - happier
- Extend building life

TEMPERATURE CONTROL

Table 1 is a little too simplistic. Pigs, like humans are homeotherms, which are able to control their body temperature providing that their environment does not exceed extremes of temperature or relative humidity (rh = the amount of moisture present in the air).

So for welfare and economic reasons we need to keep the pigs in their environmentally-influenced temperature comfort zone. The primary ventilation objectives to achieve this are:

• **In normal temperate conditions**, to provide varying air throughout so we can achieve a correct house temperature to maximise performance.

• **In temperate summers**, to keep the temperature around the pigs within 3° to 4° C (6°-9°F) of the outside conditions.

• **In very hot conditions**, have the capability to provide high air speeds over the pigs to increase their upper temperature tolerance level, often using wetting to assist evaporative cooling.

• **In cold to very cold conditions,** to provide a controllable minimum level of ventilation, but at the same time directing the air adequately across the pigs with good mixing, so avoiding draughts from cold/very cold incoming air. Supplementary heating may be called in to assist this process.

THE COST OF SUB-OPTIMAL VENTILATION

Table 2 is an attempt to quantify what incorrect ventilation can cost, based on my own experience of tackling many ventilation problems and following up subsequent performance. The final column should give considerable pause for thought! My records show that many farms have, without too much trouble, achieved improvements in FCR, ADG and MTF of at least half these huge penalties, resulting in increases of MTF to the order of 25 kg (55 lb). This is equivalent in worldwide economic terms to a reduction in all feed costs of 20%, and this is a 10 to 12% lower production cost, with one third more gross margin.

Table 2 THE COST OF POOR VENTILATION

Depression re:	Not achieving ECT or LCT	Appetite	Stress*	Health	Total
Food conversion	0.21	0.1	0.05?	0.23	0.4 to 0.5 worse
Daily gain (g/day)	45	20	10?	50	125g/day slower

* difficult to measure, minimal estimate only.
Poor ventilation alone can cost up to 50 kg (110 lb) of lean meat per tonne feed.

Table 3 summarises five case histories.

Table 3   BEFORE-AND-AFTER[1] RESULTS FROM 5 FARMS WHERE THE VENTILATION WAS EITHER RENOVATED OR RENEWED. A professional agricultural engineer was employed in each case to quantify the problem, then to rectify the design and supervise installation[2]

	Estimated penalty cost/pig of present ventilation[2]	*Estimated cost of renovation per pig[2,3]*	*Actual cost of renovation per pig[3]*	*Estimated value of improved performance after renovation per pig[2,4]*	*Actual value of improved performance measured over the first 12 months[4]*
	£	p	p	£	£
Farm 1	3.50	56	60	3.80	3.15
Farm 2	2.29	41	66	2.74	2.20
Farm 3	4.60	103	97	5.09	3.86
Farm 4	2.74	54	63	2.74	1.87
Farm 5	2.80	87	90	2.46	2.58
Average	£3.19 ($5.13)	53 ($0.85)	70 ($1.13)	£3.36 ($5.40)	£2.73 ($4.39)

Key:- [1] From individual records; 3% per annum allowance for normal year-on-year improvements.

     [2] Using a computer model.

     [3] Over 10 years at 12% interest.

     [4] Excludes Health & Reduced Overheads (Figures quoted from FCR & ADG only).

Farm 1	Complete renewal/update of fan ventilation.
Farms 2 & 4	Natural to Automatic Natural (ACNV)
Farms 3 & 5	Complete renewal including insulation. Fan ventilation.

The agricultural engineer and his service cost approximately 10% of the capital outlay

Comment: Renovation costs were usually higher than estimate.
Actual performance improvements were lower than computer-forecasted.
Nevertheless, actual economic value of the improved performance increased gross profit by 33%.

WHY DO SO MANY PIG PRODUCERS FAIL TO PROVIDE CORRECT VENTILATION?

There are two primary susceptible areas.

1. **Not understanding the physics of air movement, air exchange and the thermodynamics of incorrect (insufficient) ventilation in cold weather conditions – in winter or at night.**

    There is a minimum level of ventilation which is essential to get rid of water vapour and dangerous gases to safeguard the pig's health and performance and the health of the stockpeople. In cold conditions, if this air exchange, called the ventilation rate, is too low mainly in order to conserve heat, then the air in the house becomes stuffy and polluted.

    However many farmers become aware of this and increase the ventilation rate with rudimentary supplementary heat input when needed. Often, because it is not done correctly, air quality is insufficient and in many cases energy losses and costs rise quickly to erode profits if this extra air required is not precisely measured, calculated and controlled.

    Because most pig producers have understandably not been trained in air movement and thermodynamics, they fail to provide ventilation equipment and a control system to provide minimum ventilation rate. This is because they don't consult a ventilation engineer in the same easy way they often do a veterinarian. To compound the problem, the equipment they buy, even if satisfactory, is often poorly installed and the manufacturer is then blamed for deficiencies. Figure 1 gives an estimate of where the responsibilities may lie.

2. **Failing to use professional advice from a (preferably independent) agricultural engineer to design the correct ventilation equipment and control/monitoring facilities and supervise correct installation.**

    Ventilation expertise is definitely not an 'art', neither can it necessarily be acquired from 'experience'. Correct ventilation is mostly a matter of getting the sums right, just like designing a bridge or a high-rise building. Unfortunately trained ventilationists are nowhere near as common as structural engineers or architects for example, but they do exist and must be sought out. They have computer models which, with the correct input data, can design an installation to satisfy the circumstances encountered on 95% of the pig farms in 90% of the world.

Do not try to do it yourself; you'll get it wrong!

Read the 'How to Use a Ventilation Engineer Checklist' for what a professional ventilationist can do for you.

HOW VENTILATION PROBLEMS OCCUR

I find there are 3 main areas where mistakes are made – (1) Specification, (2) Design and Installation, (3) Operational Control and Maintenance.

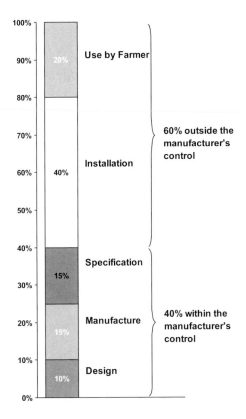

Source: Crabtree (1995)

Comment:
- Nearly two thirds of complaints on poorly functioning ventilation equipment are outside the primary manufacturer's control.
- Incorrect siting and installation by sub-contractors can be a major source of error.
- Incorrect operation by the pig producer is probably less of a problem than is often claimed, but still accounts for one-fifth of any disappointing results in my experience.

Figure 1 Fan ventilation: which factors have the most influence on reliability

(1) Specification

Correct ventilation is, first of all, mathematics. It is a numbers game! By collecting about 20 basic pieces of data on the **expected climatic (weather) changes** round the building (mostly temperature and wind speed); **the structure of the building** (including insulation values); the **pig numbers, input and output weights** (*i.e.* maximum/minimum) to be housed; and the **feed energy inputs** of the stock (*i.e.* the target physical performances), the agricultural engineer can provide a set of ventilation specifications which will cope with reasonable climatic, stocking density and feed regime variables expected on your farm. He can even design- in protection against 50 year extremes of weather – if you wish – but at a cost. Let him discuss where the trade-off is in your opinion.

Not to have this done kills the project at the start. You will be paying out for years in reduced productivity, whatever you spend on the ventilation equipment. It won't cope adequately enough. Alternatively I have seen installations which are over-specified thus wasting food energy even if the pigs have plenty of air changes.

(2) Design and installation

Having got the numbers right, the correct design of the ventilation system is equally vital. It matters very much what the air patterns are, where they go, and how fast or slow the air moves. This will vary all the time as the outside temperature changes, the wind speed and direction changes, and as the stock grow or are moved in and out. The design system needs to be able to follow these changes and react accordingly.

(3) Controls and maintenance

Now we are nearly there. The final critical sector is to select and install reliable control equipment. So many farmers alter things by hand. You *love* twiddling knobs! Don't do it! Install well-designed electronic controls and leave it to them. A good system will do it four times better than you, never sleeps or leaves the building – and saves its cost in 3 months! You've got enough to do in the piggery – leave ventilation to a (good) controller.

Lastly – maintenance. Having spent wisely, *do* look after the gear. A good fan manufacturer should offer a maintenance service contract.

The real test of any equipment is how well it is supported by the supplier.

AN ADVANCE CHECKLIST TO HAVE READY: HOW A VENTILATION ENGINEER APPROACHES VENTILATION REFURBISHMENT

He will ask many questions to ascertain the present position.

✓ **Details of buildings & furniture**

The major overall "internal" dimensions of the building should be shown on a sketch. Include all the following:

❖ **Building overall dimensions**
- Length
- Width
- Height to ridge
- Height to eaves or ceiling

❖ **Pen dimensions**
- Length
- Width
- Height
- Number

❖ **Axis of building to prevailing wind**
- Location (area and whether exposed or sheltered)
- Proximity of nearby buildings and their width/ length/height

✓ **Building construction**

Full details of the materials of construction and their thicknesses should be given on the sketch; use the following notes as a checklist or fill in the details as appropriate.

❖ **Floor**
- Solid concrete
- Slats
- Straw / other bedding *e.g.* sawdust
- Perforated
- Earth

❖ **Roof**

Details of the materials of construction and their thickness need to be provided. Indicate also the presence of any air gap and its depth or of a loft space.

- Outer skin
- Air gap
- Insulation
- Inner skin
- Material
- Thickness

❖ **Walls**

Two types of walling are normally found *ie* panel walls or masonry walls. For buildings with both types the dimensions, in particular height, of both should be detailed on the sketch. Full details of the materials of construction, their thicknesses and the presence of any air gaps should be provided.

- Panel walls
  - Outer skin
  - Air gap
  - Insulation
  - Inner skin
  - Material
  - Thickness
- Masonry walls
  - Single skin or double
  - Material
  - Thickness
- Cavity walls
  - Outer skin
  - Air gap
  - Inner skin
  - Material
  - Thickness

✓ **Fittings and obstructions**

Any internal fittings such as water tanks, feeders or feed pipelines, other obstructions *e.g.* beams or roof trusses should be indicated and their dimensions and location shown on the sketch.

✓ **Stocking details**

Details of building stocking levels should be based on a realistic appraisal of the number of animals that are being or will be housed.

❖ **Number of pigs to be housed**
- Maximum no
- Minimum no
- No of pigs/pen

❖ **Stocking Policy**
- Continuous
- All-in/all-out

❖ **Liveweight of pigs**, kg (or lb in USA)
- Weight in :
- Weight out :
- Expected mean weight:

✓ **Feeding details**
❖ *Ad lib*
❖ Restricted
❖ Pellets
❖ Meal
❖ Wet/dry
❖ Pipeline fed
❖ Energy density of food (MJ/kg):DE or (kilocals/lb, USA)

If restricted
- Food fed/pig day to start (kg)
- or *ad lib*
- Food per sow per year (tonnes)
- Weight of pig at which restriction starts (kg)
- Food fed/pig day when restriction starts (kg)
- Food fed/pig day to finish (kg)

✓ **Typical performance**
❖ FCR/ADG over liveweight range stated in 'stocking details' above

✓ **Current ventilation**
❖ Type/make
❖ Number per room
❖ Capacity – max/min speed
❖ Diameter

❖   Housing
❖   Controller details
❖   Type of sensor
❖   Sensor(s) position

Why all this information is necessary

From these data the ventilation engineer uses a computer model to calculate:

- The thermal properties and heat losses from the structure as of now.
- What this is costing in predicted performance loss.
- What this is costing in lost income/extra costs.

We now have a penalty cost figure (per pig, or per m² (ft2) or per year) to set against the cost of rectification.

- Where the current system is underachieving (*e.g.* poor control of ventilation rate; poor air inlet design; wrong airflow pattern; prone to outside wind interference etc) and whether it can be rectified by alteration or replacement.

We now have guidance on what structural alterations may be needed to allow any ventilation system (however good) to *function* properly.

- What the specifications are of a ventilation system to provide the correct max and min ventilation rates and what heat input/evaporative cooling may be further required.

We now have specifications (just like nutritional specifications for feeds) to enable us to approach any fan manufacturer to quote his best prices for equipment and installation. The specifications *must* be met. If not, refer the manufacturer to the ventilation engineer before purchasing, as he may be able to design it in. If not, go to another supplier.

The procedure recommended above allows the producer to set the cost of a correct solution against what the current system is costing him.

Many replacements / refurbishments / new installations give trouble or affect optimal performance because the producer is dissuaded by the initial capital cost of renewing/updating the ventilation system. In my experience paybacks vary from 4 months to 2 years, with a return on investment (ROI) of over

6:1 subsequently across a 6 year depreciation period. Typical costs are analysed in Table 3 on page 485.

The cost of a qualified ventilationist has been between 8% and 12% of the capital cost, and is well worth it.

THE BASIC PRINCIPLES OF FORCED VENTILATION BY FANS

Rule One   Get control of air movement.

In natural ventilation, in contrast to fan ventilation, the pigs and the outside weather conditions dictate the internal air patterns. With forced ventilation the fans and the internal shape of the room are the primary dynamic forces.

Rule Two   Control of air movement can only be achieved by sealing the building.

And I do mean seal! How can you possibly control air currents if the building 'leaks', thus distorting them? The average well-built piggery has cracks and crevices around doors and windows adding up to over a square metre of open space. Air movement cannot be efficiently controlled with a metre-square hole in the wall or roof! So such cracks have to be sealed with brushes and draught-excluders. It is not a luxury, it is a necessity, and is worth the extra cost.

Rule Three   Ventilate the dunging area first, the pigs last.

Most dirty-pen problems are due to a failure to follow this primary rule. Except in very hot conditions where the air currents can of course be reversed to cool the pigs. Less important in totally-slatted pens, but you still need to delineate between the sleeping/resting area and the voiding/ drinking area. Air positioning primarily influences this.

Rule Four   Have a long air travel.

This ensures two vital things, the colder outside air is slowed down to no more than 0.2m/sec over the pig's back, and good air mixing with warm inside air is achieved so that the pigs are never in a cold air current, or draught. In hot weather a properly designed system is reversed as control of air currents allows this to be done.

Basic rules of forced ventilation

Two dissimilar ground floor plans of a finishing piggery. By following the basic rules of fan assisted ventilation, any shape of building can be ventilated adequately. The two cardinal errors in forced ventilation practice across the world are: not to get control of air currents and insufficient ability to cope with temperature and wind velocity extremes.

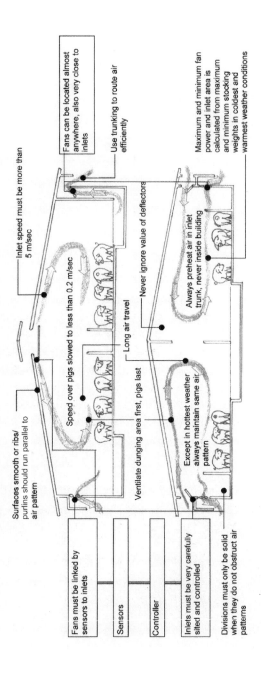

Surfaces smooth or ribs/purlins should run parallel to air pattern

Fans must be linked by sensors to inlets

Sensors

Controller

Inlets must be very carefully sited and controlled

Divisions must only be solid when they do not obstruct air patterns

Speed over pigs slowed to less than 0.2 m/sec

Ventilate dunging area first, pigs last

Long air travel

Except in hottest weather always maintain same air pattern

Inlet speed must be more than 5 m/sec

Never ignore value of deflectors

Always preheat air in inlet trunk, never inside building

Fans can be located almost anywhere, also very close to inlets

Use trunking to route air efficiently

Maximum and minimum fan power and inlet area is calculated from maximum and minimum stocking weights in coldest and warmest weather conditions

Figure 2 Basic rules of forced ventilation

Rule Five   Fans do not direct the air.

Fans are solely the source of power. Do not use fans to try to direct the air pattern. Providing the building is well sealed, fans can be placed anywhere, in the ceiling, in the wall or even under the floor. However, the correct place for the fan is where it is most easily serviced.

Rule Six   Inlets should direct the air pattern.

Unlike fans, which can be placed anywhere, inlet size and placing is extremely important. It is just as important to control the inlets so that when they are fully open a large volume of air can drop straight downwards to cool pigs when hot, and at the opposite extreme, can be closed up to a narrow slit to project a much smaller volume of cold air across the building on a long air travel to warm it up and slow it down before reaching the pigs.

Rule Seven   Inlet closure and fan speed must be interlinked by a controller.

Most climates consist of, every week, 7 "winters" during the nights and 7 "summers" during the days. At night, when outside air is colder, a smaller volume of air is needed, thus fan speed is reduced (or better, some fans are cut out) but the inlets are narrowed to maintain the same air pattern nevertheless. In daytime more air is drawn in to maintain the correct temperature, and so the inlets open wider to keep the same pattern flowing. Temperature sensors at several positions are needed, also linked to a microchip controller ('Dicam' from Farmex UK is the best I know and superior to anything I've seen so far).

Rule Eight   Preferably exhaust the air, not pressurise.

Merely because exhausting air makes it easier to stabilise incoming air. A fan exhausting air in a properly sealed building has to draw in air from the inlets, thus the inlets "pressurise" the building anyway. There is, however, one important exception to this rule, the recirculation system which blows in air through a plastic duct which must be kept inflated, thus a pressure system is created. The Recirculation Concept (see page 519) is often very useful in certain buildings – consult your agricultural engineer.

Rule Nine   Under most climatic conditions, air should enter the inlet at not less than 5 m/sec.

This is usually sufficient to ventilate most spaces in which we house pigs, and this amount of inlet *speed is needed to drive the correct air patterns, whether the volume is large or small.*

Rule Ten   Fix the maximum air requirement.

This is done by calculating maximum stocking density under the hottest summer weather expected in relation to the UCT (Upper Critical Temperature) limit of the pigs.

Rule Eleven   Fix the minimum air requirement

Only purchase fans which are capable of minimum ventilation rates. Failure to do this is a common fault. Table 4 gives the essential capabilities to achieve this.

Table 4   TYPICAL MINIMUM VENTILATION RATES REQUIRED

Housing	Minimum ventilation rate as % of maximum rate
Dry sow house	10
Service house	10
Farrowing	5*
1st stage weaning (batch stocking)	2*
2nd stage weaner (batch stocking)	2*
Fattening house (batch stocking)	5*
Fattening house (continuous stocking)	10

Where the minimum fan speed is below 10% it is wise to use 2 or more fans with stepped control or some form of recirculation.

Rule Twelve   Fail-safe devices must be built-in

Because in a power cut a sealed building becomes a coffin. Automatic drop-out panels (size and position calculated to give adequate temporary ventilation) and an alarm bell are needed.

Rule Thirteen   Beware of obstructions

Ribs in metal surface cladding, purlins, fluorescent tubes, pipes and pen divisions all obstruct airflow. But a planned obstruction can become a useful deflector to position air where it is needed or protect pigs from a draught.

Rule Fourteen   In cold and windy areas try to achieve target ventilation rates with several small fans rather than a few large ones.

This way it is easier to get down to minimum air input, as one or two small fans can be running at high speed yet only provide the low overall air changes needed. Thus they can override outside wind pressure when one or two large fans running very slowly cannot push air out against the force of the wind outside.

THE THERMODYNAMICS OF VENTILATION

Understanding LCT, ECT and UCT

The temperature at which you dictate your pigs should live has a marked effect on how they convert their food. Pigs, rather like humans, are very good at getting rid of large quantities of heat into the surrounding air. We put clothes around us to conserve that heat and build houses round our pigs to do the same thing. If our clothes are too thin then much heat escapes and we feel cold. If we move outside into an icy wind and a lot of cold air blows through them, we feel colder still.

A poorly insulated house is a thin set of clothes for the pig, and draughts or over-generous ventilation in winter (or at night at any time of the year) has just the same effect on a pig as our being out in a cold wind.

Pigs too cold

Thin clothes, icy wind, poor insulation, overdone ventilation. All these add up to exactly the same set of unconscious reactions in both human and pig.

Reaction one: preserve body temperature at all costs – humans at 37°C (98.4°F), the pig somewhat hotter at 39°C (102.2°F).

Reaction two: do this by generating more heat within our bodies (shivering, unnecessary movement).

Reaction three: secure this heat by diverting food from more productive processes – like growth in the case of adolescent pigs or humans.
These three reactions are involuntary. The pig will carry them out

anyway and he doesn't have to think about it. But he then follows this up with three conscious reactions which also burn up energy much better earmarked for growth or reproductive purposes.

Reaction four: go and look for somewhere warmer (shelter, huddling, piling in a heap).

Reaction five: go and look for more food

Reaction six: if neither can be secured, start worrying about it all and, if things get really bad, become miserable, consider getting ill and as a last resort, dying.

Pigs too hot

When pigs get too hot, productivity also suffers.

Reaction one: look for somewhere cooler, away from a warm companion, seeking out a draught or a breeze, somewhere wet, a cooler floor or wall. Energy is used up in restlessness.

Reaction two: if that doesn't work, make your own wet patch by drinking more, urinating and dunging where you decide to lie down, and cover yourself in wetness to help it along. Cut down on food intake. All this severely affects productivity.

Reaction three: if the surrounding temperature goes on rising, start panting to assist what little evaporative cooling you possess. Lie still and become very stressed. More energy is used up. Maybe stop eating altogether.

Reaction four: give up the ghost when the temperature becomes intolerable.

The comfort zone

There are three thermodynamic thresholds across the temperature range within two of which the pig finds itself quite happy and comfortable. This quite broad band of temperatures is called the 'zone of thermoneutrality' by the physiologists. This makes them sound very wise and scientific; in fact it took a meteorologist (C V Smith, in 1963) to invent a much better phrase for it – the Comfort Zone – for that is exactly what it is. However, you might as well call it the zone of maximum profitability if you want to.

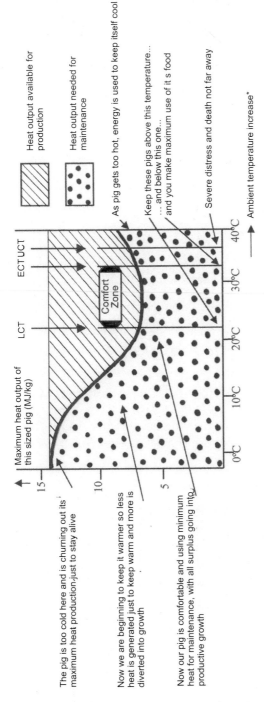

The pig is too cold here and is churning out its maximum heat production-just to stay alive

Now we are beginning to keep it warmer so less heat is generated just to keep warm and more is diverted into growth

Now our pig is comfortable and using minimum heat for maintenance, with all surplus going into productive growth

Maximum heat output of this sized pig (MJ/kg)

Heat output available for production

Heat output needed for maintenance

As pig gets too hot, energy is used to keep itself cool

Keep these pigs above this temperature... and below this one... and you make maximum use of it s food

Severe distress and death not far away

Ambient temperature increase*

This classic diagram looks a bit ferasome, but take it steady and you will see where lower critical temperature comes in. While the evaporative critical temperature threshold is important in hot summers or in hot climates, in temperate or cold zones it is the LCT threshold which is vital, as this costs the most money

*IMPORTANT NOTE: These temperatures are variable, used here as an illustration only

Figure 3 Influence of environmental temperature

LCT (Lower Critical Temperature)

This is the temperature below which the pig starts to direct food away from growth or reproduction into keeping warmer.

ECT (Evaporative Critical Temperature)

The temperature at which the pig starts calling on what little evaporative processes it has. As it virtually cannot sweat, panting is its main option. Moisture removes heat from the lungs and respiratory passage. Not much, but it helps.

UCT (Upper Critical Temperature)

The temperature above which the pig starts to lose control of its metabolism. Normal functions are impeded, and if the surrounding temperature rises only a few degrees further, death will result (see Figure 3).

UCT OR ECT?

The Comfort Zone lies between LCT and ECT, although some authorities favour LCT and UCT. I prefer LCT/ECT as ECT possesses a distinct visible warning symptom (panting) and reaching UCT is too near to the risk of disaster.

Factors affecting LCT

The story now gets a bit more difficult. You may remember we agreed that the pig, like the human, is pretty good at throwing off heat. And that a cold building or a cold draught encourages it to do so, and that it diverts food from production – growth – into maintenance as it sets about readjusting matters to keep warm. Pigs differ. Larger pigs have higher maintenance demands and so produce more heat. Also fast-growing pigs generate more heat than slower growers.

Again, fatter pigs have more body insulation in the shape of subcutaneous (under the skin) fat, so they lose less heat than leaner pigs; older pigs are similarly better-endowed in this respect than baby pigs. And so on. One can go on showing that pigs are different from others in respect of their comfort zones. Some, for example those that are younger or leaner, will have comfort zones in a hotter temperature range than those that are older and fatter.

Thus the comfort zone is a moveable area, especially towards the LCT threshold in colder climates. And it depends on a wide variety of both internal and external influences such as age, food intake, fat cover, flooring, ventilation, insulation, etc.

The same is true for ECT and UCT, but to a lesser extent. A small thin pig can get rid of excess heat more easily compared to an older or fatter one, so it will have a higher toleration for warmth, in other words its upper thresholds of ECT/UCT are further up the temperature scale.

A moveable threshold

These variations in pig size, fat cover, energy intake, and the environment it finds itself in means *that LCT is not a fixed, static* temperature and, moreover, what is satisfactory for one group of pigs can be damaging to others. You will see the extent to which these factors affect LCT in Table 5.

Table 5 TWO FACTORS INFLUENCING LCT TAKEN FROM VARIOUS PUBLICATIONS

		Bodyweight and food . . .			
		Between 20 and 100 kg there is approximately a			
		1°C fall in LCT per 10 kg bodyweight increase			
Bodyweight, kg (lb)		*Food intake, kg/day (lb/day)*		*LCT, °C (°F)*	
1.5	(3.3)	*Ad libitum* milk		30	(86)
10	(22)	*Ad libitum* creep		25	(77)
20	(44)	0.8	(1.8)	20	(68)
40	(88)	1.3	(2.9)	17	(63)
60	(132)	1.8	(4.0)	14	(57)
80	(176)	2.3	(5.1)	12	(54)
100	(221)	2.8	(6.2)	10	(50)

It explains why temperature standards have to be within variable bands, and that individual temperature ideals have to be ascertained for each house. Also *you* must recognise them on your own premises and then ensure they are reached.

Figure 4 illustrates the concept.

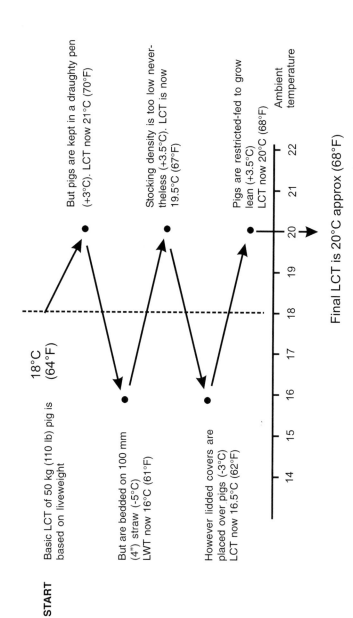

Figure 4 Many factors influence LCT - learn how to assess LCT like this .

So how do we make sense of this variation?

High LCT bad; low LCT good

Remember that. Learn it parrot-fashion if you like. Why is a high LCT a bad thing? Look at it this way – the lower critical temperature is the ambient (surrounding) temperature at which a pig starts to divert food energy from growth into just keeping warm. If this temperature threshold happens to be **high**, say 21°C (70°F) (because the pig is young, growing fast, on a perforated steel floor etc, all of which raise the LCT) **then the temperature around the pig has to be kept up to or beyond 21°C in order to get maximum growth** – and this costs money.

If you do not do this, then the pig either eats more food to compensate – assuming it can get it, or it grows more slowly for the same food intake if it is denied more food (Table 6).

Table 6 SOME MORE LCT INFLUENCES

Draughts and over-generous ventilation . . .

Air Velocity (m/sec)	LCT °C (°F)	
60 kg pig, Fed 1.9g/day/insulated house on concrete		
0.15	17.7	(64)
0.30	18.9	(66)
0.60	20.2	(69)
1.20	21.5	(71)

Imagine air taking 10 seconds to cover a metre (0.1 m/s) and then 1 second to cover 4 metres (4 m/s), much faster. This is the difference between gentle and brisk ventilation on a coldish day; it will raise the pigs LCT by 5°C.

Floor type . . .	LCT °C (°F)	
Concrete slats	19	(66)
Asphalt	15	(59)
Asphalt + 3" bedding	12	(54)

3" straw bedding lowers LCT between 3-5°C dependent on other factors.

Social behaviour . . .

Huddling reduces the drain on energy. We all know that, but here it is quantified.

Groups of	9	4	1
LCT	14°C (57°F)	16°C (61°F)	19°C (66°F)

All the same, it may cost real money to keep a building at 21°C. Now, if the pig's LCT happens to be 16°C, or *low* (because the pig is older, has a higher fat level, has more food on offer or is on a 3" straw bed) then it can be kept that much cooler without much loss of growth, and that is usually an economical thing to do. So high LCT is costly but low LCT sets a lower hurdle to clear. But whether your LCT target is high or low, you must reach it as not reaching it costs money.

Achieving LCT

The complex interaction of the various factors affecting LCT as shown in Figure 5, for example, makes it impossible for the stockperson to arrive at an *exact* LCT for a given group of pigs. All is not lost, however, as an approximate figure can usually be worked out along the lines of Table 7. There are also more detailed tables in specialist textbooks on pig housing and most countries have leaflets and monographs which give the necessary information for an owner to do his own basic but approximate calculations similar to Figure 4. The 40 page Monograph published by the UK Farm Electric Centre in 1990 "Controlled Environment for Livestock" is a model of its kind and gives many valuable tables of this nature.

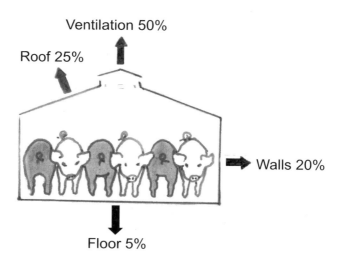

Figure 5 Where the heat goes

What I find disturbing from my on-farm work across the world is how often the LCT is underachieved in winter or alternatively during the late

Table 7 A GUIDE TO MINIMUM TEMPERATURE SELECTION

Floor Type:		Straw bedded °C (°F)	Solid insulated concrete °C (°F)	Fully slatted °C (°F)
Piglets Pre-Weaning				
(Age)				
Day 1		30°C (86°F)	31°C (88°F)	32°C (90°F)
Day 3		28°C (82°F)	30°C (86°F)	29°C (84°F)
Day 5		24°C (75°F)	26°C (79°F)	27°C (81°F)
Day 17		22°C (72F)	24°C (75°F)	25°C (77°F)
Pigs Post Weaning				
(Weight)	(lb)			
6 kg (day of weaning)	13	27°C (81°F)	28°C (82°F)	30°C (86°F)
6 kg (once started eating)	13	26°C (79°F)	27°C (81°F)	29°C (84°F)
8 kg	18	24°C (75°F)	26°C (79°F)	28°C (82°F)
10 kg	22	21°C (70°F)	23°C (73°F)	25°C (77°F)
15 kg	33	19°C (66°F)	20°C (68°F)	23°C (73°F)
20 kg	44	15°C (59°F)	17°C (63°F)	21°C (70°F)
30 kg	66	13°C (55°F)	15°C (59°F)	19°C (66°F)
45 kg	99	10°C (50°F)	12°C (54°F)	17°C (63°F)
60 kg	132	9°C (48°F)	11°C (52°F)	16°C (61°F)
90 kg	198	8°C (46°F)	10°C (50°F)	15°C (59°F)
Dry Sows (in groups)				
Service Area		10°C (50°F)	12°C (54°F)	15°C (59°F)
(assumes "Flushing")				
Gestation		15°C (59°F)	17°C (63°F)	20°C (68°F)
Farrowing Sows				
Day 1 to 3 post farrowing		18°C (64°F)	20°C (68°F)	23°C (73°F)
Day 4 to weaning		16°C (61°F)	18°C (64°F)	20°C (68°F)

NB: 1 All the above temperatures are provided as a *guide* only. They assume that the pigs perform as expected on an ad-lib feeding regime and are housed without draughts. Pigs fed on a restricted ration are likely to require higher temperatures and draughts can increase the required temperature by up to 4°C (approx 8°F).

It is important for the best feed intake, health and performance to keep pigs as cool as possible without any signs of huddling and consequently the behaviour of the animals must remain the main criterion for the actual temperature selected.

2 °F are approximate only (to the nearest degree)

Table 8 A GUIDE TO MAXIMUM TEMPERATURE THRESHOLDS

Stock	Age weeks	Weight kg (lb)	Airspeed M/sec	Airspeed Draughty	Skin Dry	Skin Wetted	ECT °C (°F)	UCT °C (°F)
Piglet	1	2 (4.4)	<0.15	<-	✓	–	35 (95)	41 (106)
11 in litter	4	5 (11.0)	<0.15	–	✓	–	33 (91)	39 (102)
Nursery on	5	7 (15.4)	0.2	–	✓	–	35 (95)	41 (106)
perforated floor.	6	10 (22.0)	0.2	–	✓	–	33 (91)	39 (102)
Groups of 50+	8	16 (35.0)	0.2	–	✓	–	30 (86)	37 (99)
(for solid floors add 1°C)								
Growers on	9	20 (44.0)		0.5	✓	–	32 (90)	39 (102)
insulated	15	50 (110.0)		0.6	✓	–	30 (86)	37 (99)
concrete.	21	90 (199.0)		0.7	✓	–	29 (84)	36 (97)
(For 100% slats drop 1°C)								
Growers	9	20 (44.0)		0.5	–	✓	34 (93)	42 (108)
100% concrete	15	50 (110.0)		0.6	–	✓	32 (90)	40 (104)
slats	21	90 (199.0)		0.7	–	✓	31 (88)	39 (102)
Dry Sow		150 (331.0)	0.2	–	✓	–	27 (75)	36 (97)
(or boar 300 kg)		150 (331.0)		0.5	✓	–	29 (81)	38 (100)
		150 (331.0)	–	0.5	–	–	33 (91)	40 (104)
Lactating sow		150 (331.0)	<0.2	–	✓	–	22 (72)	32 (90)
on solid floor		150 (331.0)	<0.2	–	–	–	23 (73)	33 (91)
with neck drip								

Comment: Based on/extrapolated from Australian recommendations (Kruger Taylor & Crosling *et al*, 1992) plus other sources
Note that the ECT levels can be considered action levels and the UCT figures as emergency levels.
°F conversions to the nearest degree. M/sec for most practical purposes can be taken as yards/sec.

night hours for as much of three-quarters of a year. 5°C below is not uncommon and 3°C is typical. 3°C below LCT for half the pig's life costs 18 kg (40 lb) MTF, equivalent to raising the cost/tonne of feed in Europe by 13%.

Ventilation is often a critical proportion of this shortfall as so much valuable heat is lost in a normal correctly-operated ventilation system (Figure 5) let alone a badly-conceived one.

Some guidelines on LCT, ECT and UCT targets are given in Tables 7 and 8.

Do we have to achieve LCT exactly?

Not necessarily. As can be seen from the tables the thermal comfort zone is quite broad. The pig should perform more or less the same within the LCT-ECT bands, but achieving a temperature 1° or 2°C above LCT or in hot conditions below ECT will give the best *economic* benefit in terms of lowered operating costs, as it costs money to keep a building warmer in cold weather and cooler in hot weather. If this *extra* warmth or coolness is not going to affect performance unduly, there is no need to spend money achieving it. Just manage the environment to keep a degree or two above LCT at all times at the lower end of the comfort zone. The important criteria are to ascertain the correct thresholds and realise that properly controlled ventilation has a major influence on whether these are achieved or not.

Here is a final checklist on what you need to do so that your pigs enjoy the benefits of good ventilation.

A REVIEW CHECKLIST ON ACHIEVING GOOD VENTILATION PRACTICE

✓　　Understand the basics of air movement.

✓　　Understand the basics of thermodynamics.

✓　　Then you will realise that good ventilation is all physics and maths. A numbers game. You will also appreciate that substandard ventilation can be extremely expensive.

✓   Thus there are at least 12 major ventilation systems which are perfectly satisfactory for pigs. The correct one for you depends on your locality, your existing farm plan and layout, and your labour load, rather than the depth of your purse.

✓   Employ a good agricultural engineer, who is also a specialist in ventilation.

✓   Be aware that installation can be critical, is sometimes substandard, and like an architect in building construction, an agricultural engineer should supervise the installation and initial running of the ventilation process, not just design or recommend the system to be used and then go away.

✓   While most equipment firms employ good ventilationists, remember that they are working within their own range of products which may or may not be the best for your circumstances. An independent ventilationist/consultant is a wise insurance against 'bending' a sales package to suit your circumstances. A manufacturer's or supplier's service contract is not a bad idea. It keeps them – and you – on your toes.

✓   Constantly measure and monitor what is going on. Check, check, check.

✓   Use simple tools to detect flat spots, draughts, diurnal (day/night) differences in room temperature and where the air is or isn't going.

✓   Sometimes it is the gases you **cannot** smell which are holding back potential performance, but use the pungent ones, like ammonia, as 'monitor' or 'marker' gases heralding trouble. These are easy to measure.

✓   You are not a good judge of a foul atmosphere, as you have become inured (habituated) to smell. Use simple instruments to forewarn you.

✓   A badly ventilated piggery can be very damaging to **your** long-term health, as medical research shows.

✓   Veterinarians are very keen on fresh air! That is fine, but after his visit and advice on keeping pigs healthy, make sure the air pattern is such that the pigs are above their LCT and not chilled by it, or use supplementary heat.

✓   It is just as important to ensure kennels and lidded covers are properly ventilated. These usually house young pigs which react acutely to poor ventilation (pneumonias, meningitis).

✓   Manage air patterns to keep lactating sows cooler and their piglets warmer.

MORE THOUGHTS ON AIR MOVEMENT

How air moves

Piggery air does not usually move in straight lines, but in a series of tumbling rotary patterns of ripples and swirls. The swirls or bubbles of air move across and up and down the enclosed spaces in patterns dictated by inlet speed, incoming air temperature, obstructions, internal surfaces and warm or cold areas inside the structure. An important source of heat being the pigs themselves – for example 10 x 50 kg (113 lb) pigs produce as much heat as a 1 kW (two-bar) fire.

Air passage across a pig removes heat from its thermal envelope – a good thing when conditions are hot, but it can be costly at other times and disease-threatening if it chills the pig.

Of course we cannot see air currents, unless we use smoke or dense vaporising fluid phials, cigarette smoke or view dust particles through a sunbeam, but a knowledge of how air affects the thermodynamics of pig metabolism is a basic of stockmanship and this is covered later.

A CHECKLIST ON AIR MOVEMENT

✓   Generally speaking, when a pig is within its comfort zone, we want the air to cross it at no more than 0.2m/sec. This is 5 seconds to cross one metre, or about one yard if you wish. Pass your hand across a distance of one metre/yard taking 5 seconds to do it – quite slow, isn't it! An airspeed of under 0.2m/sec will not constitute a draught unless the air is very cold.

✓   As the pig gets warmer, *i.e.* approaching ECT, passing the air across the thermal envelope will remove heat, and 1m/sec or more (one second to cross one metre) will do this very effectively. Less so if the air is warm and moist, but more effectively if the pig's skin surface is wet due to the evaporative cooling effect.

✓ A very rough but quite useful method of detecting where air is moving is copiously to wet the back of your hand or arm. Due to the air evaporating moisture on the skin, which lowers the surface temperature, this wetting of your arm will detect low speed air currents where a dry skin surface, such as your face, will not.

✓ Small smoke diffusion tubes (Dräger) containing crystals which emit harmless white smoke when air is blown through them via a rubber bulb or bellows, are useful for such close work – around creep areas and over surfaces prone to condensation.

✓ For larger areas, a hand-held liquid fog generator (Dynafog) is invaluable, although you need to be sparing with the smoke! Most producers possess neither of these useful aids and they will learn a lot more about ventilation if they do.

✓ Pig behaviour is a major instrument in detecting the effects of wrong air placement, especially when 'piling' or huddling is evident. Get expert at noticing unusual lying behaviour, open-space abnormal urinating and wrong-mucking and act on what the pigs are telling you about your incorrect air placement, which will be largely responsible for it.

✓ Snout cooling is a very useful ventilation aid in the farrowing house. A small 230 mm (9") fan blows outside daytime air along a central distributor pipe which falls down a vertical tube to exit over the feed trough at about 0.15m/sec. A damper butterfly valve in the downpipe controls airspeed. While this is called snout *cooling*, it is the *fresh* air from outside – even if it is hot – which can improve lactation intake by 15% or more in hot conditions – rather than any cooling effect, which is minimal or non-existent.

Table 9 shows its effect.

Table 9 THE EFFECT OF SNOUT COOLING ON LACTATION APPETITE

	With	*Without*
Average feed intake/day 1-26 day, kg (lb)	5.1 (11.25)	4.06 (8.95)
Temperature over sow at midday, °C (°F)	27.2 (81)	27.8 (82)

Source: Clients records

✓ Down draughts falling alongside an end or outside wall can materially affect performance of those pigs sleeping against the wall. Detect them at night with a smoke phial or the back of your wetted hand. The wall surface cools the air, the air thus becomes heavier and falls on to the sleeping pigs below all night. This can be remedied by nailing a triangular-section wooden batten at least 5cm (2") proud at about 1 metre (3 ft) from the floor. This deflects the falling air into the rising warm air from the pigs below. A water pipe will do the same thing and is more durable!

✓ The placing of air over the pigs is extremely important. Correctly, ventilationists, veterinarians and nutritionists stress the importance of achieving maximum and minimum ventilation rates as well as air exchange measured in cubic metres/hour or cubic feet/hour. Even if this has been successfully designed-in, many piggeries I visit in cold conditions have fallen down on the design, placing and operation of the *inlets* (not the fans) to ventilate the *foul areas first*, achieve *good mixing* of fresh and indigenous air by a *long broad route* towards the pigs and *slow it down* over the pigs themselves.

✓ Many producers (and some ventilation designers it seems!) do not realise that a different piggery floor plan can be accommodated by just changing the inlet source, *not* the placing of the fans, so as to reverse the air direction in cold or hot weather. Many cases of persistent respiratory problems have been removed permanently by alteration to air inlets, and inlets alone.

✓ Similarly, hot conditions can be accommodated by just altering inlets – not to throw the air across a long travel to the pigs – but to direct it down on to the pigs to cool them. The fans do not need altering unless the maximum ventilation rate is not being achieved (this is a simple enough calculation for a ventilation engineer) when more fan capacity is required for such conditions.

✓ Minimum ventilation rates (also easy to calculate) give most trouble. As many as 40% of piggeries are 'shut up' too tightly in cold weather because of lack of mathematical design. The system as found just couldn't cope. When this is rectified (some heat increment may be needed, a third easy-to-calculate figure), nursery pigs especially leap ahead – I have known 80g/day (2.8 oz/day) more growth between 10-25 kg (22-55 lb), with a 50% reduction in mortality.

✓ Whether a building is pressure-ventilated or the air is extracted does not particularly matter as long as the numbers are right, and done by a professional. The sensor and control devices need to be state-of-the art, the installation must be sound, and the equipment well-maintained.

Thus almost any internal space can be properly ventilated. Sure, some buildings are awkward and not very brilliant, but they can still be ventilated correctly, even if some expenditure is needed.

✓   The best forced ventilation system? There is no best system – unless it be the one which works satisfactorily at the least cost. And this depends on getting the sums and installation right for the individual circumstances.

NATURAL VENTILATION

Natural ventilation moves air both through adjustable and fixed openings. Adjustable includes windows, eave panels, ridge openings, and ventilated doors, while fixed openings use open building points and continuous ridge and eave gaps.

While manually-operated natural ventilation is cheaper than mechanical ventilation, the manual alteration of such systems is never frequent enough and so cannot match that of mechanical control.

Table 10 illustrates this dramatically when one building was part-converted to a modern fan installation under correctly-sensored control, leaving the other naturally-ventilated half until later. While the fan installation added 37p/pig (about 14%) housing and operating costs to every pig produced over a projected life 5 years, the return on extra outlay (REO) was still 5.4:1. We later kept temperature records in both structures (Figure 6).

Table 10 IMPROVEMENTS IN PERFORMANCE WHEN ONE HALF OF A PIGGERY WAS CONVERTED TO AN ENGINEER-DESIGNED FAN VENTILATION SYSTEM, LEAVING THE OTHER HALF TO MANUALLY-OPERATED NATURAL VENTILATION AS BEFORE

	Fan	*Natural*
Pigs processed	3278	5112
30-90 kg (66-198 lb) days	81	93
MTF kg (lb)	+12 (+26.5)	–
Value of better performance/pig	£1.99 (US$3.20)	–
Cost of alteration and running cost/pig	£0.37 (US$0.59)	–
REO	5.4 : 1	–

Source: Clients costings

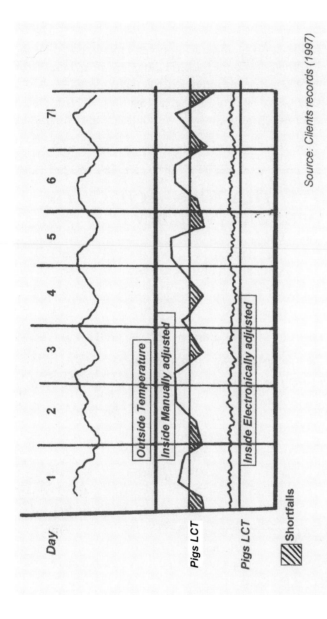

Source: Clients records (1997)

Comment:

- Manual adjustment failed to keep pigs above LCT for 40% of time, 90% of this at night. This is equivalent to 3 whole days per week below LCT and 2 whole days 2°C (approx 4°F) below LCT.

Figure 6 Natural ventilation is never altered frequently enough to compete with electronic controls

Apart from the inability to alter the ventilation frequently enough, as in the above example, producers often provide insufficient naturally ventilated inlet and outlet areas. This is particularly true in hot or airless (low outside windspeed) conditions.

The following Checklist gives an indication what these should be for cooler/temperate climates. These specifications are *insufficient* for tropical conditions where very different structural designs are essential (for example, openings are 33% of floor area) and for poorly-insulated buildings in temperate climates where ventilation improvement is required.

NATURAL VENTILATION – A CHECKLIST
OPEN PLAN BUILDING

Outlet air

✓   Calculate total liveweight of pigs in kg, at maximum stocking rate.

✓   Allow 230 cm² (35.4 in²) opening per 100 kg (220 lb).

✓   This gives the total exhaust air opening required.

✓   Divide by number of squares on floor plan – in this example divide by 4.

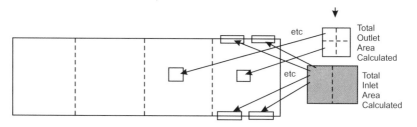

✓   Insert ventilator trunks (insulated) at centre of each square.

✓   Provide adjustable valve/close to 95%.

Inlet air

✓   Divide total liveweight by 100; allow 700 cm² (107.8 in²) per 100 kg (220 lb).

✓   Divide floor plan into squares.

✓   Divide number of squares into total allowance.

✓   Place four equal sized inlets on 2 sides of each square.

✓   Provide hinged shuttering to close upwards to 95%; but drop down completely in hot conditions outside.

✓   Inlets should open to deflect air to the ceiling.

✓   A high ceiling is desirable, sloped to the exit ventilators.

✓   Exit ventilator trunks should be insulated to a minimum 'u' value of 0.5 . U values of 0.4 in ceilings and 0.5 in side walls are advisable.

✓   To avoid wind shadow, naturally-ventilated buildings up to 5m (16.4 ft) high should be at least 15m (50 ft) apart to allow uninterrupted inlet airflow. For every 0.5m (1.6 ft) higher, allow 1m (3'3") more lateral space.

✓   Keep the system simple. This makes it easier to understand, operate, and maintain.

Note: Inlet and outlet allowances vary from source to source. The figures are a compromise and seem to work well in non-extreme conditions.

NATURAL VENTILATION FOR KENNEL/ MONOPITCH STRUCTURES

These popular traditional low cost structures, primarily for young growing pigs, need careful ventilation if stale, stuffy air is to be avoided along with the associated problems of lack of appetite, tailbiting, pneumonias and probably streptococcal meningitis.

Advice on design specifications will be found in the Tailbiting Section.

ACNV AUTOMATICALLY-CONTROLLED NATURAL VENTILATION

While manually-operated natural ventilation has advantages (cost, quietness, freedom from power failures) the disadvantages of inadequate control, bird ingress and windless conditions has evolved a compromise/

halfway solution where automatic, sensored control of natural ventilation removes the major problem of insufficient control, especially out of working hours.

ACNV is a simple system made up of a controller unit, a thermostat, adjustable flaps in the side walls of the building and a motor which operates the flaps.

The controller unit uses the thermostat to check the air temperature within the building at pre-set time intervals. If the temperature is above or below the required level the controller switches power to the mechanism which opens or closes the flaps to adjust the ventilation rate. Thus if the house is too hot the flaps are opened, if too cold they are closed.

The resulting system is one which does not continually adjust ventilation rate, but if necessary changes it in small steps at pre-set intervals.

Advantages of ACNV

- Can be used in a wide variety of housing – farrowing, nursery, growing/finishing, kennels, monopitch, and operate curtains, flaps, louvres, etc.

- Low energy needs thus very low running costs.

- Virtually silent.

- ACNV gives good temperature control, particularly at low ventilation rates.

- ACNV will operate with low air speeds within the building, even in windy conditions.

- ACNV does not create dusty conditions.

- Based on a combination of the stack effect and wind, ACNV does not depend on forced air movement to function so there is less likelihood of major problems should the power fail.

- ACNV is virtually fail-safe. If there is a power failure when the flaps are open they would remain in that position and ventilation would be provided. Unacceptable temperature and power-failing warning devices can easily be included in the system and are strongly recommended.

And the disadvantages?   How to avoid problems with ACNV

- ACNV needs some windspeed to operate it fully, so can be a problem in a sheltered hollow if a heavy weight of pig (*i.e.* finishers) is to be ventilated properly in still or oppressive conditions.

- ACNV will not give control of air direction within the building.

- In frosty, windy conditions, cold air can fall on to pigs below the flaps.

- The installation needs to fit 'design temperatures' for the locality so that flap size and operational control can be calculated. This needs expert advice.

- While running costs are much lower than for forced (fan) ventialtion, capital/installation costs are still only about 15-20% cheaper for a well-designed fan system.

A CHECKLIST FOR ACNV OPERATION

- ✓ Consult an engineer/ventilationist to agree siting, summer and winter design temperatures and from this data inlet flap sizes and design, and whether extra ridge outlets are needed – for wide buildings in a hollow, for example.

- ✓ Choose equipment only from a supplier with a track record for ACNV who can recommend an experienced installer.

- ✓ Take care in selecting a good controller giving a range of sample time between 3 and 10 minutes. Generally 5 minutes allows the interior conditions to 'settle' following what may have been a change in ventilation arising from the previous change. Too rapid sampling can give unstable conditions.

- ✓ Position the controller so it can be easily accessed.

- ✓ Automatic controls with a manual override are best, but allow the auto-controls to take precedence. Avoid 'fiddling' with knobs!

- ✓ The controller needs to be robust, waterproof and easily maintained. Keep it clean and free from pressure-washing damp.

✓ Flaps should be light and robust *and fit well*, be easily changed, and insulated to avoid condensation problems. In colder areas they should have an adjustment to hinge *upwards* inside the building.

✓ Thermostats should be placed at pig height, armoured if needs be to prevent damage by chewing. The thermostat is normally set at ±2°C (±4°F) of the LCT of the pigs present and have a range of 10°-40°C (50-104°F). Seek advice on type and placement.

✓ Fail-safe provision is essential, *e.g.* power failure and unacceptable temperature warning.

✓ Ensure all stockpeople are trained in 'what-if' emergency procedures.

✓ A simple fan and ventilator sock (*see 'Recirculation' page 519*) is a useful and not expensive back-up in localities prone to hot, still conditions.

✓ The ACNV in kennels and farrowing houses needs as much care in design as for the larger grow-out houses.

CURTAINS

Vertically rising reinforced canvas curtains or blinds are a common method of ventilating buildings especially in the tropics. Their design and operation leave much to be desired, however. They are very useful in areas with daytime temperatures between 26°C and 34°C (79°-93°F) where a reasonable flow of air is required across the pigs, along with spray cooling. They can also be used in cool/temperate climates but are less valuable if frost is commonplace.

In warm climates, insufficient thought is given to late evening and night-time temperatures which can often fall sufficiently to cause chilling. This is especially dangerous with young pigs in the 20-40 kg (44-88 lb) band. While there is a problem (growing pigs in this weight range will eat 18g (0.64 oz) food more/day for each 1°C below their LCT when growing at 400g/day (14.1oz/day) and so grow 12g/day (0.42 oz/day) slower, which results in 2 days longer to slaughter) *the real danger from slightly low temperatures is chilling*. Chilling causes stress and stress lowers the animals' immune defences to many diseases, especially respiratory ones. Chilling is an additional effect to the ambient (surrounding air) temperature, as the floors are often wet from the day's spraying to keep the pigs cool. Wet floors increase thermal conductivity so the pig loses additional heat, not only to the cooler air over it, but also to the cold surface under it. Some critical organs (liver and kidneys) are in direct contact with this too-cold surface.

Curtain alteration is rarely frequent enough

As the warm evening turns to a cooler, chilly or even cold night, the curtain is the main device to retain the correct air temperature – just above the pigs' LCT. These, being manually operated, are usually raised (to conserve heat) too late; as a result the pigs are chilled and stressed as the outside temperature falls.

I have measured such falls in the tropics between 7 pm and midnight. In each case I asked the owner what he thought the range of fall over this period was, the reply being "4-5°C" (about 9°F), when the instruments showed it was between 8° and 12°C (14°-22°F). Only one curtain adjustment meant that all pigs were way outside the correct range across 4 hours, and from midnight onwards were frequently below LCT by 2°C (older pigs) and as much as 5°C below LCT for weaners.

As curtains are primarily used in hot climates, their correct installation and operation is fully described in the Hot Weather Section.

RECIRCULATION VENTILATION

Probably an underused concept, which uses polyethelene ducts, rather like an elongated stocking, to distribute a mixture of outside air and inside air down through the piggery. The amount of air mixing is controlled by a damper from 95% recirculation to 100% extraction, which is activated by pre-set temperature sensors.

Advantages

• Fans are working at high speeds when the recirculation rate needs to be low in cold conditions, so the air placement performance is stable.

• These high fan speeds can override outside wind pressure *ie* the system is windproof, a distinct advantage over multiple fans on exposed sites in cold weather.

- The capital cost is about 35% lower than multifans and concentrated in one exchanger box and the controller/sensors. Further savings can be made if the producer makes his own ducts. For how to do this, see below.

- It is simple to install into almost any building. Wide-span buildings can use more than one unit.

- Air can be placed precisely where it is wanted.

- Maintenance is easy and accessible.

- An even simpler, non-recirculation duct can be used to place air at variable speeds exactly over the pigs to cool them in hot weather, or in extreme conditions can be used together with wetting.

Disadvantages

- Can be expensive in power as the unit runs continuously.

- The exchange box tends to clog with dust and needs frequent cleaning (but an easily-removed filter helps sanitise the air).

A CHECKLIST AND CONSTRUCTION GUIDE FOR A HOME-MADE AIR DISTRIBUTION STOCKING.

Inflatable lightweight pressurised ducts to assist ventilation in hot weather or awkwardly-shaped existing structures.

✓ Pressurised ducts can be made from heavy duty (500 gauge) polyethylene.

✓ Make the initial duct diameter match the fan diameter. Ordinary propeller fans can be used *i.e.* 15", 18", 24" etc (3800, 4570, 6096 mm), but ensure the fan speeds achieve 1,400; 1,200 and 940 rpm respectively so as to inflate the duct firmly.

✓ Taper the duct. This ensures the duct pressure and discharge air speed are constant so that the hole spacing along the duct can be uniform.

Taper to 3:1 (wide:narrow ends) for ducts 25 to 50m (82-164 ft). 1.5:1 for ducts 51-70m (167-230 ft) and no taper over 70m. Seal the end. Use the spare material to create a web hanger.

(The duct is clamped to the air straightener (see below) with a steel or wire band, held by an 'Oxford' nut.)

✓ Select the hole size required. In general diameters greater than 40mm (1.6") give directional jets and diameters of less than 30mm (1.2") tend to give a diffusing jet. But holes less than ½" diameter usually clog with dust.

✓ Calculate the number of holes required. The total *area* of holes is equal to the total cross sectional *area* at the fan.

✓ Decide where to cut the exit holes – downwards, to one or both sides etc.

✓ Cut D shaped holes with the flap of the D *facing* the fan. This will ensure that as the air ripples down the duct – and exits at different pressures – the D flap acts as a direction guide so that the air exits at right angles rather than forwards. D flaps are called sipes.

✓ Fit an air-straightener between the fan and the duct of the same diameter. An air straightener is a light metal tube 1½ times longer than the fan diameter with fins forming a simple cross welded inside, although 8 rather than 4 is better. This stops the air spinning in the polythene duct and twisting it.

✓ Hang the duct from the web at 1m (1 yard) intervals. This is the simplest idea, because alternatively, using wire hoops of the *same diameter* as the inflated duct at the point of suspension means each one has to be progessively smaller. But wire rings do allow the tube to be unclamped and rotated to vary the directional impact of the air.

Source: Extrapolated from G E Carpenter (1980)

PROBLEMS WITH GASES AND VENTILATION

When I was 23, and working on a farm, I got within 5 seconds of losing my life due to gases, so this section is written with some conviction! It happened like this. I was repainting equipment in an unoccupied dry sow house and was working about 30 metres in from the nearest door when a colleague with his tractor came to extract the slurry from the deep channel pits in the buildings. He had no idea I was inside and, as is customary, pushed air through the tanker's hose to agitate the crust into the liquid before sucking it out. When the stench (ammonia and hydrogen sulphide)hit me I put the paint-pot down and walked to the door to protest. Within a few metres of the opening my head swam and eyes misted over, but I just had sufficient momentum to stagger out and collapse unconscious outside. It could also have been a very high concentration of carbon monoxide in among the other gases which was lethal.   Had I got up from painting 5 seconds later, so the doctors eventually told me, I'd not be writing this now!

That's how dangerous slurry gases are. Across the world they kill between 10-15 stockmen every year and thousands of head of livestock and poultry. That's the dramatic aspect of air quality; ***piggery gases are dangerous things***.

Problems for humans: an insidious menace

But gases also have a stealthy, insidious side to them, dragging down both animal and human health particularly over a period. Humans first. While it takes years of exposure to ammonia (NH_3) and hydrogen sulphide (H_2S) to create permanent lung damage, by the time the patient sees a doctor the damage cannot be reversed. A survey among British people whose careers had been in intensive livestock buildings showed that their average life-span was some 3 years shorter than industrial workers living in urban surroundings - where the air quality is not all that marvellous either! And it seems that piggery workers have 40% more chest complaints than the rest of us. A good deal of this discrepancy must be due to dust, of course, but dust particles are a common carrier for various toxic gases in chemical (solid microparticle) form, so the two can be considered as interlinked.

Animal attendants should also be aware of the "Monday Morning Response". If you go to work every Monday morning feeling good and then begin to feel lousy during the day - check your air quality. The response happens because you get away from the gases during the weekend. But once back in the building, your body has to adjust all over again.

A survey of intensive animal attendants in the 1980s revealed a far higher level of ailments associated with air quality than in a normal population.

Problems with livestock

The insidious nature of gases at levels commonly found in confined buildings affecting animal production is not easy to measure and while evidence is accumulating, we could do with still more data to back up the indicative surveys so far completed. Farmers have yet to awaken to the drag on productivity caused by gases occurring in "normal" situations on their farms.

The effect of **acutely noxious** concentrations of certain air pollutants on animal production are generally well known and are recognised. However, these high levels occur relatively infrequently and are usually the result of ventilation failure or agitation of a long-term manure storage within or adjacent to the animal housing facility. Of more concern should be the continuous exposure of animals to **sublethal concentrations**.
Sainsbury (1981) claimed there is evidence that manure gases may affect the production rates of animals by reducing feed consumption, lowering growth rates, and by increasing the animals' susceptibility to certain infectious diseases. A series of results conducted at the Illinois Agricultural Experimental Station showed that NH_3 affected the growth rate of pigs as well as the course of an infectious disease. The rate of gain in young pigs was reduced by 12% during exposure to 50 ppm of NH_3 and by 30% at 100 to 150 ppm. At concentrations of 50-75 ppm the young pigs also had difficulty in clearing bacteria from their lungs. When an ascarid infection was imposed on young pigs, their growth rate was reduced by 28% compared with the control group. When both the worm infection and 100 ppm NH_3 were imposed on them, the reduction in growth was additive, *i.e.* 61%. Results also suggested that while the feed conversion efficiency was not affected, the growth rates were reduced 10 and 30% for pigs exposed to 50 and 100 ppm respectively. This would have a significant effect on raising overhead costs even if the feed cost was not inflated.

Acceptable and non-acceptable levels of gases

Table 11 surveys the main gases found in piggeries, together with their threshold levels and danger levels.

Table 11 CHARACTERISTICS AND DANGERS OF PIGGERY GASES

Gas	Symbol	Heavier or lighter than air	Safety threshold	Danger	Smell?
Marker gases					
Ammonia	NH_3	L	25-35 ppm	Irritant Stress Long-term lung damage	Pungent smell (over 12-15 ppm)
Carbon Dioxide	CO_2	H	2000 ppm Fresh air = 340 ppm Breath = 50,000 ppm (5%)	Wrong-mucking and Tailbiting at 0.5% to 0.7%? Slowed growth in weaners at 1.2%	No smell
Killer gases					
Hydrogen Sulphide	H_2S	H	5-10 ppm *Caution:* maximum smell response seems to be as low as 50 ppm *i.e.* the senses are dulled over this level	S. Meningitis? Death at 100 ppm *Dangerous.* This gas can rise to 1500 ppm on pit agitation. Exposure to 10-15 ppm over time can cause respiratory troubles.	Pungent smell (over 1 ppm)
Carbon Monoxide	CO	H	20-30 ppm but in growing pigs may be as high as 150-200 ppm	*Dangerous.* Death at 500 ppm. Stillbirths at >150 ppm Abortions at >200 ppm 21% reduction in growth rate at 300 ppm. S. Meningitis	No smell
Probably harmless gases					
Methane	CH_4	L	>80% in air	5 to 15% mixture with air can cause explosion	Little or no smell

How gaseous are our piggeries?

Probably more than we think. I've sampled many, and about 40% have been over the safety threshold in cold weather. Pigs weren't dying of course, but I suspect their performance was below best. This opinion seems to be borne out by Figure 7 which details a careful trial in the UK. While carried out more than a decade or more ago, our housing and ventilation standards today, while improved, suggest that only a slight improvement will be found on these disturbing ammonia levels, where 45% were over the standard safety level, 22% exceeded the serious excess level, and 93% were high enough to affect performance to some degree.

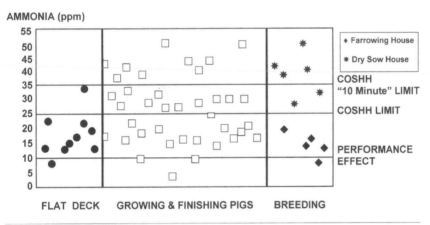

+ The survey covered the range of building types found in England, but controlled ventilation, slatted units formed the majority.

* COSHH = UK Control of Substances Hazardous to Health – Recommendations laid down by Government.

<u>Source:</u> Williams and Pitt (1991)

Figure 7 Ammonia levels measured on 60 UK pig farms

Recent visits to medium to small pig farms in other cool/temperate zones (Northern USA, Canada, Sweden, Denmark, Japan and Korea) suggest the UK figures are probably equally applicable in those countries today. Ammonia is the gas which accounts for the majority of smell pollution on our pig farms, helped onwards by hydrogen sulphide in combination with it. Table 12 gives the latest advice on the effects of rising ammonia concentration.

Table 12 AMMONIA – THE INVISIBLE PROBLEM

The effects of ammonia on human health, and the health and performance of pigs are well documented

5 ppm	–	Lowest level detectable by smell.
6 ppm	–	Eye and respiratory irritation begins.
11 ppm	–	Animal performance reduced.
25 ppm	–	Maximum level for 8 hour exposure (COSHH regulations).
35 ppm	–	Maximum allowed level for 10 minute exposure.
40 ppm	–	Headaches, nausea and appetitie loss in humans.
50 ppm	–	Severe reductions in animal performance and health
	–	Increased possibility of pneumonia
100 ppm	–	Sneezing, salivation, irritation to mucosal surfaces in pigs and humans.

WHAT CAN THE FARMER DO TO REDUCE THE HIDDEN MENACE?

1. ***Check on ventilation adequacy***: Generally speaking this is not just a matter of achieving satisfactory air changes per hour as the aim must be to sanitise the air quality ***and*** minimise heat-loss, allowing for behaviour and performance, so the complexities involved are best solved by a visit from an agricultural engineer specialising in ventilation.

2. ***Measure the gas concentration***: This, amazingly, is almost never done, despite there being simple test kits available for a low cost and done in 30 seconds or so by merely depressing a squeeze-grip. Better is to do an 8-hour crystal discoloration test, which spread of time will provide a better indication of overall gas concentration. Some noxious gases are invisible and unsmellable as Table 11 shows.

3. ***Keep things cleaner***: Routine, methodical sanitation not only lowers bacterial and viral challenge but also reduces the nuisance of smell and flies to tolerable proportions. Many slurry pits are left to build up waste material too close to the slatted floor level, so heavier-than-air toxic gases are pushed up to form a deadly layer from which the growing pigs or vulnerable sows have to breathe while at rest. Never allow slurry to reach within 150 mm (6") of the ***bottom*** of the slat.

4. ***Control the digesta within the digestive tract so that less gaseous formation occurs once elimination has taken place.*** This is an

interesting new field of biotechnology. A water-soluble extract of the *Yucca schidigera* plant, trade name 'De-Odorase', reduces the levels of ammonia released from fresh manure and during anaerobic decomposition.

Studies carried out with the Yucca extracts used in pig diets and in manure handling systems at various American universities and private research facilities have shown that adding the product to pig starter, grower and finisher diets can give the following advantages:-

- reduction in ammonia gas levels; (Table 13)

- enhanced performance for starter, grower and finishing pigs;

- elimination and prevention of further accumulation of solid manure and sludge;

- reduction of winter fuel costs;

- reduction of metal corrosion potential.

Yucca extract can also be added to slurry pits and lagoons to reduce smell. Because the Yucca extract stimulates the rate of breakdown of complex carbohydrates and water, it can be used to break down accumulations of solid manure and sludge. Once sludge is reduced, pumping becomes much faster and easier and breeding grounds for flies are also reduced. The Yucca extract not only prevents ammonia build-up, it also stimulates sludge micro-organisms to reduce hydrogen sulphide production.

Table 13 EFFECT OF SARSAPONIN ON AIR QUALITY IN THE NURSERY

		Control	*Sarsaponin*	*% reduction*
Nursery study –	Ammonia	22 ppm	14.5 ppm	34%
	Hydrogen Sulphide	50 ppm	25 ppm	50%
In vitro trial –	Ammonia (mg/d)	125.9	109.6	13%

Source: Alltech (1999)

The Yucca and its water-soluble extract have a long history of safe use as a food material for humans and livestock. As long ago as 1965, Yucca was even approved for use in human food without restriction.

But does it always work?

Usually yes, but in my experience there have been criticisms that it has been 'ineffective' in reducing complaints of noxious piggery smells. Why is this? The observations are always subjective as few if any complainants have measured the ammonia levels before and after, naturally enough.

But I did, using Dräger bellows-type crystal discoloration tubes, and found that while the 33% reduction was still there, the original ammonia level was so high (at 35 to 40 ppm) the resulting reduction to over 20 ppm (sometimes 25-27 ppm) was still so noticeable that the reduction was largely not detectable by the human nose. In other words a very bad smell just became a bad smell!

Secondly, and just as important, it takes about 3 weeks for the reduction via the feed source to reach full effectiveness – and in half the cases the users were grumbling about lack of smell reduction after a week.

Moral: If you are using yucca to control ammonia, get the NH_3 level down to around 15-20 ppm first by cleaning the place up. Then yucca in the feed will give a noticeable and better result, but give it 3 weeks to do so.

GAS EMISSION FROM SPACE HEATERS

Because of its 40% cheaper cost compared to electricity, heaters in many nurseries are fuelled by propane gas or alternatively LPG. After combustion the residual gases, (primarily carbon dioxide, CO_2, with an advised maximum level of 0.3%) high levels of carbon monoxide (CO) have occasionally been found in farrowing houses and nurseries. A maximum allowance of 0.02% CO is recommended for industrial workers and if the minimum ventilation rate is inadequate, this can be exceeded.

Problems associated with medium-high concentrations of CO (and possibly CO_2, a ratio of under 0.02 $CO:CO_2$ is advised) could give rise to stillbirths (due to the gases interfering with blood haemoglobin) and abortions at farrowing - possibly strep meningitis later on.

Heaters should be regularly cleaned and serviced and some are easier to clean than others. The poultry industry is well up on this knowledge. With many wide-span wean-to-finish sheds appearing, gas brooders during the 5-25 kg (11-55 lb) nursery stage look like making a comeback on cost grounds and new wean-to-finish operators will encounter gas problems until they are as experienced as their poultry colleagues.

PREVENTING & CURING CONDENSATION

Pigs are very wet animals. Quite apart from urine and dung, a pen of 12 midweights will exhale 5 litres (over a gallon) moisture a day. As this is in fine droplet form and warm, it rises upwards to condense on the nearest cool surface.

Water vapour always tries to move from an area of high humidity to one of lower humidity. As the vapour moves it cools down to the dewpoint and condenses inside the structure, often soaking the insulation layer. And that can reduce its heat-retention properties to almost nil. The problem is noticed as wet patches on ceiling and walls.

We all tend to put a damp proof course under our concrete floors, but a damp course is equally important inside the building superstructure, especially if the building is pressure-ventilated either by positive or negative pressure.

How can you stop internal rot? When the building is new, 5 cm (2") adhesive tape can be used to seal the gaps. If the insulation is old and greasy, or dusty, timber cover straps should be used as the adhesive will not stick.
It is much better, of course, to line the gap between facing-board and insulation layer with 1000-gauge polyethylene sheeting. This provides an effective vapour seal.

Condensation also strikes in other places:

- *Inside ducts*: Always divide the warm internal air from the colder incoming air with 25 mm (1") of insulation board. Also trunk-ventilators condense towards the upper or outer lengths and must always be well-insulated all the way up to stop dripback. Even with natural ventilation, a failure of the internal air to escape due to overcooling in the confines of the duct occurs.

- ***Drop-out ('fail-safe') panels***: These often condense on the upper surface. The panel should always be made of insulation board – but they rarely are.

- ***Internal surface condensation***: I find – in all except the very worst cases which need major redesign of the ventilation system and its specifications – that farmers make very heavy weather of curing localised condensation.   The problem is usually caused by a warm moisture-laden air moving too slowly and hitting a cooler surface.

  These situations are usually not too difficult to remedy. I set about it like this.

A CONDENSATION CHECKLIST

✓ Condensation occurs in 'dead spots' ventilation-wise, where the humid air has time to deposit its load of moisture droplets on a cold surface. It is important to know where the air bubbles are striking the surface causing the condensation, how fast the air is moving at that spot and from which direction it is coming.

✓ Know where your air currents are coming from, going to and where they are slowing down across a potentially cold surface. So use smoke tubes to ascertain this.

✓ Most stockpeople think that scouring the cold surface with a gentle airflow of 0.1 m/sec (10 seconds to cross a metre or yard) will remove condensation. It will help, but can be difficult to do, although judicious use of a 'table top' baffle board (see below) to deflect an air stream can help.

✓ Is there an air current passing fairly close to the condensation? If so, is it strong enough for some of it to be deflected to scour the area concerned? Use a simple deflector – a piece of plywood or a polyethylene sheet tacked to a light batten frame hung from plastic string at an angle sufficient to skid the air towards the desired route. Yes, it does look odd, collects dust and so forth – but it often works. And it costs nothing but a little patience in lengthening and shortening the string 'hangers' to get the angle of deflection right. Use your smoke tubes to check on what is happening.

✓ If the air pattern is too weak, you must consider increasing the fan power, sealing the building more thoroughly, reducing the inlet area nearest to the condensation or inserting an extra inlet nearer to the problem. Any of these solutions help. But at all times you should never accidentally increase the air flow over the pigs so that it causes a draught, unless it is summertime, of course.

✓   Coanda Deflection: The friction of a jet of air flowing along a smooth surface is less than the friction of the jet passing through the air. Odd, but true! Air entering a building may therefore follow or 'skid' along a smooth ceiling or wall far further than it would do across an open void. This is known as the Coanda effect. Look and see if there is a smooth flat surface near the condensation area and experiment by deflecting some incoming air onto the linking surface by using deflectors or altering the inlet configuration.

✓   Are any obstructions interfering with the Coanda effect? Can you remove them – pipes – battens – electric cables?

✓   A further extension of this can be to construct a duct, or cheaper, a polyethylene tube or tunnel, to blow cool air across a troublespot. (*See the Recirculation section for how to make one*). Be careful about creating downdraughts at pig level, however.

✓   Insulation: If warmer, humid air condenses on a colder surface then it goes without saying that the problem can be cured by making that surface warmer by insulating. I've left this until last because in most condensation cases this is either difficult or prohibitively expensive to do. It should have been done earlier, but it hasn't.

Sometimes however, surface insulation *is* the only answer. Some temporary insulation-board tacked on to a surface, spraying the troublespot with polyurethane foam and the use of "insulation paint" (for nail heads or small areas) are all solutions which have worked out where stripping, re-insulation and re-panelling is ruled out.

I am often called upon to 'teach' our stockmen here how to cure condensation by these methods who can claim a great many more successes than failure when they return to their units. However they, and most farmers are very vague about the terms used in insulation. To conclude this section, Table 14 is included for reference purposes.

Table 14 GRAPHICAL EXPLANATION OF INSULATION TERMS

K Value = Thermal Conductivity of a Material
Measured in: Watts* per metre per °C (W/m °C)
Thermal conductivity. A measure of a material's ability to conduct heat. Defined as the quantity of heat (watts) which will flow between two opposite faces of a 1m² of material with a 1°C temperature difference between those two opposite faces.

Bad Conductivity   Good Conductivity

(Different materials)
Note: The materials are the same thickness

Lowest values are best. Typical insulation materials = 0.02 to 0.2 (straw = 0.07, copper wire = 200)

R Value = Thermal Resistance of a Material
Measured in: Square metres per °C per watt (M² °C/W)
Thermal resistance. A measure of the resistance to heat flow of 1m² of a given thickness of material, or structural component made up of a number of materials.

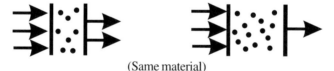

(Same material)
Note: Different thicknesses of the same material

Highest values are best: **Range**: 0.12 to 0.55 for typical piggery measurements/ materials

U Value = Thermal transmittance of a structure i.e. can be a combination of various materials
Measured in: Watts per square metre per °C (W/m² °C)

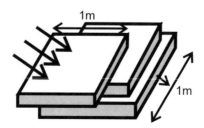

1m

1m

Note: K value (above) refers to an **individual** material

Note: U value is the amount of heat which passes through 1 square metre of the materials comprising the whole structure when the temperature difference from outside to inside is 1°C.
Lowest values are best. **Typical range**: 0.5 to 5.5

What is a Watt?

The basic metric unit of energy is the Joule (J) which is used to measure various forms of energy, including heat energy. 1 joule = 0.239 calories. Heat flow is the rate of transmission of heat. The metric unit of heat flow is the joule per second, called the watt (W).

A GLOSSARY OF TECHNICAL TERMS AS THEY AFFECT PIGS

Attention to problems on the pig farm often involve the producer reading, or being quoted, technical terms in reports from veterinarians and other advisers. These are often couched in scientific jargon (to facilitate accuracy) which may or may not help the farmer and his stockpeople take decisions. In addition, many research papers are similarly peppered with technical terms understood by the scientist but not always by a lay reader.

This glossary is not exhaustive by any means, and lists the sort of terms and definitions I may have had to look up over the past 35 years. Thus it may also be of help to the reader – to whom I apologise if I have insulted your intelligence with some of them! But we are all different, and I wish someone had done this for myself years ago! I have consulted a wide variety of sources, both printed and verbal, to whom I am grateful, but the interpretations are largely my own. If you disagree, can put it better or can suggest omissions – I'd be glad to know.

Absorption	to take in or assimilate substances eg *into* tissues. Commonly confused with adsorption (qv)
Acral	(*vet*) affecting the extremities
Acute	of recent onset – rather than depicting severity (although it is commonly used in this latter form)
Ad libitum	(*ad lib*) without restraint
Adipose	fatty
Adjuvant	a material that aids another, ie in a vaccine to increase antigen potency
Adsorption	attracting and holding other substances *onto* surfaces
Adventitious	(*vet*) acquired, not in the correct place (accidental)
Aetiology	(*vet*) the science dealing with the causes of disease
Aerobe	(*adj* aerobic) a micro-organism which needs oxygen to function fully
Agalactia	(*vet*) partial or complete lack of milk
Aitchbone	the hip bone
Algesia	(*vet*) sensitivity to pain, hence analgesic = pain-killer

Amino-acid	(*nutr*) protein building block
Ampere	(*constr*) the unit measuring the strength of an electric current
Anabolism	(*adj* anabolic) the formative stage of metabolism (qv)
Anaerobe	(*adj* anaerobic) a micro-organism which can grow in the absence of oxygen
Analogous	resembles, similar to
Anaphylaxis	(*adj* anaphylactic) severe or unusual allergic shock reaction
ANFs	Anti-Nutritional Factors. Materials present in certain feed raw materials which interfere with digestive or metabolic pathways
Androgen	male hormone
Anoestrus	no oestrus, lack of oestrus
Anoxia	(*vet*) (*adj* anoxic) interference with (lack of) oxygen supply
Anterior	towards the front
Anthropomorphism	attribution of human characteristics to animals
Antibody	specialised proteins produced by lymphocytes (white blood cells) in response to presence of an antigen (qv)
Antigen	(*adj* antigenic) any substance foreign to an organism (ie a pig) reacting with an antibody so as to produce an immune response within the organism/pig
Antioxidant	(*nutr*) material which inhibits the oxidation (qv) of compounds eg prevents rancidity of fats
Arthritis	(*vet*) inflammation of a joint
Aspirate	(*vet*) Suck out. (However, aspiration can mean inhalation)
Astringent	(*vet*) causing contraction
Ataxia	(*vet*) muscle incoordination
Atresia	(*vet*) (*adj* atresic) closure of a structure
Atrophy	(*vet*) wasting; shrinking
Attenuation	(*vet*) reduction; thinning (diluted)
Attrition	wearing away
Atypical	not typical, irregular
Audit	a systematic review
Autogenous	(*vacc*) self-generating; originating within the body
Autolysis	self destruction of a cell
Axis	a line about which a figure, curve or body is symmetrical (or about which it rotates)

The New Terminology – 'A'

AIV : Annual Investment Value The number of times per year the savings or improved income from the extra investment is turned over in relation to the investment cost per tonne, sq metre, per pig, etc.

AMF : Absolute Mortality Figure How many piglets were lost relative to those born alive. A much better figure than % mortality. Expressed as 'AMF 0.9 of 12 BA' (BA = born alive). **Target AMFs** vary from 0.6 for 8 BA to 1.2 for 14 BA

Bacterin	a vaccine made up from killed bacteria
Bacteriocide	substance which destroys bacteria
Bacteriostat	substance which inhibits but does not destroy bacteria
Batch farrowing	farrowing sows in deliberately formed groups to facilitate workload and supervision, also to make disease in the offspring easier to contain
Bentonite	a pure clay capable of absorbing moisture
Berkshire	a pig breed noted for its meat carcase quality
Beta carotene	(*nutr*) Vitamin A
Bile	fluid produced by the liver which breaks up large fat globules for digestion by enzymes. Stored in the gall bladder
Billion	one thousand million and is commonly shown as 10^9. (Not one million million)
Bioactive	secures a response from living tissue
Bioassay	testing the power of a drug on a living organism
Biopsy	(*vet*) removal (for microscopic examination and testing) tissue from the living body
Biosecurity	all the measures taken to preserve health and defend against disease
Biotechnology	the application of scientific biological principles for industrial purposes (eg genetic engineering, pharmaceuticals, etc but also from wholly natural sources, eg yeast by-products)
Biotin	(Vitamin H) Vitamin B complex, involved in hoof strength
Birthweights	target on at least 15 to 17 kg of living neonates (qv) per litter. *Target at weaning 1.5 kg piglet. Action level <1.2 kg*
Blastocyst	early stage of embryo formation (from 'blast', a bud)

Blind teat   milk may be present in the mammary gland but the teat canal is blocked. In gilts can be a genetic defect

Bloat   (*gastric*) distension with gas, common with feeding (hot) whey

Blood poisoning   common term for blood infected with bacteria or their toxins

BOD   Biochemical Oxygen Demand, used to measure the potency of effluent and is the amount of oxygen needed over a specific time to decompose organic matter at 20°C

(Body) condition score   a subjective method of estimating the fat cover of (usually) a sow, across a range of 1 (emaciated) to 5 (obese)

Born alives   those piglets which drew at least one breath, confirmed by the bucket test – did the lungs float or sink quickly? *Target: born deads < 5%*

Breech presentation   foetus buttock-first at parturition

Brewer's grains   feedstuff residue after starch fermentation

Brewer's yeast   brewing by-product after harvesting and drying saccharomyces cerevisiae yeast

Brooder   substantial cover over heat source used early-on in wean-to-finish housing

Brown fat   fatty tissue which gives off heat. Pigs with higher (genetic) levels of brown fat 'burn off' food – a useful effect in us greedy humans but which raises FCR in pigs

BTU   British Thermal Unit

Buffer   material in solution which increases the amount of acid or alkali needed to produce a unit change in pH (qv)

Bulk density   the density of a granular substance (eg animal feed) calculated as the unit volume of the substance including the spaces between the particles/grain (see density)

Bursa   (*vet*) small fluid-filled cavity the body produces to protect against friction

Bursitis   inflammation of a bursa

Caecum   a small pouch between the small intestine and colon containing cellulose-splitting bacteria. Poorly developed in pigs and humans compared to ruminants

Calculi   (*vet*) accretal stores (as in kidney stones)

Calcaemia	*(vet)* excessive blood calcium
Calorie	*(nutr)* the amount of heat needed to raise 1 gramme of water by 1°C (1 calorie = 4.187 joules)
Calpain	an enzyme which breaks down muscle structure thus improving tenderness. Calpostatin is an inhibitor and increases with stress.
Capacitor	instrument for storing charges of electricity
Capsid	shell which protects a virus nucleus. Made up of capsomes
Carbohydrate	the simplest carbohydrates are the sugars (saccharides). More complex are the polysaccharides (eg starch and cellulose). Sugars (eg glucose) are intermediates in the conversion of food to energy; polysaccharides serve as energy stores in plants and seeds, potatoes, etc. Cellulose, lignin etc provide supporting cell walls and woody tissue in plants thus are not very digestible.
Carcinogen	*(vet)* cancer-causing substance
Cardiovascular	*(vet)* to do with the heart blood vessels
Caries	*(vet)* decay
Casualty	any pig slaughtered in an emergency due to disease, injury or distress. Casualties must be distinguished from culls
Catabolism	*(adj* catabolic) procedure within the body where complex structures are broken down into simpler compounds with the release of energy
Catalyst	*(n* catalysis) a substance which assists/speeds up a reaction but which is not used up during the process.
Cathartic	causing bowel evacuation
Caudal	*(vet)* towards the tail
Cell-mediated	affected by the cells in the body rather than chemicals
Cellulose	*see carbohydrate*
Celsius	0°C freezing, 100°C boiling. Centigrade = 100 steps. Celsius to Fahrenheit (qv) °F = (°C x 9/5)+32 ie °C x 9 ÷ 5 + 32 = °F.
Centimetre	(cm) 100th of a metre, 0.3937 ins
Cervix	*(vet) (adj* cervical) neck, or narrow part of an organ. In the female between the uterus and the vagina. A safety valve to protect the uterus from foreign bodies.
Chelate	*(nutr)* claw. Inert substance which holds a trace element (mineral) until the right digestive conditions release it for digestion.

Chemotherapy   *(vet)* medication

Chitterlings   *(nutr)* deep-fried delicacy; made from sections of the pigs large intestine

Chromosome   contains coiled DNA. In animal cells, determines sex and transmits genetic information

Chronic   *(vet)* in existence for a long time and causing less of a reaction than acute (qv)

Cilia   tiny hair-like substances which move the cell or move mucus over it

Circadian (-rhythm)   body activities which occur at regular intervals irrespective of light or dark influences ('biological clock')

Clinical   obvious disease (sub-clinical; less obvious or undercurrent)

Coander (Coanda) effect   'skidding' (water or) air along a flat surface thus lessening resistance and helping the direction of flow.

Coefficient   (of variation) *(stats)* the change between the variation of certain factors expressed as a ratio

Cohort   *(stats)* a group of animals with similar characteristics used in a research trial

Colitis   *(vet)* inflammation of the colon

Colon   the large intestine between the caecum and rectum

Colonisation   *(bact)* the ability of bacteria to adhere to a living surface and then multiply

Commensal   (usually *bact*) living on or inside another organism but causing no harm to it

Condensation   (vaporous) the change of a vapour into a liquid (see Dewpoint)

Conduction   the movement of energy (sound, heat or electricity) by the agitation of molecules inter alia

Congenital disease   present at birth

Congenital   evident from birth

Consultant   an ordinary guy a long way from home, or who has left salaried employment on grounds of economics or age, and whose quality of life improves markedly thereafter.

Convection   the movement of heat through a liquid or gas. Heat expands portions of the material, they become less dense and rise; their place is taken by colder portions, thus setting up convection currents.

Convex   curving outwards (concave is curving inwards)

Correlation	*(stats)* the degree of association of variables. Linkage, ie age of the pig can be correlated or linked to increased fat cover
Cortex	an outer layer
COSHH	Control of Substances Hazardous to Health. UK regulations (1989) providing one set of stipulations for all occupational health risks
Cost/benefit analysis	taking account of social costs/benefits as well as purely financial ones
Costive	constipated
Critical temperature(s)	see Temperature
Crude fibre	*(nutr)* the non-digestible cellulose, hemi-cellulose and lignin portions of a feed (see also Fibre)
Cycle	not the same as parity (qv). Cycle denotes a time from event to event eg birth to birth or breeding to breeding
Cystitis	*(vet)* inflammation of the bladder
Deadweight	carcase weight dressed to a specific standard
Deamination	*(nutr)* processing of surplus protein to waste material
Deliquescent	absorbing moisture from the air, eg copper sulphate is deliquescent
Denature	*(nutr)* to produce a structural change in a protein which causes it to reduce its biological properties
Density	the ratio of the mass (weight) of a substance to its volume
Dermal	*(vet)* to do with the skin. Dermatitis = skin disease
Dessication	drying
Deviation Standard deviation	*(stats)* a measure indicating variability from the average. Data which have a normal distribution of 66% of the data points are within 1 standard deviation and 95% fall within 2 standard deviations.
Dewpoint	the temperature at which water vapour present in the air saturates the air and begins to condense to form water deposits, ie dew begins to form
Diagnosis	the identification of disease. Clinical diagnosis is the identification from clinical signs during life backed by by laboratory tests (see also prognosis)
Dietetic	*(nutr)* to do with the diet
Differential diagnosis	*(vet)* using the differences in diseases derived from symptoms backed up by epidemiological (qv) tests to select a diagnosis most suited to the evidence

Digestible energy (DE)	the gross energy eaten less that voided in the faeces. 1 MJ (megajoule) DE = 239 Kcals. (See also Metabolisable Energy and Net Energy)
Dilation	*(vet)* stretching
Discrete	separate. (Note: discreet = tactful)
Disseminated	scattered
Distal	*(vet)* remote; far end of
Distribution	*(stats)* 'normal distribution' is a graphical curve appearing like a bell, symmetrical on both sides of the vertical axis, increasing to peak incidence on the left side and decreasing to zero incidence on the right side
Diuretic	increasing urine amounts also a product to do the same
Dose-response curve	*(stats)* shows how a drug responds in its effects to increased dosage
Dräger tube	hand held gas detector
Dressing percentage	(Killing Out % : Yield, USA) deadweight (qv) expressed as a percentage of liveweight shortly before slaughter.
Ductile	drawing out without breaking
Duodenum	first organ leading out of the stomach which primarily digests fats. The bile and pancreas empty into it
Dynamics	(eg thermodynamics) the reaction of bodies to force (in this case heat)
Dysplasia	*(vet)* abnormal development
Dyspnoea	*(vet)* difficult breathing
Dystocia	*(vet)* abnormally laboured farrowing (foetal d. = due to the foetus; maternal d. = due to the sow)
Dystrophy	disorder due to faulty nutrition
Eclampsia	*(vet)* post-natal convulsions
Ectoparasite	a parasite living on the host's body, eg fleas (endo = inside the body, eg worms)
Electrolyte	a substance, normally a mineral salt, which allows the intestine to insorb water at the same time it may be exsorbing it (ie during diarrhoea) thus deterring dehydration.
Elisa Test	enzyme-based test for degree of immunity to detect and measure either an antigen or the antibody to it (qv). Useful in specific circumstances
Embryo	from the time the organism develops a long axis to the time the major limbs etc have started to develop, when it becomes a foetus

Emetic	*(vet)* used to induce vomiting
Empirical	simple, basic
Empty days	(Open Days; Non-productive days) the number of days per year or per litter the sow is not carrying or feeding piglets. *Targets* per year 28-30; per litter 12-13
Emulsion	two unmixable liquids where one is dispersed with the other both in small droplet form. These can settle out and many emulsions need shaking before use
Encephalomyelitis	*(vet)* inflammation of both brain and spinal cord
Endemic	present at all times
Endocrine	hormonal
Endogenous	*(vet)* produced by the body or organism itself
Endometrium	the lining of the womb
Enteritis	*(vet)* inflammation of the intestine
Entire	both testicles descended. (As distinct from cryptorchid where neither have descended)
Entopic	occurring in the correct or expected location
Enzyme	*(nutr)* a protein which acts as a catalyst (qv). A chemical 'go-between' facilitating metabolism. The pig may contain 13,000 different enzymes. Addition of some to the feed can improve performance/reduce pollution
Epidemic – epizootic	*(vet)* disease attacking many subjects at the same time
Epidemiological	the study of diseases and their causes
Epidermis	outermost skin layer(s)
Epithelial	*(vet)* to do with cell formation in the body (see epithelium)
Epithelium	the cell covering of external and internal body surfaces
Epizootic	[1] widely diffused and rapidly spreading [2] an epidemic
Erythrocyte	red blood cell (corpuscle)
Ethology	study of animal behaviour
Excipient	adding a filler or carrier
Exogenous	outside the body
Exponential	(growth) *(stats)* ever-increasing
Extrapolation	(extrapolated from) *(stats)* inferred or deduced from the data presented
Extrinsic	outside (opp: intrinsic)
Exudate	*(vet)* fluid emanating from a wound or irritant
Eyponym	name including a person's name eg Aujeszky's disease

F1	First filial generation or first cross, terms used in genetics
F2	Second filial generation, ibid
Fahrenheit scale	Freezing point is at 32°F and boiling point of water at 212°F. Fahrenheit to Celsius °C = (°F – 32) x 5/9
Falciform	*(bact)* sickle shaped
Farrowing index	number of farrowings a sow achieves per year. *Target* 2.4. Can be used on a herd basis ie

$$\frac{Total\ number\ of\ farrowings\ in\ a\ given\ year}{Average\ sow\ inventory\ for\ that\ year}$$

Farrowing interval	*Target* 152 days in normal circumstances
Farrowing rate	number of farrowings to a given number of services. *Target*: 87-92% (indoors)
Farrowing fever	*(vet)* MMA syndrome (qv)
Fermentation	enzyme conversion of carbohydrates etc to simpler substances (like lactic acid). Done artificially, ie for the animal, it helps digestive efficiency
Fibre	crude fibre (qv) is today considered a largely meaningless term. Neutral Detergent Fibre is a better term (qv) but still has limitations as it does not quantify Non-Starch Polysaccharides (NSP) (qv)which themselves can compromise pig performance. Levels, and treatment of, NSP can however improve the digestibility of fibre. Fibre quality and amounts offered can be deficient in modern sow diets (constipation; lack of gutfill; stress)
Fibroma	*(vet)* fibrous tumour or swelling
Filamentous	a long threadlike structure (as in a filamentous blastocyst whose 'arms' attach themselves to the womb wall at implantation (qv)
Fimbriate	*(bact)* fringed border
First litter sow	female pig between the date of the first effective service and the date of the next effective service (following the successful completion of the first pregnancy)
Fixed costs	*(econ)* labour, contractors' costs, buildings and rent, machinery and equipment, finance charges, stock leasing, feeds, insurance, sundries
FLF	*(nutr)* Fermented Liquid Feeding. A process where either the complete feed or critical starch ingredients (such as wheat) are fermented by the addition of a

	starter culture (and possibly enzymes) plus heat to help predigest the feed.
Flocculation	settling out of particles in a liquid. Usually soft particles as distinct from harder ones (sedimentation). Can happen in wet-feed circuits between feeds.
Fomite	inanimate material which carries disease – bedding, dust, faeces, etc
Fructose	a sugar found in fruits and honey (lactose = milk sugar; mannose = yeast sugar)
Futures	*(econ)* specified quantity of products guaranteed for delivering at a specific future date at a contracted price
Galacto	pertaining to milk
Gastritis	*(vet)* inflammation of the stomach lining
Gastroenteritis	*(vet)* inflammation of the stomach lining *and intestine*
Gene	unit of heredity comprising a simple segment of DNA molecule that makes up a chromosome. Two copies of each gene, one from each parent, are present in every cell.
Generic	*(name)* the name of a drug not protected by trademark, usually describing the drug's chemical makeup
Genome	the complete inventory of hereditary traits contained in a half-set of chromosomes
Genotype	the entire genetic makeup of an animal (see also phenotype)
Gestation	110-116 days in the sow, from fertilisation to birth
Gilt	correctly, a young female pig which has not yet had a litter of pigs, rather than one before first conception. She then becomes, *after parturition*, a first-litter sow
Glässers disease	*(vet)* contagious disease of young pigs caused by the haemophilus organism
Glucosinolates	found in brassica crops, interfere with iodine uptake
Glutaraldehyde	a popular disinfectant, superseded by Peracetic Acid and Peroxygens (qv) against viruses
Gnotobiotic	an animal which has been born germ-free (see also SPF)
Grain	(high moisture) contains 22% to 40% moisture but must be ensiled anaerobically
Gravid	pregnant
Gutfill	[1]correctly, the amount of the pig's total body weight

	which is comprised of food, digesta and faeces [2]Also describes satiation
Gross margin	*(econ)* net output minus total feed costs and other variable costs (qv)
Habituation	gradual acceptance and adoption of a certain form of behaviour caused by an outside stimulus which is often unpleasant
HACCP	(pron. 'Hassap') Hazard Analysis Critical Control Points - a structured approach to identifying food production safety problems and controlling those problems
Haemoglobin	*(nutr)* component of blood which transports oxygen to and from muscles. (Neutralised by too much carbon monoxide being breathed in)
Halothane	an anaesthetic (Halothane Test on pigs identifies individuals susceptible to certain stressors)
Hepatitis	*(vet)* inflammation of the liver
Heterogeneous	not uniform: dissimilar *Note: pronounced heterogeneous*
Heterogenous	from another source; not originating in the body eg 'foreign body', *Note:pronounced heterogenous*
Heterosis	*(gen)* a first generation offspring showing greater vigour than either parent
Histology	*(vet)* science involving the structure of tissues
Homeostasis	stabilising mechanism within the body akin subjected to changing conditions (eg disease / stress / hunger) so as to mitigate their effect
Homeotherm	warm-blooded
Hormones	a wide variety of chemical messengers with specific functions. Work via the bloodstream
Humidity	the amount of moisture in the air (see Dewpoint)
Humoral	processes carried out by the body's fluids
Hybrid vigour	*(gen)* heterosis (qv) better performance and viability in the first generation from matings of parents of different breeds. The advantage is quickly lost of hybrids are then interbred
Hydatid	*(vet)* Cyst-like
Hydrochloric acid	secreted by cells lining the stomach, vital (especially in the weaned pig) to 'sanitise' the ingesta prior to it being passed on to the duodenum (qv)

Hyper- and **hypo-**	'hyper-' indicates extremeness, excessivity, eg Hyperactive; while 'hypo' indicates deficiency, beneath or under, eg Hypoglycaemia (low blood sugar) and hypodermic
Hypertrophy	*(vet)* increase in size of an organ due to excessive cell production
–itis	*(vet)* words ending in –itis refer to the inflammation of a particular organ (Enteritis: inflammation of the intestines)
Ig	prefix denoting any one of the 5 immunoglobulins (qv), IgA, IgD, IgE, IgG, IgM
Ileitis	inflammation of the ileum (qv)
Ileum	latter part of the small intestine
Immunoglobulins	a specialised protein usually produced following exposure to an antigen (qv)
Implantation	the attachment of the fertilized egg to the womb wall (endometrium) usually between day 10 to 30 after service
In vitro	seen in a test-tube or artificial environment
In vivo	in the living body
Incremental	(costs) *(econ)* added or increasing costs
Infarct	*(vet)* area interrupting the blood supply
Inflammation	*(vet)* the normal healing reaction of the body to an injury ie protective walling off of an injured area from the cause
Ingestion	eating/swallowing material (ingesta)
Inguinal	*(vet)* the groin area
Intumescence	*(vet)* swelling
Iodophor	skin disinfectant
Ishigami system	specifically designed very cheap double skinned polypropylene growout housing based on 300-400 mm deep, re-used sawdust. Called 'pipehouses' in Japan from the arched supports. Unexploited in Europe.

The New Terminology – 'I'

ILR : Income to Life Ratio a variant of PLR (qv). Preferred by some producers as there are many forms of ***profit*** while ***income*** is a singular, finite figure and thus easier to use.

Joule unit of energy. The energy needed to move one newton over 1 metre. 1 joule = 0.2388 calories. (A newton is a measure of applied force)

Juxta- (prefix) near or next to

K value (*Constr*) Thermal Conductivity of a Material (used in insulation calculations). Measured in Watts per metre per °C (W/M°C). Range of typical insulation materials is 0.02 to 0.2 with straw 0.07 (and copper wire 200). In insulation terms lowest is best; in electrical conductivity highest is best

Kcal (*nutr*) Kilocalorie. 1000 calories (qv) or one Calorie

Keratin (*nutr*) primary constituent of horn, hoof, hair, nails and secondary constituent of tooth enamel and skin outer layer

Label Rouge French farm food quality designation – 'excellent'

Labrum (*vet*) edge

Lactogenic stimulating milk production

Lacunae (*vet*) small cavities / hollows

Laminar / laminated (*constr*) layered

Lard commercially rendered pig fat

Latent concealed, not obvious

Laxative mild = aperient, strong = purgative/cathartic

LD50 poison indicator. The dose which will kill 50% of those tested

Lesion (*vet*) wound, ulcer, sore, tumour, bite, scratch, etc. A deviation from the normal in a body

Lights butcher's name for lungs

Lipase (*nutr*) fat-splitting enzyme

Lipids fats, greases, oily and waxy compounds

Lipoproteins how fats are moved in the blood eg HDL = high density lipoproteins and conversely LDLs

Litter scatter (*econ*) the percentage of litters with more or less than a specified number of liveborn piglets. A litter scatter average of <8 is usually taken as a performance warning. *Target* 10% <8

Livid correctly *discoloured*, not necessarily red. Also black/blue

Lochia (*vet*) discharge after parturition

Lumen correctly, the inside of any tube, eg the lumen of the

	gut/intestinal lining. Also a measure of the flow (flux) of light
Lux	Measure of illumination. One lumen (q.v) distributed evenly over 1m². E.g. darkness = 10-15 lux, bright sunshine outdoors = 500 lux+. Breeding units need at least 2/3 300-350 lux (author's opinion) intermittent with 1/3 diurnal darkness for best results
Lymphocyte	white blood cell, T cell
Macerate	soften by wetting
Macrolide	a specific type of antibiotic eg tylosin
Mandatory	required by law, essential
Marker gene	*(repr)* many genes are extremely difficult to find on the chromosome but can be associated with others found in greater numbers – marker genes
Mastication	chewing
ME *(nutr)*	Metabolizable Energy. The gross energy less that used in the faeces (DE) and the urine and gases. Gross to DE losses in the pig are about 16% and DE to ME about 5%
Mean	*(stats)* average, midway between two extremes
Medial	mid position, midline
Medulla	the inner portion of any organ. Core
Meningitis	*(vet)* inflammation of the lining of the brain, the meninges
Mesentery	*(vet)* membrane(s) attaching various organs to the body structure
Metabolism	*(nutr)* all the processes which lead to the build-up of the body (anabolism) and the breakdown of body molecules to provide energy (catabolism)
Metritis	*(vet)* inflammation of the uterus
Microgram	one millionth of a gram or one thousandth of a milligram. Written as µg
Microingredient	*(nutr)* a nutrient only needed in ppm, milligrams, micrograms/tonne
Micron	one thousandth of a millimetre
Micronutrient	element *(nutr)* a trace element, eg Se, Fe, Cu, Zn, etc
Milligram	one thousandth of a gram. Written as mg
Millimetre	one thousandth of a metre, written as mm
MMA	*(vet)* Mastitis-Metritis-Agalactia syndrome, known also as Farrowing Fever

Modulation	how a cell adapts to its environment
Morbidity	diseased
Morphology	the form and structure of an organism, cell etc
MOS	Mannan oligosaccharide. A naturally sourced (yeast) cell used as a feed growth enhancer and immunity modulator
Motility	movement
Mummified	degenerate (discoloured and shrivelled) piglets which died before farrowing
Mutagenic	*(gen)* inducing mutation
Mutation	*(gen)* the structural alteration of the DNA strand giving rise to a different genotype (qv)
Mycosis	(bact) any disease caused by a fungus

NT *The New Terminology – 'M'*

MTF (Saleable): Meat produced per Tonne (Ton) of Feed fed. A more useful measure for the working farmer than FCR as it relates primary income (from lean meat after dressing out) against the primary cost of feed. It is also easier for the working farmer to secure the necessary data, and simple to convert to an equivalent cost per tonne (ton) of feed basis (see PPTE)

Mortality – AMF (Actual Mortality Figure). Better than % mortality as it more accurately expresses losses in relation to litter size, where % mortality can mislead.

MSC: More for the Same Cost. There are two practical ways of making profit – either produce More for the Same Cost (MSC) or produce the Same for Less Cost (SLC). Theoretically SLC is better because if all producers produced more at the same cost the pig price may fall due to oversupply, while producing the same for less cost secures profit without overloading demand, thus stabilising pig price. The ideal of producing More for Less Cost (MLC) is usually unattainable in pig production!

Nano	one thousandth millionth
Nares	openings of the nose
Nascent	(just born) more commonly just emerging from a chemical reaction
Necrosis	*(vet)* cell death (*adj* necrotic)
Nematode	roundworm
Neonatal	just born (usually up to one week). Neonate (n)
Neoplasm	a tumour
Nephritis	*(vet)* inflammation of a kidney

Net or **nett**	*(econ & nutr)* total amount (remaining) eg net energy. No further deductions made
Net margin	*(econ)* gross margin (qv) less fixed costs (qv)
Net output	*(econ)* sales plus credits, less purchases plus valuation charge (closing valuation minus opening valuation)
Neuritis	*(vet)* inflammation of a nerve
Neutral detergent fibre NDF	*(nutr)* the amount of cellulose, hemi-cellulose and lignins in the diet, all indigestible components, but useful in sow diets, less so in others.
Nervous system	Central Nervous System involves the brain and spinal cord Autonomic Nervous System is not subject to voluntary control
Non-starch polysaccharides NSP	*(nutr)* a constituent of NDF fibre which may have anti-nutrient properties, especially for the young pig
Nostrum	*(vet)* a quack remedy
NVQ	National Vocational Qualification. UK award for stockmanship skills etc
Nucleotide	*(nutr)* building block of DNA (as distinct from amino-acid = building block of a protein)
Obstetrics	*(med)* science of pregnancy and birth
Occluded	*(vet)* closed (or sometimes, severely obstructed)
Oedema	*(vet)* build up of fluids in a body (eg Bowel Oedema)
Oesophagus	the throat to the stomach passage (ie gullet)
Ohm	*(constr)* the unit of electrical resistance
Oocyst	the highly resistant stage of a coccidia's life cycle
Opportunity cost	*(econ)* earmarking money put by for one project to be be spent on another should it appear opportunely
Orchitis	*(vet)* inflammation of a testis
Organic	produced only with assistance from materials harvested from living organisms and/or vegetable or animal fertilisers rather than those coming from synthetic chemicals
Orthopaedics	*(med)* the practice of muscular / skeletal surgery
Osteitis	*(vet)* bone inflammation
Overheads	*(econ)* fixed costs (qv)
Oxytocin	hormone which acts as a stimulant towards pregnancy and releases milk (together with the suckling stimulus) in lactation

Oxidation (nutr) replacement of negative charges (electrons) on a molecule by positive charges (protons). The opposite of reduction. (See also anti-oxidant)

P1 / P3 optical probe backfat measurements in mm at two fixed points over the loin. 4.5 cm (P_1) and 8 cm (P_3) from the midline of the back of the last rib. The two added together describe the degree of fatness. Generally however, P_2 is the most commonly used, at 6½ cm

Palpate examine by touch

Pancreas organ which produces enzymes to break down proteins, carbohydrates and fats (pancreatic 'juice')

Papilloma wart

Paradox quite different to what is expected

Parakeratosis *(vet)* thickening and cracking of the skin, in pigs due to Zn deficiency

Parameter a measurement which can be expressed numerically

Parenteral *(vet)* administered not through the alimentary canal ie by injection

Parietal *(vet)* referring to the walls of an organ

Parity [1] similarity;
[2] in the sow, the number of times a sow has farrowed eg a gilt is in parity 0 and a sow which has farrowed 4 times is in her 4th parity

Parturient giving birth or related to birth

Passive external stimulus (as distinct from 'active' where the animal responds spontaneously / originates the response)

Pathogenic *(bact)* disease producing

Pathology the study of disease

Pectoral *(vet)* the chest region

Pellucid translucent

Peracetic acid / new and powerful virucide disinfectants (eg Virkon
peroxygen S) capable of very quickly penetrating the various protective layers of many viruses to destroy the nucleus

Peracute *(vet)* very acute but shortlived

Peri- *prefix*; around or close to, eg perinatal (as distinct from neonatal = just after)

Peripheral near to the edge (of). (Periphery = outer edge or outside the central object)

Peristalsis	*(nutr)* the involuntary wavelike motion on down the gut
Permeable	permitting passage of a substance
Petecheal	*(vet)* tiny blood-blistering
pH	measure of alkalinity / acidity. <7 = acid. >7 = alkaline. 7 = neutral (Range 1-14).
Phagocyte	cell which eats micro-organisms (pathogens) and other foreign particles
Phenotype	the outward appearance of an animal in expressing its inheritance (as distinct from genotype = its whole genetic make-up)
Photoperiod	the time of exposure to daylight, or artificial light
Pili	hair-like structures found on the surface of bacteria helping them adhere to internal surfaces eg a gutwall. Also called fimbria
Pint	*(imperial)* = 586 ml (American 473 ml). Both approx
Pipeline feeding	*(nutr)* food mixed to a gruel and piped to a trough or feeder
Plasma	blood fluid containing the corpuscles
Plasma protein	*(nutr)* protein-rich fraction of blood also containing immunoglobulins, especially IgG
Polypeptide	*(nutr)* a protein substance containing two (dipeptide) three (tripeptide) or more (polypeptide) linked amino-acids
Polysaccharide	see carbohydrate
Postpartum	after farrowing. Also 'postparturient'
Potentiation	the effects of two combinations being greater than the sum of either two alone
Power	(to the power of) mathematical symbol to simplify the display of very large (or very small) numbers, eg 1000 = 10^3 or 1 x 10 x 10 x 10 (kilogram, kilowatt). Small numbers have a minus prefix eg 0.001 = 0^{-3} or 1 ÷ 10 ÷ 10 ÷ 10 (millilitre, milliamp). Thus one billion (one thousand million, 1,000,000,000) is simplified as 10^9
Prebiotics	act on gut conditions or precondition nutrients or capture hostile organisms eg oligosaccharides. Different from probiotics (qv)
Precursor	forerunner, usually leading to another more active result
Premature farrowing	before day 110 of pregnancy but where some foetuses have survived, nevertheless, for 24 hours
Primiparous	a sow which has had at least one pregnancy resulting in viable offspring

Probability	*(stats)* P = a measure of likely reoccurrence. The number of times an event did occur divided by the number of times it might have occurred. (The number of times it might have occurred is defined as the adding together of all the positive and negative outcomes)
Probe	[1] a fat measuring instrument; [2] the measurement itself (see P_2)
Probiotics	beneficial organisms which colonise the gut surface rather than pathogens eg lactobacilli. Competitive exclusion
Prognosis	*(vet)* a forecast of the likely effects of a disease and its cure prospects. Diagnosis is the identification of a disease
Proliferation	increase, multiplication
Prophylaxis	*(vet)* fending off disease, prevention *(adj* prophylactic)
Prosthesis	*(vet)* an artificial body part replacement
Protocol	an action plan, set of guidelines
Provitamin	a substance from which an animal can form a vitamin
Pseudorabies	American name for Aujeszky's Disease
PSS	Porcine Stress Syndrome. Sudden death especially after transportation, fighting etc, associated with PSE (Pale Soft Exudative pork)
Pyrexia	*(vet)* elevated body temperature, in the case of the pig above 40ºC (103.5ºF)

NT *The New Terminology – 'P'*

PPTE : Price Per Tonne (Ton) Equivalent. Rightly or wrongly pig producers still make econometric judgements on price per tonne (ton) of feed. PPTE is a simple calculation which can convert economic advantages into a per tonne of feed equivalent figure ie how much the advantage would reduce the per tonne cost of feed across the feeding period.

PLR : Profit to Life Ratio further refines REO (qv) incorporating the *time* it takes to obtain the 'Return' part of the REO figure. PLR quantifies payback. Used principally in longer-term transactions, *ie* use of equipment and building refurbishment.

Quadrant	one quarter of the circumference of a circle
Quadrate	square, four-sided *(adj* quadratic)
Qualitative	non-numerical description, eg colourful, small, etc
Quantitative	numerical description eg fourth, two kilometre, etc

Quartile	*(constr)* one fourth of a dimension plan or structure. Mainly used in ventilation design
Replacement cost	*(econ)* value of breeding stock purchased together with their valuation charge (closing valuation – opening valuation)
R-factor	*(constr)* thermal resistance of a material (as distinct from thermal conductivity of a material – k value, qv). Measured in m² per °C per Watt (M² °C/W). Range 0.12-0.55 with highest values best
Radiant	emitting heat from a surface
Ratio	the relationship between two quantities
Reagent	*(chem)* material used to produce a chemical response so as to detect and measure other materials
Recessive	*(gen)* a gene which only functions when it is provided by both parents
Reduction	*(nutr)* see 'oxidation', its opposite reaction
Reed-bed	an effective and underused method of small-scale effluent disposal
Replication	*(stats)* the repetition of an experiment to improve statistical accuracy
Resorption	reabsorption
Retroactive	(response) requiring stimulation to act. The opposite of proactive, which is to initiate an action
Return to service	see 'Service'
Rhinitis	inflammation of nasal lining
Rideal-Walker N°	effectiveness of a disinfectant compared to phenol

The New Terminology – 'R'   **NT**

REO : Return to Extra Outlay Ratio a useful measurement of value for money (ie added value) enabling the producer to prioritize use of his capital and the vendor to justify good quality in a product

Saline	salty
Sandcrack	a crack in the claw of a pig running in the direction from the coronet to the toe
Sanitizer	*(bact)* correctly, a combined detergent *and* disinfectant product (less good than using them separately)
Saprophyte	*(bact)* an organism which lives on dead tissue
Sarcoma	*(vet)* a malignant tumour capable of very fast growth
Satiety	*(nutr)* complete hunger satisfaction

Sebaceous gland	secretes sebum, an oily substance, around the hair follicles (over-secretion dries as dandruff)
Sedentary	inactive, lazy
Semi-sternum	sitting/lying upright on the chest as distinct from lying supine (qv)
Sensitivity	*(bact)* susceptibility of an organism to a compound eg an antibiotic
Sensitivity analysis	*(stats)* the comparison of the performance of one or more actions to a common mean
Septicaemia	*(vet)* blood poisoning
Septum	the partition in the pig's snout
Serum	usually involves the clear portion of blood plasma that does not contain blood cells. Blood serum from pigs containing antibodies to a disease is called antiserum, and is used to provide temporary immunity to that disease when injected
Service	Normal or regular return to oestrus evident 18-24 days after previous service, measured from the first day of mating. Irregular return to as above but oestrus occurs after 24 days
Sesqui-	*prefix* meaning one and a half
Shedding	*(bact)* releasing (pathogenic) bacteria
Sib	*(gen)* correctly a blood relative but can be an abbreviation of sibling
Sibling	*(gen)* brother or sister
Sigmoid	*(stats)* S-shaped
Significance	*(stats)* when the relationship between two variables is greater than by random chance. There are various degrees of significance denoted by 'p' (probability qv) expressed as a percentage
Skatole	a chemical constituting part of boar taint odour, along with androstenone
SMEDI	(Stillbirths, mummifications, embryonic deaths, infertility) *(vet)* due to Enteroviruses causing high piglet mortality
Somatic	*(vet)* whole body tissue rather than the cells which make it up ie muscles, skin etc
SPF	Specific Pathogen Free pigs reared disease-free (gnotobiotic) for pig trial research purposes. Also used to indicate high quality pork in Japan
Specific gravity	the weight of a liquid related to the specific gravity of water, which is 1.0

Stable	Scandinavian term for a piggery
Standard deviation	see deviation
Stasis	Cessation or slowing
Stenosis	narrowing
Sterotypies / sterotypic	abnormal behaviour(s)
Sternum	the breast bone
Stillborn	correctly, piglets which did not draw breath once expelled, as with born-deads. Can be confirmed by the 'bucket-test'. (see born-alives)
Straw-flow	bedded flooring design, steeply sloped, where gravity and the pig's feet gradually move the fouled straw to a collecting channel outside the pen
Stress	conditions and reactions affecting the wellbeing, mental and physiological, of the pig
Striated	streaked
Subclinical	(see clinical)
Subcutaneous	under the skin
Subjective	an unconfirmed, personal, opinion. The opposite of objective
Supernatant	the liquid lying above a layer of deposited insoluble material
Supine	lying flat on its side
Surfactant	substance which reduces surface tension, thus releasing bound particles eg soap, detergent
Symptomatic	indicating a symptom (of)
Synchrony	(*adj* synchronous; synchronism) occurring at the same time
Syndrome	(*vet*) a pattern or total of clinical signs constituting a whole picture
Synergy	(*adj* synergistic; synergism) combined action so that the total effect is greater than that of the two separately
Syntax	(*econ*) the rules of a language or computer program
Systemic	affecting the whole body system. Comprehensive
Single Space Feeding (SSF)	see wet/dry Feeding

The New Terminology – 'S'   **NT**

SLC : Producing the Same for Less Cost see MSC

T Cell	*(bact)* lymphocyte, white blood cell
T$_2$ toxin	a mycotoxin
Tare	*(weight)* the weight of a vehicle less its load
TDN	**Total Digestible Nutrients** *(nutr)* Now discontinued measure of energy
Temperature; **Lower,** **Evaporative** and **Upper Critical** **Temperatures.**	*Lower* is the ambient temperature below which the pig needs to divert food energy into keeping warm. *Evaporative* is the temperature at which panting occurs and signals the onset of heat stress and the serious diversion of energy into keeping cool. *Upper* is the temperature which when reached, the animal's life is in danger
Terminal **crossbreeding**	*(gen)* continuing breeding from a first cross without crossing further
Therapeutic	*(vet)* treating disease, curing, alleviating (*n* therapy)
Therm	heat required to raise 1000 kg of water 1°C. 1 therm = 1000 Kcal = 106 megajoules (MJ)
Titre	*(bact)* the amount of one substance required to react with another, used in determining antibody levels present
Tomography	used in radiology to visualise fat/lean deposition by scanning a cross section through the pig's carcase
Topical	*(vet)* a localised area
Torsion	twisting . As in gastric torsion (bloat)
Transducer	*(constr)* a device which converts pressure, temperature etc to an electric pulse
Transverse	*(constr)* side to side, across
Type	*(bact)* to identify an organism or blood group, etc
Ultrasound	used in pregnancy diagnosis by equipment capable of emitting radiant energy at over 20,000 cycles per second
U-Value	*(constr)* measure of thermal transmittance of a material, used in insulation calculations. The amount of heat which passes through 1 m^2 of a structure where the outside/inside temperature differs by 1°C. Expressed as Watts per sq metre per °C (W/m^2 °C). Typical range 0.5-5.5 with lowest values best.
Variable costs	*(econ)* costs which are likely to vary frequently
Vascular	(system) to do with the blood vessels / blood supply

Vector	*(constr)* an effluent sprayer covering a defined area
Vegetative	its most common meaning is resting ie vegetative state
Vein	blood vessel leading from various organs back to the heart in contrast to an artery which carries blood from the heart to various organs and the extremities
Velocity	speed (air movement) vital in correct air placement in a piggery
Ventral	*(vet)* abdominal area; *(constr)* towards
Vermifuge	expels worms. A vermicide is a substance that kills them
Viable	correct (rational, acceptable)
Villus	*(pl* villi) a microscopic, very sensitive thread-like growth covering the intestinal surface which absorbs nutrients from the digesta, thus increasing the absorption area many thousandfold.
Viraemia	*(vet)* blood infected by a virus
Virulence	the degree of pathogenicity of an organism
Viscera	large internal organs
Viscous	*(adj* viscosity) sticky, thick liquid
Volatile	evaporates easily and quickly
Volt	*(constr)* the unit of electric movement or force (one ampere of current versus one ohm of resistance)
Vomitoxin	a mycotoxin causing vomiting, especially in young pigs
Watery pork	PSE pork (see PSS)
Watt	*(constr)* a measure of electric force ie the work done at 1 joule per second. Equal to 1 ampere under the pressure of 1 volt
Wet-dry feeding	*(nutr)* technique where a small amount of meal is nudged by the pig into a receptacle which is then moistened once the pig activates a drinker nozzle over it. Also called (inaccurately) Single Space Feeding
Wet feeding	*(nutr)* pipeline feeding (qv). Also known as Liquid Feeding (see also FLF)
Wiltshire cure	keeping bacon in a dry cool environment after a 3 to 4 day soak in a brine solution
Withholding period	mandatory period of withdrawal of a drug from animal treatment before that animal can be used for food
Working capital	*(econ)* the capital required for the daily operation of a business
Weaning to service interval	the time between the date of weaning and date of first mating. Date of weaning is day 0.

NT   *The New Terminology – 'W'*

WWSY : (Weaner Weight per Sow per Year) The commonly-used term of pigs (weaners) per sow per year (PPS) is a less indicative figure than WWSY because it gives no indication of the weight of weaner achieved. Much better to express this as weaner **weight**, not weaner **numbers**, produced over a period when used in the critical economic assessment of a breeding farm – the sow inventory

Yield	dressed carcase weight
York Ham	ham first pickled, then stored over a long period in dry salt
Yorkshire	*(gen)* term for the Large White breed (in Europe) used elsewhere in the world
Zearalenone	an oestrogenic mycotoxin (qv) particularly dangerous to pigs, especially gilts and breeding sows
Zinc oxide	*(bact)* a useful anti-diarrhoeal used at high levels in the food of weaned pigs for a short period, but causes Zn build-up in soil
Zoonose	an animal disease transmissible to man
Zygote	*(gen)* the fertilised ovum just before first cleavage

FURTHER READING

I give below some suggestions on the more useful journals and books which anyone wishing to improve his knowledge of the rapidly-moving field of pig technology might consider for his bookshelf. They are not in any order of importance, but are the ones – among over 50 possible sources – where interesting, groundbreaking and up-to-date information on pigs will be found.

MAGAZINES

The doyen of them all is *PORC MAGAZINE* (French). This is a quite remarkable publication, but only for those of us who can read French! Between 100-150 pages each month, with 100 or more colour photos and diagrams. No pig journal in the world can touch it; the French and French-Canadians are so lucky. Other pig journals – please copy!

PIG LETTER (USA) is first class, if rather heavily weighted towards the veterinary aspects of pig production. Some articles, written by brilliant young scientists, are a little too obtuse/complicated for the average producer but they do stretch our brains and are well worth both the challenge and the private subscription.

PIG PROGRESS (Dutch, in English) is an excellent monthly, with good, solid, informative guest-written articles.

NATIONAL HOG FARMER (USA). The best of the American monthlies. The magazine is weighted towards US economics and internal strategies, but it is the only magazine to publish, two or three times a year, a series of 'Blueprint' editions where one subject – disease, nutrition, management – is covered in depth. Worth subscribing just for these authoritative statements of where we are at these days.

WESTERN HOG JOURNAL (Canadian). This bi-monthly has developed into a first-class general pig journal over the past two years, supported by the excellent practical academics at the agricultural and university centres in Alberta and Saskatchewan in particular.

ANIMAL TALK (British). Despite its title, this two-sided monthly on pig nutrition is written by two well-known and highly respected pig consultants, Drs Des Cole and Bill Close. Essential reading – concise and informative.

PIG INTERNATIONAL. Good articles on world-wide pig production. You need to see what the rest of the world is doing: this journal does it well.

PIG WORLD. Now that the 50 year old 'Pig Farming' is no more as a separate entity, it's successor is 'Pig World'. This is a much-improved monthly source of important information on the British pig scene now it has been adopted as the official organ of the UK's National Pig Association, and hopefully heralds the regeneration of the much-battered British pig industry at the time this book goes to press.

TECHNICAL JOURNALS

There is a wide range of these to choose from. All are expensive and the majority publish research papers about other livestock species too. Most are biennial or quarterly.

PIG VETERINARY SOCIETY JOURNAL (British) is really for pig veterinarians but it does contain good material on management/disease-associated items. Twice-yearly.

It is difficult to recommend one technical journal for the pig producer, but both the *JOURNAL OF ANIMAL SCIENCE* and *ANIMAL PRODUCTION* provide ground-breaking research papers, the former primarily nutrition and the latter on many aspects of farm animal technology.

BOUND PROCEEDINGS

There are two important international pig conferences (outside the large annual American circuit) whose annual proceedings are worthy of study. The proceedings of the Banff (Canada) Pork Seminars "*ADVANCES IN PORK PRODUCTION*" and the Australian Pig Science Association "*MANIPULATING PIG PRODUCTION*" are important updates on current pig technology. So too is (I believe the best) of the American yearly conferences '*THE ALLEN D LEMAN SWINE CONFERENCE PROCEEDINGS*' published by that bubbling cauldron of pig research, the University of Minnesota. Alltech's annual volume on biotechnology from naturally sourced products (see later) is also an essential for anybody's library.

MY TOP TEN BOOKS ON PIGS

Arguably the best book on pigs ever written is Mike Muirhead and Tom Alexander's 600-page volume "*Managing Pig Health and the Treatment of Disease*". While specialising in veterinary aspects of pig production, it engages itself deeply into practical management, as it should. While expensive at over £95 (US$130) no pig farm's office should be without a copy.

Another first class book, this time on nutrition, is Close and Cole's '*The Nutrition of Sows and Boars*' (Nottingham University Press). Easy to read and understand, with clear directions on diets and feed levels.

David Taylor's regularly updated book '*Pig Diseases*', now in its 7th edition, is a valuable reference source, and not expensive. Published by Dr Taylor.

For the slaughter pig, English, Fowler, Baxter and Smith's '*The Growing and Finishing Pig*' (Farming Press) is still a valuable source, if a little dated now (1988).

Prof Colin Whittemore's majestic work '*The Science & Practice of Pig Production*' (Longman Scientific & Technical) is authoritative, but personally I prefer his shorter '*Elements of Pig Science*' (Longman's) concisely written by a real word-craftsman, rare among leading academics.

Paul Hughes and Mike Varley's '*Reproduction in the Pig*' (Butterworths) gives an excellent grounding in this critical and easily 'got-wrong' area.

On housing, Gerry Brent's '*Housing the Pig*' (Farming Press) while somewhat aged now (1986) is still a remarkably good reference source on getting dimensions and construction basics right.

There is no good reference work I know of on the important subject of ventilating piggeries, but a remarkably good monograph is the now out-of-print UK Farm Electric Centre booklet '*Controlled Environments for Livestock.*' If you ever find a copy, grab it! It is absolutely first class and desperately needs a reprint, but the FEC has been disbanded on cost grounds along with many of our vital establishments in Britain. So short-sighted of our politicians.

Co-products (by-products) can play a part in reducing feed costs so '*Co-Product Feeds*' by Robin Crawshaw (Nottingham University Press) is the most comprehensive I've seen.

I have not come across a really good book on baby pig management and weaner nutrition, but I'm told a couple are in the offing. Badly needed and the information is there, but has tended to be commercially guarded.

Good textbooks on outdoor pig-keeping, stockmanship/man management and slurry disposal also exist. I buy these off the Landsman's Bookshop list.

For 30 years I've been looking for a book on interpreting statistics which didn't lose me after the first 10 pages. At last I've found one – Derek Rowntree's '*Statistics Without Tears*' (Penguin). This is a gem, and even the academics don't sniff at its simple language.

MISCELLANEOUS

Students and practitioners of pig production should keep their eyes open for good commercial monographs and booklets. The UK Meat and Livestock Commission (MLC) for example and the advisory branches of Government organisations (Universities in the USA) like ADAS in the UK from time to time produce important monographs. For example the 6 ADAS monographs on group housing of sows referred to in this book and the very recent MLC one (2002) on PMWS/PDNS. The superb Annual Reports of the Danish Ministry of Agriculture (National Committee for Pig Production) show what this vibrant nation is doing in its pig research, and are helpfully available in English. Several commercial firms [Farmweld (USA); Eli Lilly/Elanco (USA/Europe); Antec International (UK); Intervet (UK); JSR Genetics (UK); Alltech (USA) etc] have excellent information and well-balanced and not overtly commercial booklets on their spheres of interest.

In the field of nutritional biotechnology, i.e. the use of naturally-sourced feed additive products, Alltech stands supreme in its literature. '*Nutritional Biotechnology in the Feed and Food Industries*' is an annual volume incorporating the annual Pig Science Symposium which shows all of us in pigs the cutting edge of this area of growing interest in, and importance to, global pig production.

'*THE PIG PEN*'

Alltech also kindly sponsor my own 3 times a year 8-page magazine '*The Pig Pen*' (now in its 8th year), where I am allowed to discuss, without fear or favour, recent developments in pig production both good and not so good. Available free from Alltech (Ireland) Ltd, 'Sarney', Summerhill Road, Dunboyne, Co Meath, Ireland, or from the author.

Finally, there are a few excellent internet sources of pig data, if you can find the time to hack through the jungle of garbage that surrounds them these days. Also some of the commercially produced training videos/disks are good value, although I must admit, leave out a good deal which I would include – hence my writing this book! So, I confine this review to the printed word, still happily far from dead.

Porc magazine	Editions Boisbaudry
	CS 77711.35577
	Cesson-Sevigné cedex
	France
	E-mail: mmontjole@editionsduboisbaudry.fr
International Pigletter	Pig World Inc
	PO Box 8031
	St Paul
	MN 55108
	USA
	E-mail: pigworld@mindspring.com

Pig Progress

Elsevier International
PO Box 4
7000 BA
Doetinchem
The Netherlands

Fax: +31 314 340516

National Hog Farmer

Primedia Business Magazines Inc
9600 Metcalfe Avenue
Overland Park
KS 6612-2215
USA

E-mail: primediabusiness.com
Editor's E-mail:
dpmiller@primediabusiness.com
Website: www.nationalhogfarmer.com

Western Hog Journal

Alberta/BC/Sask/Man Pork Boards
4828-89 Street
Edmonton
Alberta
Canada TG3 SK1

E-mail: pwpm@telusplanet.net

Animal Talk

Nottingham Nutrition International
14 Potters Lane
East Leake
Loughborough
Leicestershire LE12 6NQ, UK

E-mail: des.cole@talk21.com

Pig World

UK National Pig Association
PO Box 29072
London
WC2H 8QS, UK

E-mail: sam@pigworld.org
Website: www.pigworld.org

The Pig Pen

Published for the author by:
Alltech Ireland Ltd
Sarney
Summerhill Road
Dunboyne
Co Meath
Ireland

Fax: +353 1 825 2251
Editor's E-mail: jngadd@aol.com

Animal Production

British Society for Animal Science
PO Box 3
Penicuik
Midlothian
EH26 0RZ   UK

E-mail: bsas@ed.sac.ac.uk

INDEX

I tend to use a textbook's index as a better guide to the scope of the subject matter inside, rather than the conventionally-abbreviated list of contents in the opening pages. With over 550 pages of text, the index is a long one so to further help speed up your search for information, I include an Index of Checklists (page 587) and also an Index of Performances (page 589). This latter gives a quick reference to the research and field trial evidence which indicates either a reduction in performance and economics from negative practices, or how positive ones have improved them.

INDEX OF CHECKLISTS

PERFORMANCE INDEX

Scattered throughout this book are many tables, mostly compiled from published research and field trial sources, of side-by-side and 'before and after' comparisons. I think it helpful if these are collected together as two lists:

1. Demonstrating improvements in performance, profit and/or lower costs from some good ideas and products/additives
2. Showing the penalties likely from some less-than-ideal actions or omissions in day-to-day work.

The two lists, based on reliable trial work, comprise a revealing source of improved profitability for the typical pig farm. Running your eye down them provides a fast-track route to profit which you may find interesting.

Performance reduction from:

Foot and leg troubles (ADG, MTF, gross margin, income/pig), 368
High immune system activation in the sow (VFI, weaner weight, milk yield), 404
Not matching feed specifications to disease status (MTF, PPTE, REO), 403, 263
A high disease challenge (VFI, ADG, FCR, lean gain), 397
Seasonal infertility (various), 91
Stress (author's estimates only; WSY, FRCR, ADG, MTF), 420
Every 1°C fall below LCT (VFI, ADG), 501
Fluctuating temperatures (VFI, ADG, FCR), 48
Gases (ADG), 523
Inadequate temperature (FCR, MTF), 277
Inadequate ventilation (FCR, ADG), 277
Suboptimal ventilation (FCR, ADG, MTF), 484
Too restricted ventilation (ADG, % mortality), 511
15% overstocking rate (ADG, FCR, extra costs/pig), 355
Increased costs of overstocking (costs/pig, REO), 291
Increased costs of understocking (sales/m²), 297
Inaccurate FCR measurement (FCR), 119
Incorrect feeder operation/adjustment (ADG, MTF, grading), 283
Mixing 10 days before shipping (ADG, AFI, FCR), 340
Post weaning check ADG, grading, MTF), 475
Short herd life (cost per weaner sold), 408
Action levels v top targets for growth rate (FCR, MTF, PPTE), 239
Cost of slower growth (ADG, FCR, cost/pig, income per tonne of feed fed), 240
General overheads in relation to growth rate (FCR, ADG, overhead costs/pig), 232
Increase in empty days (cost/sow/year), 56

Performance improvement from: